Nonlinear Effects in Fluids and Solids

MATHEMATICAL CONCEPTS AND METHODS IN SCIENCE AND ENGINEERING

Series Editor: Angelo Miele
George R. Brown School of Engineering
Rice University

Recent volumes in this series:

32 **PRINCIPLES OF ENGINEERING MECHANICS**
Volume 1: Kinematics – The Geometry of Motion • *Millard F. Beatty, Jr.*

33 **PRINCIPLES OF ENGINEERING MECHANICS**
Volume 2: Dynamics – The Analysis of Motic • *Millard F. Beatty, Jr.*

34 **STRUCTURAL OPTIMIZATION**
Volume 1: Optimality Criteria • *Edited by M. Save and W. Prager*

35 **OPTIMAL CONTROL APPLICATIONS IN ELECTRIC POWER SYSTEMS**
• *G. S. Christensen, M. E. El-Hawary, and S. A. Soliman*

36 **GENERALIZED CONCAVITY**
• *Mordecai Avriel, Walter E. Diewert, Siegfried Schaible, and Israel Zang*

37 **MULTICRITERIA OPTIMIZATION IN ENGINEERING AND IN THE SCIENCES**
• *Edited by Wolfram Stadler*

38 **OPTIMAL LONG-TERM OPERATION OF ELECTRIC POWER SYSTEMS**
• *G. S. Christensen and S. A. Soliman*

39 **INTRODUCTION TO CONTINUUM MECHANICS FOR ENGINEERS**
• *Ray M. Bowen*

40 **STRUCTURAL OPTIMIZATION**
Volume 2: Mathematical Programming • *Edited by M. Save and W. Prager*

41 **OPTIMAL CONTROL OF DISTRIBUTED NUCLEAR REACTORS**
• *G. S. Christensen, S. A. Soliman, and R. Nieva*

42 **NUMERICAL SOLUTION OF INTEGRAL EQUATIONS**
• *Edited by Michael A. Golberg*

43 **APPLIED OPTIMAL CONTROL THEORY OF DISTRIBUTED SYSTEMS**
• *K. A. Lurie*

44 **APPLIED MATHEMATICS IN AEROSPACE SCIENCE AND ENGINEERING**
• *Edited by Angelo Miele and Attilio Salvetti*

45 **NONLINEAR EFFECTS IN FLUIDS AND SOLIDS**
• *Edited by Michael M. Carroll and Michael A. Hayes*

A Complete Series Listing of Volumes appears at the back of this volume.

A Continuation Order Plan is available for this series. A continuation order will bring delivery of each new volume immediately upon publication. Volumes are billed only upon actual shipment. For further information please contact the publisher.

Nonlinear Effects in Fluids and Solids

Edited by

Michael M. Carroll

Rice University
Houston, Texas

and

Michael A. Hayes

University College Dublin
Dublin, Ireland

Plenum Press • New York and London

Library of Congress Cataloging-in-Publication Data

Nonlinear effects in fluids and solids / edited by Michael M. Carroll and Michael A. Hayes.
p. cm. -- (Mathematical concepts and methods in science and engineering ; 45)
Includes bibliographical references and index.
ISBN-13:978-1-4613-8000-9 e-ISBN-13:978-1-4613-0329-9
DOI: 10.1007/978-1-4613-0329-9

1. Continuum mechanics. 2. Elasticity. 3. Nonlinear mechanics. I. Carroll, Michael M. II. Hayes, M. A. (Michael A.) III. Series.
QA808.2.N64 1996
531--dc20 96-15818
CIP

ISBN-13:978-1-4613-8000-9

Softcover reprint of the hardcover 1st edition 1996

A Division of Plenum Publishing Corporation
233 Spring Street, New York, N. Y. 10013

10 9 8 7 6 5 4 3 2 1

Professor Ronald S. Rivlin

Contributors

D. R. Axelrad, Micromechanics Research Laboratory, McGill University, Montreal, Quebec, Canada H3A2K6

G. Bao, Department of Mechanics, Peking University, Beijing, People's Republic of China 100871; *present address*: Department of Mechanical Engineering, Johns Hopkins University, Baltimore, Maryland 21218

M. F. Beatty, Department of Engineering Mechanics, University of Nebraska at Lincoln, Lincoln, Nebraska 68588-0347

Ph. Boulanger, Département de Mathématique, Université Libre de Bruxelles, Brussels, 1050 Belgium

M. M. Carroll, Brown School of Engineering, Rice University, Houston, Texas 77005

J. L. Ericksen, Department of Aerospace Engineering and Mechanics and School of Mathematics, University of Minnesota, Minneapolis, Minnesota 55455; *present address*: Florence, Oregon 97439

G. Fichera, Department of Mathematics, University of Rome-1, 00199 Rome, Italy

A. E. Green, Mathematical Institute, University of Oxford, Oxford OX1 3LB, England

M. Hayes, Department of Mathematical Physics, University College, Dublin 4, Ireland

W. E. Langlois, IBM Almaden Research Laboratory, San Jose, California 95120

E. H. Lee, Division of Applied Mechanics, Stanford University, Stanford, California 94305-4040, and Department of Mechanical Engineering, Aeronautical Engineering and Mechanics, Rensselaer Polytechnic Institute, Troy, New York 12180

M. F. McCarthy, University College Galway, Galway, Ireland

G. B. McKenna, Polymers Division, NIST, Gaithersburg, Maryland 20899

P. M. Naghdi, Department of Mechanical Engineering, University of California, Berkeley, Berkeley, California 94720†

J. M. O'Neill, Department of Theoretical Mechanics, University of Nottingham, Nottingham, NG7 2RD, England

A. C. Pipkin, Division of Applied Mathematics, Brown University, Providence, Rhode Island 02912†

K. R. Rajagopal, Department of Mechanical Engineering, University of Pittsburgh, Pittsburgh, Pennsylvania 15261

K. N. Sawyers, Department of Mechanical Engineering and Mechanics, Lehigh University, Bethlehem, Pennsylvania 18015

G. F. Smith, Center for the Application of Mathematics, Lehigh University, Bethlehem, Pennsylvania 18015

A. J. M. Spencer, Department of Theoretical Mechanics, University of Nottingham, Nottingham NG7 2RD, England

A. S. Wineman, Department of Mechanical Engineering and Applied Mechanics, University of Michigan, Ann Arbor, Michigan 48109

---

†Deceased.

Preface

This volume of scientific papers is dedicated with gratitude and esteem to Ronald Rivlin and is offered as a token of appreciation by former students, collaborators, and friends.

Ronald Rivlin's name is synonymous with modern developments in continuum mechanics. His outstanding pioneering theoretical and experimental research in finite elasticity is a landmark. From his work there has followed a spate of developments in which he played the leading role—the theory of fiber-reinforced materials, the developments of the theory of constitutive equations, the theory of materials with memory, the theory of the fracture of elastomers, the theory of viscoelastic fluids and solids, the development of nonlinear crystal physics, the theory of small deformations superimposed on large, and the effect of large initial strain on wave propagation. It is in Rivlin's work that universal relations were first recognized. Here also are to be found lucid explanations of physical phenomena such as the Poynting effect for elastic rods in torsion. Additionally, he and his co-workers predicted the presence of secondary flows for viscoelastic fluids in straight pipes of noncircular cross section under a uniform pressure head. While some others may have displayed a cavalier lack of concern for physical reality and an intoxication with mathematical idiom, Rivlin has always been concerned with genuine mathematical and physical content. All of his papers contain interesting and illuminating material—and may be read with profit by anyone interested in continuum mechanics.

Ronald Samuel Rivlin was born on May 6, 1915, in London. He entered St. John's College, University of Cambridge in 1933 where he obtained his B.A. in Physics and Mathematics in 1937. Subsequently he obtained the degrees of M.A. (1939) and Sc.D. (1952) of Cambridge University.

His professional career began with his appointment in 1937 as a Research Physicist with the General Electric Company, Malvern, where he remained for five years. This was followed by two years as a Scientific Officer with the Ministry of Aircraft Production and nine years with the British Rubber Producers Research Association where he was Superintendent of Research. He spent

1946–47 as a guest worker with the National Bureau of Standards, Washington, D. C. In 1952–53 he was a consultant with the Naval Research Laboratory, Washington, D. C., where he and J. L. Ericksen began their very fruitful collaboration. He was appointed Professor of Applied Mathematics at Brown University in 1953, where he was a chairman of the Division of Applied Mathematics from 1958–63. He left Brown in 1967 for Lehigh University where he was Professor of Mathematics and Mechanics, Centennial University Professor, and Director of the Center for the Application of Mathematics, until his official retirement in 1987.

Rivlin combined his very active research activities with heavy administrative duties and contributions to his profession. He was Head of the Applied Mathematics Division at Brown, and later Director of the Center for the Application of Mathematics at Lehigh. He organized conferences at Brown and at Lehigh. He has been Vice President and President of the Society of Rheology, Chairman of the Society for Natural Philosophy, on the Board of Governors of the Institute of Physics, on the Executive Committee of the Society for the Interaction of Mechanics and Mathematics, and Chairman of the U. S. National Committee on Theoretical and Applied Mathematics. He has done extensive editorial work on many journals including the *Archive for Rational Mechanics and Analysis*, *Rheologica Acta*, and *Journal of Mathematical Physics*.

Among Rivlin's many awards are the Bingham Award Medal of the Society of Rheology, the Panetti Prize and Medal, and the Timoshenko Medal of the ASME. He has received honorary doctorates from the National University of Ireland, Tulane University, and the Universities of Nottingham and Thessaloniki. He is a member of the U. S. National Academy of Engineers and the American Academy of Arts and Sciences. He is an honorary foreign member of the Accademia dei Lincei and of the Royal Irish Academy.

Rivlin has published more than two hundred papers in various scientific periodicals. His first paper (with N. R. Campbell) appeared in *Proceedings of the Physical Society*, Vol. 49 (1937).

To put his major scientific work in perspective it is first necessary to give some historical background.

Following the development of the theory of stress by Cauchy in the 1820s and his theory of strain, there was renewed interest in elasticity and developments in classical elasticity theory by Navier, Boussinesq, Voigt, Lamé, Stokes, and Kelvin. In the course of these developments of classical elasticity throughout the 19th century, many results were obtained which did not necessarily involve the basic assumption of classical elasticity theory that the strains are infinitesimally small. In particular, there are results due to Kelvin and Tait (1879) which are valid for large strains and, of course, in his development of the theory of strain Cauchy (1826) obtained results also valid for large strains.

At the beginning of this century there was available a framework for the development of a theory of finite elasticity, in which no restrictions are placed on the magnitudes of the elastic deformations. It was earlier realized (e.g., by Green in 1840) that the physical properties of a material could be characterized by a strain-energy function. Indeed, within the context of infinitesimal strain, much effort had been expended on the construction of a strain-energy function which would be adequate to model the ether—an elastic material through which only purely transverse waves could propagate. MacCullagh (1939) solved this problem, but it was overtaken by events in the development of Maxwell's theory of light (1865). Even so, mathematical physicists were conscious of the need to construct particular models within the context of the infinitesimal strain theory. That consciousness pervaded the thinking in seeking a model valid for finite strain. It was recognized that the stored-energy function cannot depend on the displacement gradients in a purely arbitrary fashion; rather, it depends on six strain components. Also late in the 19th century, particularly through the work of Voigt, it was recognized that material symmetry must play a role in the form of the strain-energy function. Voigt showed that for the thirty-two crystal classes there are just nine forms of the strain-energy within the context of infinitesimal strain. (A century later Smith and Rivlin solved the same problem within the context of finite elasticity theory.)

For finite strain it was recognized that for an isotropic compressible elastic material, the strain-energy function must depend on just three functions—the three strain invariants. It was also realized that if the particular form of this dependence is known and postulated, the corresponding constitutive equations could be derived and the equations of motion or equilibrium could be written down together with appropriate boundary conditions. Even though forms of the strain-energy density had been derived and also postulated within the context of linearized elasticity, within the context of finite elasticity no specific form of the strain-energy density had been obtained for any particular elastic material. Various special forms were suggested on the grounds of their alleged simplicity but there was no coherent theory. Murnaghan in 1937 departed from the notion of simplicity and took instead a general function of the three strain invariants as his starting point. He considered successive approximations to the strain-energy function when the strains are "small." Murnaghan considered the extensions to be small, but the rotations could be arbitrarily large. Murnaghan's strain-energy function for isotropic compressible elastic materials was a polynomial in the strain invariants. He solved some boundary value problems using his strain-energy function. What is now called the Mooney–Rivlin strain-energy function was advanced by Mooney in 1940 (on the basis of a kinetic theory approach) and later confirmed (theoretically and experimentally) by Rivlin to describe adequately the behavior of rubber when subjected to large strains. This strain-energy is the analogue for incompressible isotropic materials to Murnagahn's strain-energy for compressible materials.

In 1944, when Rivlin joined the British Rubber Producers Research Association, he was asked to look into the adhesion mechanism involved in pressure-sensitive adhesive tapes which use rubberlike adhesives. It was this task which prompted him to consider the work done when thin filaments of rubber are stretched. Employing what he has called the "neo-Hookean" model he was able to solve some problems concerning the forces necessary to maintain simple extension, simple shear, and simple torsion. He saw that to maintain simple shear it was necessary not only to provide a shearing force, as in the classical linearized theory, but also normal forces on the shearing planes and forces normal to the plane of shear had to be provided. In the case of torsion of a circular cylindrical rod he found that not only must a torque be supplied, as in the classical linearized theory, but also a thrust distributed over the ends of the rod. This was essentially the Poynting effect discovered experimentally (c. 1910) in rods of both metal and vulcanized rubber.

It was these discoveries, using the relatively simple neo-Hookean model, that led Rivlin to his revolutionary work. He recognized that by exploiting the incompressibility of the material, these same problems of extension, shear, and torsion could be solved without assuming any particular form of the strain-energy function. He proceeded to solve these problems and others. The results were reported in a remarkable series of ten papers, beginning in 1948 with "Large elastic deformations of isotropic materials. I. Fundamental concepts; II. Some uniqueness theorems for pure, homogeneous deformations," *Philos. Trans. R. Soc. London Ser. A* Vol. 240, 459–508 (1948); and continuing until 1955 with "Reinforcement by inextensible cords" (with J. E. Adkins) *ibid*, Vol. 248, 201–223 (1955). These papers were remarkable for the range of problems considered and the novel way in which the problems were formulated, treated, and solved. Rivlin was quick to recognize universal relations—results which did not directly involve the stored-energy density. In textbooks today, these problems are treated exactly as Rivlin treated them. Rivlin exploited the fact that for incompressible materials the stresses are determined by the strains to within a hydrostatic pressure. He recognized that the presence of an arbitrary hydrostatic pressure made solving problems easier, not harder. This recognition has continued to be used in solving similar problems for any materials which are subject to internal constraints, such as incompressibility or inextensibility in one or more directions.

It is interesting to note in passing that these earliest discoveries were made in the area of rubber elasticity, a subject which could be traced back to the early 1800s. In 1806 the blind John Gough discovered the curious thermal behavior of rubber when repeatedly stretched. In the 1840s there was a further surge of interest in the subject when, as Bell (p. 728) has pointed out, vulcanized rubber railway bumpers had come into use. Late in the 1850s Joule did some classic experiments on the thermodynamic properties of rubber. Even so, a form of the strain-energy function for rubber had not been postulated.

In his study of the large deformations of rubber, Rivlin was not content with theoretical results. He set out also to do experimental work to verify the theoretical results and to determine the form of the stored-energy function. He was remarkably successful. Again, to quote Bell (p. 734): "Certainly the most important 20th century experimental development in the finite elasticity of rubber was the experiments of Ronald S. Rivlin and D. W. Saunders in 1951 . . . ." "The experiments considered by Rivlin and Saunders were: (1) the pure homogeneous deformation of a thin sheet of rubber in which the deformation was varied in such a manner that one of the invariants of the strain I_1, or I_2, was maintained constant; (2) pure shear of a thin piece of rubber; (3) simultaneous simple tension and pure shear of a thin sheet; (4) simple extension of a strip; (5) simple compression; (6) simple torsion of a right-circular cylinder; and (7) superposed axial extension and torsion of a right-circular cylindrical rod." Bell went on: "These experiments of Rivlin and Saunders are a landmark in the history of experimental mechanics. . . ." He remarks on "the successful confluence of experimental observation and theoretical explanation which this study achieved." Bell concluded: "Rivlin and Saunders described with thoroughness and clarity the details of specimen preparation, the apparatus, the method of performing the experiments, and commentary on the important aspects of the correlations and the experimental limitations of certain of the measurements and calculations." Praise indeed from a master experimentalist.

One of the most striking outcomes of the work of Rivlin on finite elasticity has been the quantitative prediction of a variety of new effects which are evident in elastic materials undergoing finite deformations. Among these are the Poynting effect, mentioned previously. Rivlin also showed how to calculate the change in length of a rod of arbitrary cross section when the rod is subject to torque. He predicted from theoretical considerations and demonstrated experimentally that when a tube of isotropic elastic material is subjected to a torque over its ends and is prevented from elongating, it undergoes a contraction in diameter. Such effects are contrary to the predictions of the classical linearized elasticity theory.

The main elements of finite elasticity theory and of its applications to rubberlike materials were understood by 1950 or so, mainly through the work of Rivlin and his co-workers. From that time it has formed the basis for a great deal of research by Rivlin and many others into the solution of problems involving finite elastic deformations and has given rise to a distinct discipline in the mechanics of continua.

The idea of developing a continuum theory to describe the mechanics of elastic materials reinforced by inextensible cords—such as rubber tires—seems to have originated in a series of papers by Adkins and Rivlin in the mid-1950s. In recent years, with the advent of very strong man-made fibers there has been an upsurge of interest in the theory of fiber-reinforced materials. Much of the recent theoretical developments are due to collaborators of Rivlin—A. J. M. Spencer and his group at Nottingham, and to the late A. C. Pipkin, at Brown, a former student of Rivlin.

At the time when Rivlin was conducting his pioneering researches on finite elasticity theory, he became aware of some unpublished qualitative experiments which had been conducted during the Second World War in connection with the study of the saponified hydrocarbon gels that were used as flamethrower fuels. These effects showed a good deal of similarity to the normal stress effects which emerged from his work on finite elasticity theory. He accordingly embarked on the development of a continuum-mechanical theory for viscoelastic fluids along the same lines that had proven so effective in his theory of finite elasticity. However, his earlier attempts at such a theory were based on assumptions which were too restrictive to provide the definite theory that was his objective. That this was the case was pointed out by Oldroyd in 1950. In 1952–53, Rivlin, while working at the Naval Research Laboratory in Washington, D. C., took advantage of the points raised by Oldroyd to construct, in collaboration with J. L. Ericksen, a continuum-mechanical theory for viscoelastic materials, both solid and fluid. Materials described by such constitutive equations are now called Rivlin–Ericksen materials.

In its application to viscoelastic fluids the Rivlin–Ericksen theory was used to provide a correct theory for some of the effects which had been observed a decade earlier in flamethrower fuels and since then in a wide variety of viscoelastic materials. The Rivlin–Ericksen theory has been used since then, both by Rivlin and his collaborators and by others, to provide explanations for other flow effects that have been observed in viscoelastic fluids and to predict hitherto unobserved effects. Of the latter, the first was concerned with the flow of a viscoelastic fluid in a straight pipe of noncircular cross section under a constant pressure head. For a Newtonian fluid, the particles of the fluid move down the pipe in rectilinear paths. It was shown by Green and Rivlin (1956) and by Langlois and Rivlin (1959, 1963) that if the fluid is viscoelastic, there is superimposed on the rectilinear flow a steady secondary flow in transverse planes that depends on the cross section of the pipe, but generally consists of one or more eddies. This effect was subsequently observed by a number of workers. It was also shown, by Pipkin and Rivlin (1963), that the thrust exerted by the viscoelastic fluid on the tube wall is not constant over the periphery of a cross section, as it is for a Newtonian fluid. These results of Rivlin and his collaborators appear to have been the first of many in which the Rivlin–Ericksen constitutive equations have been used to predict, in various experimental situations, secondary flows and force distributions that are either nonexistent in Newtonian fluids or qualitatively quite different from those obtained with Newtonian fluids. Rivlin and Ericksen initiated the whole research field now called non-Newtonian fluid mechanics.

Rivlin made other significant contributions to the general theory of the mechanics of viscoelastic materials. Beginning in 1957, together with various collaborators, he published a number of papers in which the theory of constitutive equations for materials possessing memory were studied. In these studies, the basic assumption is made that the stress in a material element at a time *t* (say) depends on the history of the displacement gradients existing in the element for all times up to and including *t*. In mathematical language, the stress is a tensor functional of the displacement gradient history. In their 1957 paper, Green and Rivlin were the first to show how theorems in functional analysis could be used to give concrete representation to the expression for the stress as the sum of multiple integrals of the strain history. They also showed how the restrictions imposed by material symmetry on constitutive equations of the functional type could be made explicit.

Also, in their 1957 paper and in later papers, Green and Rivlin established the connection between constitutive equations of the memory type and the earlier Rivlin–Ericksen constitutive equation, and they showed that the latter arises naturally from the former if the flow fields are sufficiently smooth.

In the Rivlin–Ericksen continuum-mechanical theory for viscoelastic materials, it is shown that, subject to certain smoothness conditions on the deformation, the stress, which is itself a symmetric second-order tensor, may be expressed as a function of a number of symmetric second-order tensors that are defined in terms of the deformation undergone by the material. Symmetry of the material, such as isotropy, enables one to obtain an expression for the stress in terms of these tensors in a closed canonical form. It was realized very soon by Rivlin that, more generally, in any continuum theory, whether in mechanics, in electromagnetic theory, or in some other area of the physics of continua, a similar principle could be applied. Once the vector or tensor variables that are related in the constitutive equations for the material (i.e., in the equations describing the relevant material behavior) are chosen, any symmetry that the material may possess can be used to obtain these constitutive equations in canonical form, without making any further assumptions regarding the characters of the relationship. The mathematical problem involved is in many cases far from simple. However, it was shown by Rivlin that the problem can always be reduced to a problem in the classical theory of invariants, and, together with various collaborators,—mainly G. F. Smith, M. M. Smith, A. J. M Spencer, and J. E. Adkins—he embarked on a program to obtain the canonical form appropriate to virtually any constitutive assumption which is likely to be encountered in the physics of continua regarding the nature of the related variables and of the material symmetry. Apart from the importance of these results in the mechanics of materials, the program has resulted in the creation of nonlinear crystal physics paralleling the classical linear crystal physics of Voigt. Although the theory has been applied by Rivlin and his collaborators, and by others, to a few areas of continuum physics, the potential of the methods and viewpoint that Rivlin pioneered appear to be far from exhausted.

It was not only in continuum mechanics that Rivlin made fundamental contributions. During his time with the General Electric Company (1937–42) he worked mainly on problems connected with the design of electrical filters for carrier telephony and on related problems in the design and fabrication of piezoelectric elements. He is the holder or coholder of seven patents and the author or coauthor of a dozen papers in this general area.

He has retained a great interest in electromagnetism and has published basic papers on wave propagation. In his classic 1961 paper with R. A. Toupin, they consider the effect on the propagation of light of weak intensity of the presence of an applied large static electric (or large magnetic induction) field. They explain the presence of the Faraday effect and the Cotton–Mouton effect. This work was continued in a series of papers with Carroll on electro-magneto-optical effects.

In much the same spirit, Rivlin and Hayes examined the effect the presence of a large initial static homogeneous deformation has on the properties of infinitesimal plane waves propagating in isotropic elastic and viscoelastic materials. In this work is to be found the first general treatment of inhomogeneous plane waves in deformed materials. In dealing with wave propagation Rivlin was led to consider allied matters of energy flux and stability questions. Here too he made fundamental contributions.

In the early 1970s Rivlin had "the policy of not publishing criticism of the work of others" which he "thought was erroneous but rather of ignoring it." Fortunately, he later changed his policy and through his (occasionally controversial) reviews and lectures gave a lead in pointing out the fallacies of some views that had been presented as if written on tablets of stone. In this he has done his subject a great service. He has been unyielding in maintaining high standards in mathematics, mechanics, and physics. His work combines in a truly unique way, mathematical flair, physical insight, and experimental skill.

Ronald Rivlin is a marvelous raconteur, a generous host, and a loyal friend. Many of us are greatly indebted to him for his many personal and professional kindnesses.

Ronald Rivlin's first paper appeared in 1937. His latest is in press. The story of his research is unfinished.

Our wish for Violet and Ronald is that they continue to enjoy good health and the company of John and his family.

Michael M. Carroll
Houston, Texas

Michael A. Hayes
Dublin, Ireland

Contents

1

On Structural Changes in Two-Phase Materials

D. R. AXELRAD

Abstract. This paper is concerned with the stochastic analysis of the structural changes in a two-phase material. A stochastic state-space and corresponding operators in that space are introduced. The occurring structural changes of the medium are seen as a random phenomenon, which can be represented by a two-component random process in that space. One of the components refers to the "unobservables" of the process. The analysis is then illustrated by the experimentally obtained results regarding the application of a uniaxial constant stress to a material sample consisting of aluminum particles embedded in a resin matrix (fluid phase).

Key Words. Probabilistic mechanics, stochastic state-space, two-component process (stochastic), evaluation of states, structural changes of an aluminum–resin compound.

1. Introduction

Structural changes in solids usually occur during a relatively short transient period of response on application of a constant load to a material sample. The present analysis is based on more recent investigations (Refs. 1, 2) and is concerned with a class of material known as two-phase structures. They consist of an α-phase (solid) which is embedded in a β-phase (fluid) matrix of different physical characteristics.

D. R. Axelrad • Thomas Workman Professor of Mechanical Engineering, Micromechanics Research Laboratory, McGill University, Montreal, Quebec, Canada H3A2K6.

Nonlinear Effects in Fluids and Solids, edited by M. M. Carroll and M. Hayes, Plenum Press, New York, 1996.

In general, the evolution of structural changes can be regarded as a random phenomenon and can be characterized by a family of random variables, i.e., a set of $\{\mathbf{x}_t\}$, $t \geq 0$, usually taking values in $\mathbb{R}^n$. Such a family can also be considered as a stochastic process $\mathbf{z}_t \in \mathbf{Z}$ on a stochastic state-space contained in a more general probabilistic function space X. The state vector $\mathbf{z} \in \mathbf{Z}$ corresponds to a set of internal variables $\{\mathbf{x}_i\}$ or is formed by certain componets of $\mathbf{z}$ only. Generally, it is defined by the set of r-parameters, $i = 1, \ldots, r$, representing the thermomechanical states of an element of the microstructure. In two-phase materials the definition of a structural element is often difficult since the α-phase has the tendency to form a collection of individual particles or clusters that are surrounded by the β-phase (fluid) matrix. In this case, it may be convenient for the simplification of the analysis to use structural units, the random shapes of which can be idealized to be spherical. The diameter of the cross-sectional area of such units must be chosen, however, to be much smaller than the corresponding diameter of the macroscopic material sample.

From a probabilistic mechanics point of view the state vector ${}^{\nu}\mathbf{z} \in \mathbf{Z}$ pertaining to an element of the structure is seen as an outcome or elementary event E in $\mathbf{Z}$ as a result of the statistical experiment ν. However, due to experimental constraints and the accuracy with which relevant observations can be carried out, one can define such an event only within a certain range $\Delta^{\nu}\mathbf{z}$ of ${}^{\nu}\mathbf{z}$. Hence, the state of an element is characterized by $\mathbf{z}^n < {}^{\nu}\mathbf{z} < \mathbf{z}^n + \Delta\mathbf{z}^n$, where ${}^{\nu}\mathbf{z}$ is a specific value for the element ν of the structure, $\nu = 1, \ldots, n$. One obtains, therefore, subsets $E_n \subset \mathbf{Z}$ that include the states within $\Delta\mathbf{z}^n$ only and which can be regarded as open spheres, i.e.,

$$E_n = \{\mathbf{z}^n < {}^{\nu}\mathbf{z} < \mathbf{z}^n + \Delta\mathbf{z}^n\}, \tag{1a}$$

$$\bigcup_n E_n = \mathbf{Z}, \qquad E_n \cap E_k = \phi, \qquad n \neq k. \tag{1b}$$

Assuming that the state-space $\mathbf{Z}$ or if extended to the probabilistic function space X is locally compact the subsets $\mathbf{E}_n$ are also compact and bounded under closure. Hence, a σ-algebra on $\mathbf{Z}$ can be defined with the following properties:

$$\text{(i)} \quad E_n \in F^z, \qquad E'_n \in F^z, \qquad E'_n - \text{complement of } E; \tag{2a}$$

$$\text{(ii)} \quad E_n \in F^z, \qquad \bigcup^{\infty} E_n \in F^z; \tag{2b}$$

$$\text{(iii)} \quad Z \in F^z. \tag{2c}$$

The elements E_n of F^z are Borel sets and Z together with F^z so defined forms a measurable space $[Z, F^z]$. An appropriate measure on the subsets of Z is therefore

$$0 \leq P^z\{E_n\} \leq 1, \qquad P^z\{E_n\} = 0, \qquad \text{if } E_n = \phi \text{ and } P^z\{Z\} = 1. \quad (3)$$

This measure from probability theory represents the distribution of the relevant field quantities and the triple $[Z, F^z, P^z]$ characterizes then the behavior of the structure by the set $\mathbf{z} \in Z$.

2. Stochastic State-Space and Operators in Z

To formulate evolution equations for the response of the two-phase structure during the short transient in which structural changes occur by the use of the abstract dynamical system $[Z, F^z, P^z]$, it is convenient to use operators. It is assumed that these operators are linear and bounded with an algebra that corresponds to the topological structure of the state-space Z. A formal representation of Z and the mappings between the chosen subspaces is given in Fig. 1 and are discussed in some detail in Ref. 8. The most important of the operators in Fig. 1 is the mapping **M** between the subspace E, which is generated by the set of observables (strains or strain rates) and the stress-space Σ formed by the unobservable components of the state-vector $\mathbf{z} \in Z$.

In general, the operator M can be expressed by

$$M = M\{\mathbf{A}^m, \mathbf{B}^m, P, t, T\}, \tag{4}$$

where $\mathbf{A}^m$, $\mathbf{B}^m$ denote stochastic integrodifferential operators characterizing the response of a statistical number of structural elements and their interactions within a particular observable domain D^m of the macroscopic material sample, P the distribution of the corresponding field variables, t the time, and T the temper-

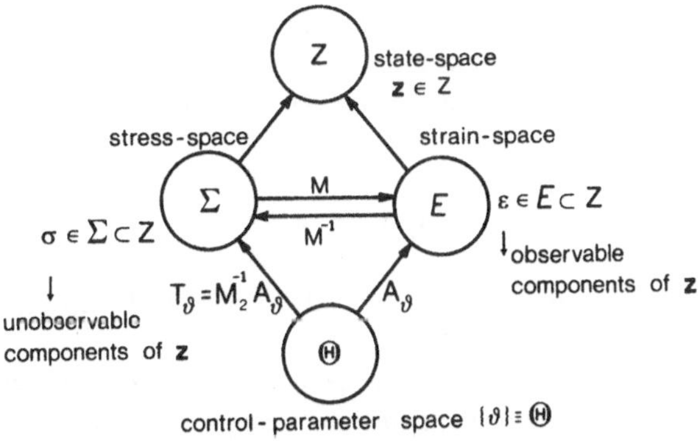

Fig. 1. Stochastic state-space representation.

ature. There are, however, certain restrictions concerning the invertibility of M, which largely depend on either the strong or strict monotonicity of the operator on some dense subsets Σ_d, E_d, respectively, that belong to Σ, $E \subset Z$. Generally, a wider class of materials can be represented by considering the weaker condition, i.e., that of strong monotonicity of M. The concept of material operators as constitutive functionals and their strict definitions are given in Ref. 3. It may be noticed from Fig. 1 that another subspace of Z, i.e., the control parameter-space Θ, has been introduced, which is linked to the subspace Σ, E via the operators A_θ and T_θ. The latter is the composition operator $[M^{-1} \circ \mathbf{A}_\theta]$. This subspace becomes important, if certain internal mechanisms are to be considered that greatly affect the stochastic field variables $\boldsymbol{\sigma}$ and $\boldsymbol{\epsilon}$.

Such mechanisms are known to occur in many microstructures. In polycrystalline solids, for instance, the dislocation densities in single crystals as well as other defects are fundamentally involved in the transition from the elastic to the plastic response of the material. In two-phase structures, experiments under the conditions of simple uniaxial confined compression testing, reveal an increase of the number of contacts between elements of the α-phase and hence an increase in contact forces and, associated with it, a decreasing effect of the mean pressure of the β-phase (fluid).

Hence, a particular state $\mathbf{z}$ of the structure during the short period of structural change will be specified generally by a probabilistic state-function or system functional of the form

$$\mathbf{f} = \mathbf{f}(\mathbf{z}_t, t, \theta_t), \tag{5}$$

in which $\mathbf{f}$ is a vector functional that contains in its argument the evolution process $\mathbf{z}(t)$ or the stochastic variable $\mathbf{z}_t \in Z$ that is directly influenced by a control parameter $\theta_t \in \{\theta_t\} = \Theta \subset Z$.

In view of the stochastic state-space representation (Fig. 1), the Euclidean-$\mathbb{R}^n$ space can be regarded as the product space of the subspaces Σ, E such that with probability one, $\mathbf{z}_t$ has right continuous piecewise paths in Z. In this case the process on the state-space will have values in $\mathbf{R}^n = \mathbf{R}^{n_1+n_2}$ or $\Sigma^{n_1} \otimes \mathbf{E}^{n_2} = \mathbf{Z}$. Thus, the stochastic process $\mathbf{z}_t$ can also be seen as two-component processes, i.e.,

$$\mathbf{z}_t = (\boldsymbol{\sigma}_t, \boldsymbol{\epsilon}_t), \tag{6}$$

where $\boldsymbol{\sigma}_t \in \Sigma^{n_1}$ refers to the "unobservable" and $\boldsymbol{\epsilon}_t \in E^{n_2}$ to the "observable" component process of $\mathbf{z}_t$.

It is evident that a sequence of successive states, which corresponds to observations made at various instants of time, i.e., $\boldsymbol{\epsilon}_i \rightarrow \mathbf{t}_i \in \mathbf{T}$, the process $\mathbf{z}_t$ will jump from the state $\mathbf{z}_{t_{i-1}} \rightarrow \mathbf{z}_{t_i}$ and requires the analysis in terms of the class of stochastic processes known as partially observed jump process (Ref. 4). This type of process for the description of structural changes has been considered in previous work (Ref. 1). The formulation of the evolution of states involves then the state jump distribution that also depends on the unobserved component process of $\mathbf{z}_t$. It becomes necessary, therefore, to consider an increasing family of sub-σ-algebras of the algebra F^z. In general, if $F^z_{\mathbf{t}}$, $t \in T$ denotes the sub-σ-algebra for a continuous time parameter set with $t_0 < t_1 < t_2 \ . \ . \ .$ and $F^z_{t_0} \subset F^z_{t_1} \subset \ . \ . \ .$, the family $(F^z_{\mathbf{t}})$ is called a filtration of the space $[Z, F^z]$. For the abstract dynamical system $[Z, F^z, P^z]$ with filtration $\{F^z_{\mathbf{t}}\}$, $\mathbf{t} \in \mathbf{T}$, a real-valued stochastic process or a sequence of random variables $\{\mathbf{z}_t\}$ is called a super or a submartingale on $[Z, F^z, P^z]$, respectively, with respect to $\{F^z_{\mathbf{t}}\}$, if the following properties are satisfied (Ref. 5):

(i) $\mathbf{z}_t$ is $F^z_{\mathbf{t}}$-measurable, i.e., $\mathbf{z}_t$ is adapted to $\{F^z_{\mathbf{t}}\}$; (7a)

(ii) the expectation $E\{|\mathbf{z}_t|\} < \infty$, $t \in T$; (7b)

(iii) $E\{\mathbf{z}_t | F^z_{\mathbf{s}}\} \leq \mathbf{z}_s$ a.s., if $s \leq t$. (7c)

If the sequence $\{\mathbf{z}_t\}$ is both a super and submartingale, it is called a martingale. Hence, denoting the σ-algebra generated by the past process $\mathbf{z}_t$ as follows:

$$F^z_{\mathbf{t}} = F^z\{\mathbf{z(s)} : t_0 \leq s \leq t\}, \qquad t \in T, \tag{8}$$

and the corresponding algebra of the observations by

$$G^z_{\mathbf{t}} = G^z\{\boldsymbol{\epsilon}(s) : t_0 \leq s \leq t\}, \qquad t \in T, \tag{9}$$

one can define a system functional $\mathbf{f}(t)$ on the past process $\mathbf{z}_t$ as a real-valued stochastic process, which for each $t \in T$ is $F^z_{\mathbf{t}}$-measurable and for which $E\{|\mathbf{f}_t|\} < \infty$.

In particular, if $\mathbf{f}(t)$ belongs to the class of bounded, square-integrable functions, given the history of observations G_t optimal filtering will occur, when $\mathbf{f}(t)$ is determined by the conditional expectation $E\{\mathbf{f}_t | G_t\}$ (see also Refs. 2 and 4).

3. Evolution of States during the Transient Period

The introduction of a probabilistic state-function or system functional and the recognition of the stochastic process $\mathbf{z}_t \in \mathbf{Z}$ by two-component processes leads to a description of the evolving states in terms of this functional and its expected value at any given time during the transiency. It remains, however, to employ the set of control variables that directly effect the changes of the structure within this time interval. For this purpose, the set of control parameters can be regarded as additional "unobservables," which influence the instantaneous equilibrium states of the structure and thus are random disturbances. Hence, the evolution of the stochastic dynamical system with the inclusion of such disturbances in terms of the system functional (5) will be characterized by the nonlinear stochastic differential equation, or

$$d\mathbf{z}_t = \mathbf{f}[\mathbf{z}_t, t, \theta_t]dt\,, \qquad t \geq t_0, \qquad t \in T. \tag{10}$$

If the disturbances are sufficiently small, they can be considered to be equivalent to white Gaussian noise so that equation (10) takes the form

$$\frac{d\mathbf{z}_t}{dt} = \mathbf{f}(\mathbf{z}_t, t) + \mathbf{g}(\mathbf{z}_t, t)|_t\, \theta_t, \qquad t \geq t_0, \tag{11}$$

in which the first term represents the generally nonlinear system functional and the second a disturbance function (g being an $n \times n$ matrix function), with the initial condition $\mathbf{z}(0) = \mathbf{z}_{t_0}$, $t = 0 = t_0$, which is assumed to be independent of the noise process $\{\boldsymbol{\theta}_t, t \geq t_0\}$. It is well known that the letter is a delta-correlated random process and hence not integrable in the mean square sense (Ref. 10). It is therefore necessary to formulate equation (11) differently. Thus, considering the **n**-dimensional random disturbance induced by the unknown interactions in the two-phase structure as an **n**-dimensional Brownian motion or Wiener process one can replace $\boldsymbol{\theta}_t$ by the formal derivative of the Brownian motion. If $\{\boldsymbol{\beta}_t, t \geq t_0\}$ denotes a vector process of independent Brownian motion, the formal derivative is given by

$$\boldsymbol{\theta}_t \cong \frac{d\boldsymbol{\beta}_t}{dt}, \tag{12}$$

and relation (11) can be written as

$$d\mathbf{z}_t = \mathbf{f}(\mathbf{z}_t, t)dt + g(\mathbf{z}_t, t)d\boldsymbol{\beta}_t\,, \qquad t \geq t_0. \tag{13}$$

The solution of this stochastic differential equation must be interpreted, however, in the sense of Itô's integral equation, i.e.,

$$\mathbf{z}_t = \mathbf{z}_{t_0} + \int_{t_0}^{t} f(\mathbf{z}(\tau), \tau)d\tau + \int_{t_0}^{t} \mathbf{g}(\mathbf{z}(\tau), \tau)d\boldsymbol{\beta}(\tau), \quad t_0 \leq \tau \leq t. \tag{14}$$

The existence, uniqueness, and solution of equation (13) is comprehensively dealt with in system control theory. On the basis of the above assumption, i.e., that the initial condition of $\mathbf{z}_t$ is independent of the Brownian motion process $\{\boldsymbol{\beta}_t, t \geq t_0\}$, the solution for $\mathbf{z}_t$ is a strong Markov or diffusion process in Z. This is readily recognized by recalling the definition of the state of a structural element in terms of the events $E \in E$. Thus, the possible events for which a probability is well-defined are the elements of the Borel field F^z of subsets of Z. If $\mathbf{z}_t$ is characterized by the observations $\boldsymbol{\epsilon} \in E$ (event-space), a probability transition function $\mathbf{P}(\boldsymbol{\epsilon}, \mathbf{E})$ is then an F^z-measurable function of $\boldsymbol{\epsilon}$ for each event $E \in F^z$ and the probability measure P^z on F^z for each observation $\boldsymbol{\epsilon} \in E \subset Z$. The evolution of the states of the structure is related to a sequence of observations, at any instant of time, by the fundamental condition that the probability measure is given by

$$P^z\{\mathbf{z}(t_i) \in E \,|\, \mathbf{z}(t_{i-1}) = \boldsymbol{\epsilon}_{t_i}\}, \tag{15}$$

or by the well-known Markov principle. Hence, the function $\mathbf{P}(\boldsymbol{\epsilon}, \mathbf{E})$ is a transition function for the Markov process $\mathbf{z}_t \in Z$ with probability one, if

$$\mathbf{P}(t_i, \boldsymbol{\epsilon}_i, t_{i-1}, E) = P^z\{\mathbf{z}(t_i) \in E \,|\, \mathbf{z}(t_{i-1}) = \boldsymbol{\epsilon}(t_i)\} \quad \text{for } t_n > t_{n-1} \cdots t_i > \cdots t_0, \tag{16}$$

satisfying the condition $\mathbf{P}(\cdot) = 1$ for $\boldsymbol{\epsilon} \in E$, and zero for $\boldsymbol{\epsilon} \notin E$ with $\mathbf{P}(\cdot) \leq 1$.

Assuming that the probability density function is continuous and bounded in the finite time interval $\{t_i\} \in T$ and that partial derivatives up to the third order exist, the evolution of the probability density will satisfy the Kolmogorov forward or the Fokker–Planck equation. For the vector-valued process $\{\mathbf{z}_t \in Z, t \in T\}$ generated by the Itô stochastic differential equation (13), where $\mathbf{z}$, $\mathbf{f}$ are r-dimensional vectors, $\mathbf{g}$ an $n \times n$ matrix function, and $\{\boldsymbol{\beta}_t\}$ an m-dimensional Brownian motion process, the evolution of the probability density can be expressed by

$$\frac{\partial p}{\partial t} = -\sum_{i=1}^{n} \frac{\partial [pf_i]}{\partial z_i} + \sum_{i,j}^{n} \frac{\partial^2 [p(\mathbf{g}q\mathbf{g}^T)i, j]}{\partial z_i\, \partial z_j}, \tag{17}$$

in which q is a positive semidefinite matrix defined by $E(d\boldsymbol{\beta} d\boldsymbol{\beta}^T) = q(t)dt$. Equation (17) can be expressed in a shorter form as

$$\frac{\partial p(\mathbf{z}, t)}{\partial t} = L\, p(\mathbf{z}, t), \tag{18}$$

where L is the forward diffusion operator of $\{\mathbf{z}_t\}$.

4. Structural Changes of the Aluminum–Resin Composite

In order to simplify the above general formulation and to validate experimental observations, the one-dimensional case of the strong Markov process (18) will be considered. For this purpose, it is convenient to regard the state-space Z as a Hilbert space, H, with the inner product

$$\langle \mathbf{f}(z), P^z \rangle = \int_{\mathbf{R}_n} \mathbf{f}(z) dP^z, \tag{19}$$

where $P^z \in M(\mathbf{R}^n)$ is a set of measures and $\mathbf{f}(z) \in B_0(\mathbf{R}^n)$ the class of bounded functions on B_0. The latter denotes the space of all real-bounded Borel measurable functions on $\mathbb{R}_n$. One can also define a family of transition operators (Ts^t) for the process $\mathbf{z}_t$ in the B_0-space so that

$$T_s^t f(z) = \int_{\mathbf{R}^n} \mathbf{f}(\eta)\mathbf{P}(s, z, t, d\eta), \tag{20}$$

in which $\mathrm{T}_s^s = \mathbf{I}$ (Identity operator) and $\mathbf{P}(\cdot)$ is the transition probability function from before (see also Ref. 3). As an approximation to the Markov process $\mathbf{z}_t \in Z$, one can consider the most probable path of $\mathbf{z}_t$ to be given by its mean, i.e., $\bar{\mathbf{z}}_t \in Z$, so that the probability density becomes

$$\mathbf{p}(\bar{\mathbf{z}}, t) = \int_{t_0=0}^{t} \mathbf{p}(\mathbf{z}, t) dt, \tag{21}$$

where t denotes the end of the transiency. Substituting for $\bar{\mathbf{z}}_t$ the two-component processes, i.e., their mean values $\bar{\boldsymbol{\sigma}}_t$, $\bar{\boldsymbol{\epsilon}}_t$ from equation (6), respectively, one obtains the probability density

$$p(\bar{\boldsymbol{\sigma}}, \bar{\boldsymbol{\epsilon}}) = \int_{t_0=0}^{t} p(\boldsymbol{\sigma}, \boldsymbol{\epsilon}) dt. \tag{22}$$

By use of a phenomenological model for the interaction between the α- and β-phases and the corresponding balance equations, an extended constitutive relation can be established involving the change with time of the observable volumetric strain ϵ_v of the material sample and the concentration of the α-phase only, so that under the condition of a constant stress (uniaxial compression),

$$\frac{d\epsilon_\nu}{dt} = -\theta_\alpha \frac{d}{dt}\left(\frac{\sigma}{\theta_\alpha}\right). \tag{23}$$

In the special case where a single observable $\epsilon \in \{\underline{\epsilon}\}$ is identified with the measurable control parameter $\theta_\alpha \in (\underline{\boldsymbol{\theta}})$, relation (22) can be expressed by

$$p(T_\theta \bar{\theta}_\alpha, \bar{\theta}_\alpha) = \int_{t_0=0}^{t} p(T_\theta \theta_\alpha, \theta_\alpha) dt, \tag{24}$$

in which the composition operator $T_\theta = M^{-1} \circ A_\theta$ of Fig. 1 has been used.

The approximation to the Markov process $\mathbf{z}_t$ and the evolution of the probability density is then characterized by

$$\frac{d}{dt} p(T_\theta \bar{\theta}_\alpha, \bar{\theta}_\alpha) = L_\theta\, p(T_\theta \bar{\theta}_\alpha, \bar{\theta}_\alpha). \tag{25}$$

Realizing that there is a one-to-one correspondence between $T_\theta \bar{\theta}_\alpha$ and $\bar{\theta}_\alpha$ given by the composition operator, the above relation can be reduced to

$$\frac{d}{dt} p(\bar{\theta}_\alpha) = L_\theta p(\bar{\theta}_\alpha). \tag{26}$$

Since for a constant applied stress σ, the condition that $\sigma = T_\theta \theta_\alpha$ and $\epsilon_v = \epsilon_v(\theta_\alpha)$ holds, the state $\mathbf{z}$ of the structure at any given time within the parameter set $\{t_i\} \in T$ can be specified by a function $\mathbf{f}(z)$ or correspondingly by a functional $F_0^t(\theta_\alpha)$ which is assumed to be linear and bounded on a dense set of the control space $\Theta \subset Z$.

This can also be expressed on the basis of the Riesz-representation theorem (Ref. 6) in the form of an integral Kernel function, i.e.,

$$z = f(\epsilon) \Rightarrow f(\theta) = \int_{t_0=0}^{t} G(t - \tau)(\dot{\theta}_\alpha) d\tau. \tag{27}$$

It is to be noted that the details of the composition operator T_θ require the determination of the material operator M and its inverse for the given two-phase structure by means of the computational procedure discussed in detail in Ref. 7. Hence, the present formulation is restricted to an ϵ_v-projection of $\mathbf{z} \in \mathbf{Z} = \Sigma \otimes E$.

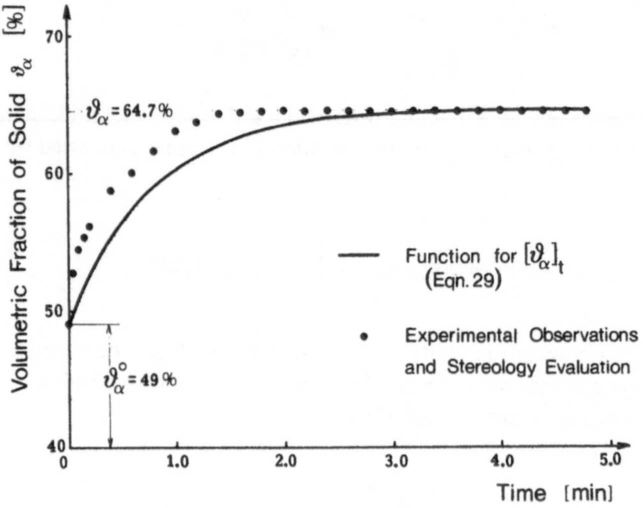

Fig. 2. Evolution of volume fractions of the α-phase.

Since experimentally the time evolution of both the volumetric strain, ϵ_v, and the volume fraction, θ_α, can be monitored, the evolution of the latter can be written as

$$\theta_\alpha = \int_{t_0=0}^{t} K(t - \tau)(\dot{\theta}_\alpha)d\tau, \tag{28}$$

where the Kernel is obtained from the relation

$$\theta_\alpha = \theta_\alpha^0 + \alpha e^{-\beta t}, \tag{29}$$

and the constants α,β are determined from a set of differential equations for two different loads applied to the material specimen. This approximation is illustrated in Fig. 2, showing some experimental results for a particular two-phase structure, i.è., an aluminum–resin composite. This structure has been chosen for its suitability in scanning electron microscopy and the required stereological evaluation of distributions and mean values. It consists of aluminum particles (diameter ~ 7 μm) representing the α-phase, which are embedded in a resin matrix (ethylene-glycol) as the β-phase (fluid). The numerical values of the constants in equation (29) are as follows: $\theta_\alpha^0 = 0.647$, $\alpha = 0.157$, and $\pi = 1.27$ for the constant stress $\sigma = 157 \pm 6.8$ kPa.

References

1. AXELRAD, D. R., *The Transient Behavior of Structured Solids*, Lecture Notes in Physics, Springer-Verlag, Berlin, Germany, Vol. 249, pp. 56–64, 1985.
2. AXELRAD, D. R., *Stochastic Analysis of Structural Changes in Solids*, Constitutive Laws and Microstructures, Proceedings of the Seminar at the Institute for Advanced Study, Berlin, Germany, pp. 23–24, 1987.
3. AXELRAD, D. R., *Foundations of the Probabilistic Mechanics of Discrete Media*, Pergamon Press, Oxford, England, 1984.
4. GIHMAN, I. I. and SKOROHOD, A.V., *The Theory of Stochastic Processes*, Vols. 1 and 2, Springer-Verlag, Berlin, Germany, 1975.
5. DOOB, D. L., *Classical Potential Theory and Its Probabilistic Counterpart*, Springer-Verlag, Berlin, Germany, 1984.
6. YOSIDA, K., *Functional Analysis*, Springer-Verlag, Berlin, Germany, 1978.
7. AXELRAD, D. R., REZAI, K., and ATACK, D., *Probabilistic Mechanics of Fibrous Structures*, Zeitschrift fur Angewandte Mathematik und Physik, ZAMP, Vol. 35, pp. 497–513, 1984, Birkhäuser Verlag, Basel, Switzerland.
8. AXELRAD, D. R., *Stochastic Mechanics of Discrete Media*, Springer-Verlag, Berlin–Heidelberg, 1993.
9. DOLEZAL, V., *Monotone Operators and Applications in Control and Network Theory*, Elsevier Scientific Publ. Co., Amsterdam–New York, 1979.
10. JAZWINSKI, A. H., *Stochastic Processes and Filtering Theory, Vol. 64: Mathematics in Science and Engineering*, Academic Press, New York, 1970.

2

Introduction to Nonlinear Elasticity

M. F. BEATTY

Abstract. This is an introductory survey of some selected topics in finite elasticity. Virtually no previous experience with the subject is assumed. The kinematics of finite deformation is characterized by the polar decomposition theorem; and Euler's laws of balance and the local field equations of continuum mechanics are described. The general constitutive equation of hyperelasticity theory is deduced from a mechanical energy principle; and the implications of frame invariance and of material symmetry are presented. This leads to constitutive equations for compressible and incompressible, isotropic hyperelastic materials. Constitutive equations studied in experiments by Rivlin and Saunders (Ref. 1) for incompressible rubber materials and by Blatz and Ko (Ref. 2) for certain compressible elastomers are derived; and an equation characteristic of a class of biological tissues studied in primary experiments by Fung (Ref. 3) is discussed. Sample applications are presented for these materials. A balloon inflation experiment is described, and the physical nature of the inflation phenomenon is examined analytically in detail. Results for the different materials are compared. Two major problems of finite elasticity theory are discussed. Some results concerning Ericksen's problem on controllable deformations possible in every isotropic hyperelastic material are outlined; and examples are presented in illustration of Truesdell's problem concerning analytical restrictions imposed on constitutive equations. Universal relations valid for all compressible and incompressible, isotropic materials are discussed. The nonuniversal, antiplane shear problem and related theorems are presented. Some examples of nonuniqueness, including that of a neo-Hookean cube subject to uniform loads over its faces, are described. Elastic stability criteria and their connection with uniqueness in the theory of small deformations superimposed on large deformations are introduced, and a few applications are mentioned.

M. F. Beatty • Professor and Chairman, Department of Engineering Mechanics, University of Nebraska at Lincoln, Lincoln, Nebraska 68588-0347.

Nonlinear Effects in Fluids and Solids, edited by M. M. Carroll and M. Hayes, Plenum Press, New York, 1996.

Key Words. Finite elasticity, constitutive equations, deformation, hyperelasticity, Rivlin's cube, elasticity, universal relations, shear, nonuniqueness, neo-Hookean material.

Preface. The theory of elastic materials subjected to large deformations has advanced considerably since its inception during the early 1940s. Significant theoretical results, many confirmed by experiments, have provided substantial insight into the physical behavior of rubberlike materials such as synthetic elastomers and polymers, in addition to natural rubber. This remarkable success stimulated numerous interdisciplinary studies in other important areas of engineering science. These include thermomechanics, electromechanics, mixture theory, wave propagation, granular soil mechanics, elastic stability, rods and shells, and viscoelasticity of solids and fluids, to name a few. And the foundations laid for finite elasticity now support important new fields of study that include the elasticity of biological materials, the mechanics of constrained continua, director theories of rods and shells, microstructural mechanics, the mechanics of liquid crystals, and the mechanics of phase transition phenomena, for example. Consequently, finite elasticity theory has attracted the interest of a great variety of applied mathematicians, engineering scientists, chemists, and physicists.

The mathematical theory of elasticity of materials subjected to large deformations is inherently nonlinear; and the mathematical difficulties encountered in the theory and its applications are considerable. Therefore, we find in the last decade that increasing numbers of applied mathematicians and numerical analysts have expressed interest in the kinds of nonlinear problems and technical difficulties encountered in the theory. Questions regarding stability and existence of solutions under various mathematical conditions set down as criteria for characterizing stable material response have attracted particular interest among mathematicians. Though eager to learn the basic problems, their own research interests and education in applied mathematics and numerical analysis have removed them from the mainstream of literature resources essential for work in this area. It was with this problem in mind that the present paper was written originally for an introductory lecture at the opening *Workshop on Equilibrium and Stability Questions in Continuum Physics and Partial Differential Equations* conducted by the Institute for Mathematics and its Applications at the University of Minnesota in September, 1984 (see Beatty, Ref. 4). The presentation was intended for a broad audience of nonexperts interested in an overview of special topics in nonlinear elasticity and elastic stability theory. The content focused on material that, in my opinion, one ought to know something about before embarking on a more thorough course of self-study of major works cited in the reading list. The lecture was well-received by the audience. When subsequently I was invited by Professor Arthur W. Leissa to submit a review article to the *Applied Mechanics Reviews (AMR)*, it occurred to me that perhaps students of mechanics and others in engineering and applied mathematics also may find the outline helpful. It was with this objective in mind that I prepared the expanded didactic review article. So, it is especially gratifying to learn from several colleagues that the essay, Beatty (Ref. 5), has since been used for several years as supplementary reading material in advanced courses on continuum mechanics and finite elasticity.

Before turning to the substance of the present article, I want to emphasize that I dedicated my *AMR* essay to Professor Ronald S. Rivlin in reflection of my personal feelings of the importance of his countless contributions to the foundations and applications of finite elasticity that I wished to sketch in that report. At the time of its appearance, Professors Michael M. Carroll and Michael A. Hayes, subject to approval by the ASME through Professor A. W. Leissa, invited me to submit the *AMR* manuscript for publication in a special book to honor Ronald S. Rivlin. Although several years passed unnoticed, as years often do, their objective remained steadfast and is now realized in this book. It is indeed my privilege to contribute this slightly abridged version of my original *AMR* article. I could not resist the temptation, however, to update the bibliography and to interject some recent research results; otherwise, the article is much the same.

1. Introduction

This essay is an introductory presentation of selected topics in nonlinear elasticity theory. The basic equations of the theory are outlined and several easy illustrative examples are sketched without going into details that may be found in sources given in the bibliography. The presentation is written mainly for engineers and applied mathematicians.

We begin in the next section with a sketch of the principal kinematical relations used to describe the finite deformation of a continuum. The Cauchy stress principle and equations of motion follow in Section 3, and the engineering stress also is introduced there. The theory of elasticity of materials for which there exists an elastic potential energy function is known as hyperelasticity. This presentation emphasizes hyperelasticity theory, but occasional annotations concerning the general theory of elasticity are included here and there.

The mechanical energy principle and the constitutive equation for a hyperelastic solid are presented in Section 4. A change of frame and the principle of material frame indifference are reviewed in Section 5. Material symmetry transformations are discussed in Section 6. Afterwards, the general constitutive equation for an isotropic hyperelastic solid is derived in Section 7. The effect of internal constraints on the form of the constitutive equation is demonstrated in Section 9 for incompressible materials. Some special kinds of compressible and incompressible, hyperelastic materials are exhibited, and their application is illustrated in some examples.

The Blatz–Ko constitutive equation for a class of compressible elastomers is derived in Section 8, its reduction to special forms having experimental support is demonstrated, and the constitutive equation more commonly identified in the literature as the Blatz–Ko material is investigated at the end. Some technical aspects of the mechanical response of these materials that may be of engineering interest are illustrated. The Blatz–Ko models are used in special applications to study the behavior typical of compressible, isotropic hyperelastic materials under finite strain. Much of this presentation, including development of the Blatz–Ko constitutive equation, appeared for the first time in my *AMR* article.

The general constitutive equation for an incompressible, isotropic hyperelastic material is obtained in Section 9. This theory is applied in Section 10 to deduce the Rivlin—Saunders constitutive equation for incompressible rubber-like materials. The special constitutive relations for the classical Mooney–Rivlin and neo-Hookean materials, and the constitutive equation for a class of biological tissues studied in primary experiments by Fung are presented there.

The inflation of a spherical membrane is described in Section 11. The inflation pressure for an arbitrary compressible or incompressible, isotropic hyperelastic membrane is derived from the general work-energy principle. The result is then illustrated for the neo-Hookean model, which fails to exhibit the entire inflation phenomenon described by data obtained in a typical balloon inflation experiment. It is shown that the Mooney–Rivlin and tissue materials are able to capture more of the overall physical effect, and the importance of the moduli is illustrated. It is shown that the neo-Hookean model yields the lower bound solution for both the Mooney–Rivlin and the biological membrane materials. Inelastic effects also are mentioned briefly. Some additional results that demonstrate the effect of compressibility are examined for a Blatz–Ko balloon. It is shown that the Mooney–Rivlin model provides an upper bound solution for the inflation pressure for any balloon in the Blatz–Ko class. The variety of results obtained in this simple problem for the various constitutive models commonly encountered in applications underscores the richness of nonlinear elasticity theory.

Some other kinds of internal constraints are mentioned in Section 12. The influence of the constraint on the Poisson function commonly associated with the simple tension problem is demonstrated. It is shown that a Poisson ratio $\nu_0 = 1/2$ in the natural state of an isotropic material does not imply that the material need be incompressible. The constitutive equation for a class of constrained, compressible materials, known as Bell constrained materials, is described.

The nonuniqueness of solutions of various boundary value problems is illustrated by several heuristic examples in Section 13. Rivlin's problem demonstrating nonuniqueness in the traction problem for a neo-Hookean cube subjected to pairs of equal and opposite forces on its six faces also is outlined there.

The problem of determining all deformations that can be produced by application of surface tractions alone in every compressible and incompressible, homogeneous and isotropic hyperelastic material is known as Ericksen's problem. The determination of all such universal deformations is important because these are the deformations around which an experimental program may be designed to determine the elastic response functions for specific kinds of homogeneous and isotropic materials. Ericksen's problem on universal inverse solutions and its importance in an experimental program are discussed in Section 14. Although only homogeneous deformations are possible in every compressible, isotropic material, in addition to these, universal solutions for incompressible materials include several families of nonhomogeneous deformations. These are described in Section 14; and the literature directed toward completion of the solution of Ericksen's problem for incompressible materials is summarized there.

Of course, it is sometimes possible that a nonhomogeneous deformation may be controlled without body force in a specific type of compressible or incompressible, isotropic material, but the same deformation generally cannot be maintained by surface tractions alone in every isotropic material. Hence, this particular nonhomogeneous deformation is not universal. A nonuniversal inverse solution is illustrated in Section 15.

The simple shear of a block, described in Section 14, is an important example of a universal deformation possible in every compressible and incompressible isotropic material. The simple shear is characterized by a formula relating the shear stress to the normal stress difference, a rule which is independent of the material elasticities. Hence, this relation is universal. A class of universal relations that includes the rule for simple shear is discussed in Section 16. The role of universal relations in experiments also is described there.

Examples presented in Section 17 illustrate Truesdell's problem concerning restrictions to be imposed on constitutive equations as tools for the objective evaluation of the physical content of theoretical results. The problem questions what restrictions are to be imposed on the strain energy function in order to ensure in analysis meaningful characterization of physical response, and to guarantee appropriate smoothness and existence of problem solutions. For an example, the empirical inequalities and their application in the physical interpretation of theoretical results are described. In fact, these inequalities are applied several times prior to our recognizing their formal connection with Truesdell's problem in Section 17. The Baker–Ericksen inequalities and the strong ellipticity condition are additional examples of restrictions that are discussed there.

Antiplane shear is not a universal, controllable deformation. A hyperelastic material may be capable of sustaining antiplane shear deformations if and only if certain conditions on the form of the strain energy function may be satisfied. Conditions necessary and sufficient for both compressible and incompressible, isotropic and homogeneous hyperelastic materials to support general and axisymmetric, antiplane shear deformations are presented, and the results are illustrated in an important example.

The reduction from the general theory to the classical linear theory of isotropic elasticity is illustrated in my *AMR* review. Caution necessary in relating the response functions to the classical Lamé moduli is demonstrated there. We shall not revisit this topic here.

An important feature of the theory, noted earlier, is that uniqueness of solution of general boundary value problems is not to be expected in the large. But uniqueness in the sense of small superimposed deformations may be established by introduction of a suitable stability criterion. Some basic criteria found in elastic stability theory are outlined in Section 18, and the relation between uniqueness and stability is discussed there. Two deficiencies of Euler's criterion are noted. The snap-through instability problem of a spherical rubber shell is discussed; and the unsolved Willis instability problem for a short, thick-walled tube is described. The review concludes with mention of some additional references to related topics for further study.

2. Kinematics of Finite Deformation

A body $\mathcal{B} = \{P_k\}$ is a set of material points P_k called particles. A reference frame is a set $\psi = \{O; \mathbf{e}_i\}$ consisting of an origin point O and an orthonormal vector basis $\mathbf{e}_i$ in an Euclidean space of three dimensions. The motion of a particle P relative to ψ is described by the time locus of its position vector $\mathbf{x}(P,t)$ relative to ψ. This locus is the trajectory or path of P in ψ. When the choice of reference frame is clear, as it is when only one frame is being used, its special mention may be omitted.

A typical particle P may be identified by its position vector $\mathbf{X}(P)$ in ψ at some reference time t_R, say. The domain κ_R of $\mathbf{X}$, the region in Euclidean space occupied by $\mathcal{B}$ at the time t_R, is called a reference configuration of $\mathcal{B}$. Then, relative to ψ, the motion of a typical particle P from κ_R is described by the vector function

$$\mathbf{x} = \boldsymbol{\chi}(\mathbf{X},t). \tag{1}$$

The domain κ of $\mathbf{x}$, the region in Euclidean space occupied by $\mathcal{B}$ at the time t, is called the current configuration of $\mathcal{B}$. Hence, as shown in Fig. 1, $\mathbf{x}$ denotes the place at time t in the current configuration κ which is occupied by the particle P whose place was $\mathbf{X}$ in the reference configuration κ_R. When no confusion may result, we shall write $\mathbf{x}$ for the function $\boldsymbol{\chi}$.

The velocity and acceleration of a particle P relative to ψ are defined by

$$\mathbf{v}(\mathbf{X},t) \equiv \dot{\mathbf{x}}(\mathbf{X},t), \tag{2a}$$

$$\mathbf{a}(\mathbf{X},\mathbf{t}) \equiv \dot{\mathbf{v}}(\mathbf{X},t) = \ddot{\mathbf{x}}(\mathbf{X},t), \tag{2b}$$

respectively. As usual, $\cdot \equiv \partial/\partial t$ denotes the material time derivative, the time rate of change following the particle P.

We shall assume henceforward that the body is a contiguous collection of

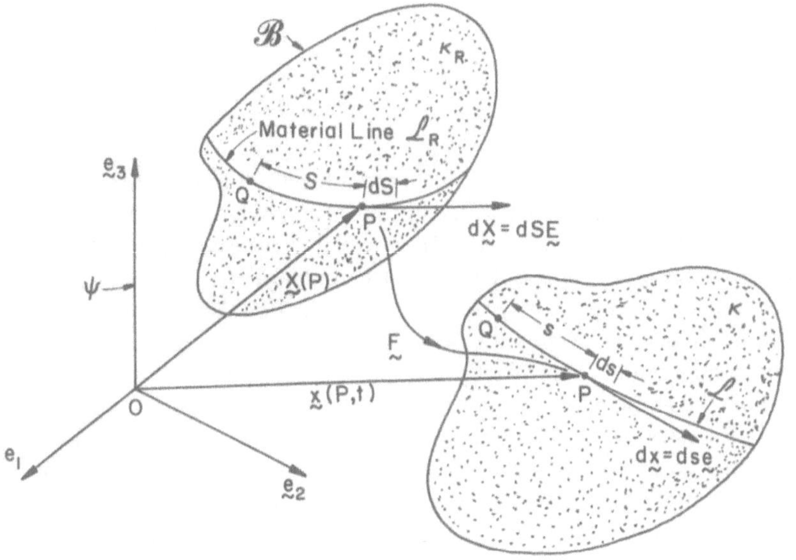

Fig. 1. Deformation **F** of a material line $\mathscr{L}_R$ in the reference configuration κ_R into its image $\mathscr{L}$ in the current configuration κ.

particles; we call this body a continuum. It is assumed that $\boldsymbol{\chi}$ is a smooth one-to-one map of every material point of $\kappa_R \rightarrow \kappa$ with

$$J \equiv \det \mathbf{F} > 0, \tag{3}$$

in which

$$\mathbf{F} \equiv \partial \mathbf{x}/\partial \mathbf{X} = \operatorname{grad} \mathbf{x} \tag{4}$$

is called the deformation gradient. This tensor transforms the tangent element $d\mathbf{X}$ of a material line $\mathscr{L}_R$ in κ_R into the tangent element $d\mathbf{x}$ of its deformed image line $\mathscr{L}$ in κ, as shown in Fig. 1. Hence,

$$d\mathbf{x} = \mathbf{F}d\mathbf{X}. \tag{5}$$

Let $|d\mathbf{x}| = ds$ and $|d\mathbf{X}| = dS$, where s and S are the arc length parameters for $\mathscr{L}$ and $\mathscr{L}_R$, respectively. Then (5) may be written as

$$\lambda \mathbf{e} = \mathbf{F}\mathbf{E}, \tag{6}$$

in which $\mathbf{e} \equiv d\mathbf{x}/ds$ and $\mathbf{E} \equiv d\mathbf{X}/dS$ are unit vectors tangent to $\mathscr{L}$ and $\mathscr{L}_R$ at $\mathbf{x}$

and **X**, as shown in Fig. 1; and

$$\lambda = \frac{ds}{dS} \tag{7}$$

is named the stretch, the ratio of the current length ds to the reference length dS of the material element. These lengths are commonly called the deformed and undeformed lengths, respectively. However, it is not essential that the reference configuration be an undistorted reference configuration, nor one that the body need actually occupy at any time during its motion. The natural, undeformed state of an automobile tire at ease on the store rack may be chosen as a reference configuration. This plainly is a possible reference configuration for the tire; but when mounted on a car wheel, it may never again occupy this special reference state. Of course, the inflated, toroidal state of the tire on the wheel also may be named as a reference configuration. But after the tire has been loaded against the road surface, it may never again occupy the toroidal configuration at any time during its motion. Indeed, the reference configuration must be one that the body can actually occupy, but it is not necessary that it ever do so.

It is seen that the relation (6) expresses the physical result that **F** rotates **E** into the direction **e** and stretches it by an amount $0 < \lambda < \infty$. This is essentially the substance of the more general and physically useful polar decomposition theorem of linear algebra (cf. Ogden, Ref. 6, p. 92; Bowen and Wang, Ref. 7, p. 140) applied pointwise to the nonsingular tensor **F**,

$$\mathbf{F} = \mathbf{RU} = \mathbf{VR}. \tag{8}$$

The proper orthogonal tensor **R** characterizes the local rigid body rotation of a material element. The positive, symmetric tensors **U** and **V** describe the local deformation of the element; they are called the right and left stretch tensors, respectively. The decomposition in (8) of the deformation gradient **F** into a pure stretch **U** at **X** followed by a rigid body rotation **R**, or by the same rigid rotation followed by a pure stretch **V** at **x**, is unique.

Because **U** and **V** usually are tedious to compute, it is customary to use their squares,

$$\mathbf{C} \equiv \mathbf{F}^T\mathbf{F} = \mathbf{U}^2, \tag{9a}$$

$$\mathbf{B} \equiv \mathbf{F}\mathbf{F}^T = \mathbf{V}^2. \tag{9b}$$

These corresponding positive, symmetric tensors are respectively known as the right and left Cauchy–Green deformation tensors. By (8) and (9), it is seen that **U** and **V**, hence also **C** and **B**, are similar tensors, that is,

$$\mathbf{V} = \mathbf{R}\mathbf{U}\mathbf{R}^{T}, \tag{10a}$$

$$\mathbf{B} = \mathbf{R}\mathbf{C}\mathbf{R}^{T}. \tag{10b}$$

It follows that **U** and **V** {**C** and **B**} have the same principal values λ_k $\{\lambda_k^2\}$ and respective principal directions $\boldsymbol{\mu}$ and $\boldsymbol{\nu}$ related by the rotation **R**,

$$\boldsymbol{\nu} = \mathbf{R}\boldsymbol{\mu}. \tag{11}$$

The λ_k are the stretches of the three principal material lines; they are called the principal stretches.

Formulas relating the respective material surface area and volume elements $d\mathbf{a}$ and dv in κ to their reference images $d\mathbf{A}$ and dV in κ_R may be easily derived by application of (5). Recalling (3), we find that

$$d\mathbf{a} = J\mathbf{F}^{-T}d\mathbf{A}, \tag{12a}$$

$$dv = JdV. \tag{12b}$$

The last relation shows that det **F** in (3) is the ratio of the current (deformed) volume to the reference (usually undeformed) volume of a material element. Therefore, the deformation is isochoric if and only if $J = 1$. It is evident on physical grounds that $0 < \det \mathbf{F} < \infty$.

The material time rate of change of the deformation of a continuum is described by the velocity gradient tensor **L**,

$$\mathbf{L} \equiv \operatorname{grad} \dot{\mathbf{x}} = \dot{\mathbf{F}}\mathbf{F}^{-1}, \tag{13}$$

where $\operatorname{grad} \equiv \partial/\partial\mathbf{x}$.The symmetric part **D** and antisymmetric part **W** of **L** are known as the stretching and spin tensors, respectively.

This completes the sketch of basic kinematics essential for our study of finite deformations of an elastic solid. Additional details may be found in Atkin and Fox (Ref. 8), Marsden and Hughes (Ref. 9), Ogden (Ref. 6), Truesdell (Ref. 10), Truesdell and Noll (Ref. 11, Sections 21–23), and Wang and Truesdell (Ref. 12). The balance principles for continua will be outlined next. We begin with Euler's laws of motion.

3. Cauchy Stress Principle and Equations of Motion

The forces that act on any part $\mathcal{P} \subset \mathcal{B}$ of a continuum $\mathcal{B}$ are of two kinds: a distribution of contact force $\mathbf{t}_n$ per unit area of the boundary $\partial\mathcal{P}$ of $\mathcal{P}$ in κ, and a distribution of body force $\mathbf{b}$ per unit volume of $\mathcal{P}$ in κ. The total force $\mathcal{F}(\mathcal{P}, t)$ and the total torque $\mathcal{T}(\mathcal{P}, t)$ acting on the part $\mathcal{P}$ are related to the momentum and the moment of momentum of the material points of $\mathcal{B}$ in an inertial frame ψ in accordance with Euler's laws of motion,

$$\mathcal{F}(\mathcal{P}, t) = \int_{\partial\mathcal{P}} \mathbf{t}_n da + \int_{\mathcal{P}} \mathbf{b} dv = \frac{d}{dt} \int_{\mathcal{P}} \mathbf{v} dm, \tag{14}$$

$$\mathcal{T}(\mathcal{P}, t) = \int_{\partial\mathcal{P}} \mathbf{x} \times \mathbf{t}_n da + \int_{\mathcal{P}} \mathbf{x} \times \mathbf{b} dv = \frac{d}{dt} \int_{\mathcal{P}} \mathbf{x} \times \mathbf{v} dm. \tag{15}$$

The moments in (15) are to be computed with respect to the origin in ψ. Herein we recall (2a), and note that $dm = \rho dv$ is the material element of mass with density ρ per unit volume in κ.

The principle of balance of mass requires also that $dm = \rho_R dV$, where ρ_R is the density of mass per unit volume V in κ_R. Therefore, recalling (12b), one finds that the respective mass densities are related by the local equation of continuity,

$$\rho_R = J\rho. \tag{16}$$

Application of the first law (14) to an arbitrary tetrahedral element leads to Cauchy's stress principle,

$$\mathbf{t}_n = \mathbf{T}\mathbf{n}, \tag{17}$$

where $\mathbf{n}$ is the exterior unit normal vector to $\partial\mathcal{P}$ in κ. Hence, the traction or stress vector $\mathbf{t}_n$ is a linear transformation of the unit normal $\mathbf{n}$ by the Cauchy stress tensor $\mathbf{T}$. Use of (2b), (17), and the divergence theorem in (14) yields Cauchy's first law of motion,

$$\operatorname{div} \mathbf{T} + \mathbf{b} = \rho \mathbf{a}. \tag{18}$$

The second law (15) together with (17) and (18) yields the equivalent local moment balance condition restricting the Cauchy stress $\mathbf{T}$ to the space of symmetric tensors,

$$\mathbf{T} = \mathbf{T}^T. \tag{19}$$

This is known as Cauchy's second law. See Ogden (Ref. 6, Chapter 3) and Atkin and Fox (Ref. 8, Sections 1.10 and 1.11) for particulars.

The Cauchy stress characterizes the contact force distribution $\mathbf{t}_n$ in κ per unit current area in κ; but this often is inconvenient in solid mechanics because the deformed configuration generally is not known *a priori*. Therefore, the engineering stress tensor $\mathbf{T}_R$, also known as the first Piola–Kirchhoff stress tensor, is introduced to define the contact force distribution $\mathbf{t}_N \equiv \mathbf{T}_R\mathbf{N}$ in κ per unit reference area in κ_R, where $\mathbf{N}$ is the exterior unit normal vector to $\partial\mathcal{P}$ in κ_R whose image in κ is $\mathbf{n}$. Thus, for the same contact force $d\mathcal{F}_c(\mathcal{P}, t)$, we must have

$$d\mathcal{F}_c(\mathcal{P}, t) \equiv \mathbf{t}_n da = \mathbf{T}\mathbf{n} da = \mathbf{T}_R\mathbf{N} dA = \mathbf{t}_N dA.$$

The vector $\mathbf{t}_N$ is named the engineering stress vector. We thus obtain with (12a) the rule

$$\mathbf{T}_R = J\mathbf{T}\mathbf{F}^{-T}, \tag{20}$$

relating the engineering and Cauchy stress tensors.

The stress principle and balance laws corresponding to (17), (18), and (19) become

$$\mathbf{t}_N = \mathbf{T}_R\mathbf{N}, \tag{21}$$

$$\operatorname{Div} \mathbf{T}_R + \mathbf{b}_R = \rho_R \mathbf{a}, \tag{22}$$

$$\mathbf{T}_R\mathbf{F}^T = \mathbf{F}\mathbf{T}_R^{\,T}. \tag{23}$$

Hence, the engineering stress $\mathbf{T}_R$ generally is not symmetric. In (22) the body force per unit volume in κ_R is defined by $\mathbf{b}_R \equiv J\mathbf{b}$; and Div denotes the divergence operator with respect to $\mathbf{X}$ in κ_R, whereas in (18) div is with respect to $\mathbf{x}$ in κ.

Thus far, the deformation of a continuum and the actions that produce it have been treated separately without mention of any specific material characteristics the body may possess. Of course, the inherent constitutive nature of the material dictates its deformation response to action by forces and torques. For a specific class of materials, the inherent relationship between the deformation $\mathbf{F}$, the rate of deformation $\dot{\mathbf{F}}$, and the stress $\mathbf{T}$ or $\mathbf{T}_R$, say, is described by an equation known as a constitutive equation. In the next section, the principle of balance of mechanical energy will be applied to derive the constitutive equation for a special class of perfectly elastic materials called hyperelastic solids.

4. Mechanical Energy Principle and Hyperelasticity

We shall ignore for the sake of simplicity all thermal effects and adopt the following mechanical energy principle: The time rate of change of the total mechanical energy $E(\mathcal{P}, t)$ for any part $\mathcal{P}\subset\mathcal{B}$ of a body $\mathcal{B}$ is balanced by the total mechanical power $\prod(\mathcal{P}, t)$,

$$\dot{E}(\mathcal{P}, t) = \prod(\mathcal{P}, t). \tag{24}$$

The mechanical power is the rate of working of the applied forces; and the total mechanical energy consists of the total kinetic energy of $\mathcal{P}$ and the total elastic potential energy with density[1] $\Sigma(\mathbf{X}, t)$ per unit volume in κ_R. Hence, in these terms, (24) may be written as

$$\frac{d}{dt}\left[\int_{\mathcal{P}} \frac{1}{2}\mathbf{v}\cdot\mathbf{v}dm + \int_{\mathcal{P}} \Sigma dV\right] = \int_{\partial\mathcal{P}} \mathbf{t_n}\cdot\mathbf{v}da + \int_{\mathcal{P}} \mathbf{b}\cdot\mathbf{v}dv. \tag{25}$$

Then, for conserved mass and with the use of (12b), (17), the divergence theorem, (18), (20), and (13), it turns out that the application of the global relation (25) to an arbitrary part $\mathcal{P}\subset\mathcal{B}$ yields the following differential equation of mechanical energy balance:

$$\dot{\Sigma}(\mathbf{X}, t) = \operatorname{tr}(\mathbf{T}_R\dot{\mathbf{F}}^T) \equiv \mathbf{T}_R\cdot\dot{\mathbf{F}}. \tag{26}$$

A hyperelastic solid is a material whose elastic potential energy is given by the following strain energy function:

$$\Sigma(\mathbf{X}, t) = \Sigma(\mathbf{F}(\mathbf{X}, t), \mathbf{X}). \tag{27}$$

Then use of (27) in (26) yields the identity

$$\left[\mathbf{T}_R - \frac{\partial\Sigma}{\partial\mathbf{F}}\right]\cdot\dot{\mathbf{F}} = 0, \tag{28}$$

which must hold for all $\dot{\mathbf{F}}$. Hence, the principle of mechanical energy balance and the constitutive assumption (27) yield the following general constitutive

[1]The elastic potential energy with density ϵ per unit mass also is used often. In this case, $\Sigma = \rho_R\epsilon$ everywhere below, and $\Sigma dV = \epsilon dm$ in (25). See also (16).

equation for a hyperelastic solid:

$$\mathbf{T}_{\mathrm{R}} = \frac{\partial \Sigma(\mathbf{F})}{\partial \mathbf{F}}. \tag{29}$$

This rule relates the engineering stress and the deformation. Here and in subsequent equations explicit indication of the possible dependence of the strain energy on the material point $\mathbf{X}$ will be suppressed. Of course, (23) also must be respected.

Use of (20) in (29) provides an alternative form of the constitutive equation that relates the symmetric Cauchy stress to the deformation of a hyperelastic solid,

$$\mathbf{T} = J^{-1} \frac{\partial \Sigma}{\partial \mathbf{F}} \mathbf{F}^{T}. \tag{30}$$

5. Change of Frame and Material Frame Indifference

A change of frame from $\psi = \{O; \mathbf{e}_k\}$ into $\hat{\psi} = \{\hat{O}; \hat{\mathbf{e}}_k\}$ is a linear transformation defined by

$$\hat{\mathbf{x}} = \mathbf{c}(t) + \mathbf{Q}(t)\mathbf{x}, \tag{31}$$

where $\mathbf{c}(t)$ is the position vector of O from $\hat{O}$ and $\mathbf{Q}(t)$ is a (proper orthogonal) rigid body rotation of ψ relative to $\hat{\psi}$, as illustrated in Fig. 2. We recall that $\hat{\mathbf{x}} = \hat{\chi}(\mathbf{X}, t)$ and $\mathbf{x} = \chi(\mathbf{X}, t)$ describe the same motion of the material point $\mathbf{X}$ but referred to $\hat{\psi}$ and ψ, respectively. For the purpose here, we shall lose no generality by supposing that the reference configuration is the same for both observers whose frames may be chosen to coincide at the reference instant t_{R}, say, and that both observers use the same clock so that $\hat{t} = t$. See Beatty (Ref. 13, pp. 308–317) for further details. Relations for the velocity and acceleration under a change of frame also are presented there.

The deformation gradient (4) under the change of frame (31) transforms as

$$\hat{\mathbf{F}} = \mathbf{QF}. \tag{32}$$

Hence, in obvious notation, the Cauchy–Green deformation tensors in (9) transform as

$$\hat{\mathbf{C}} = \mathbf{C}, \tag{33a}$$

$$\hat{\mathbf{B}} = \mathbf{QBQ}^{T}. \tag{33b}$$

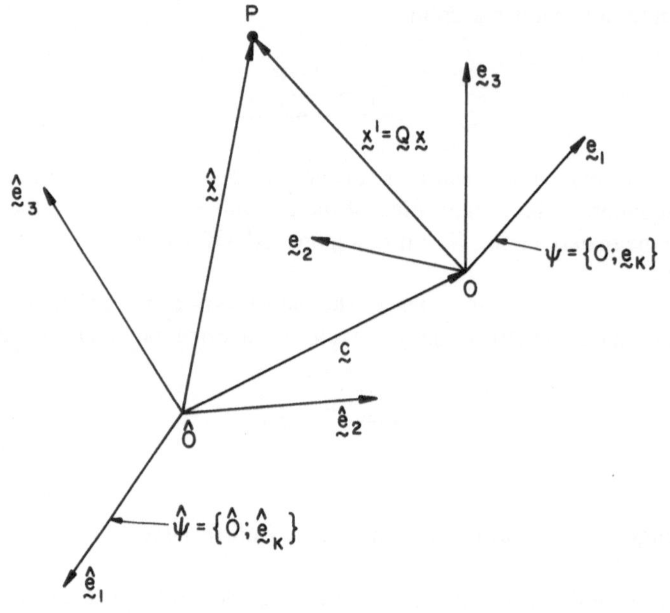

Fig. 2. Change of reference frame $\psi \to \hat{\psi} : \hat{\mathbf{x}}(\mathbf{X}, t) = \mathbf{c}(t) + \mathbf{Q}(t)\mathbf{x}(\mathbf{X}, t)$.

The first of (33) reflects the use of the same reference configuration to which both $\mathbf{C}$ and $\hat{\mathbf{C}}$ are referred.

It is an axiom that the strain energy of $\mathcal{B}$ is indifferent to the observer. This principle of material frame indifference (Truesdell and Noll, Ref. 11, Section 19) applied to the strain energy function (27) implies that

$$\Sigma(\hat{\mathbf{F}}) = \Sigma(\mathbf{QF}) = \Sigma(\mathbf{F}) \tag{34}$$

must hold for all proper orthogonal $\mathbf{Q}$ and for all $\mathbf{F}$. Therefore, we may choose $\mathbf{Q} = \mathbf{R}^T$ in the polar decomposition $\mathbf{F} = \mathbf{RU}$ to deduce the necessary condition

$$\Sigma(\mathbf{F}) = \Sigma(\mathbf{U}) = \tilde{\Sigma}(\mathbf{C}), \tag{35}$$

wherein $\mathbf{U} = \mathbf{C}^{1/2}$ follows from (9). Conversely, if the strain energy has the form (35), on replacing $\mathbf{F}$ by $\hat{\mathbf{F}}$ and recalling (33a), we may recover (34) for all changes of frame. Thus, the strain energy function for a hyperelastic material is frame indifferent if and only if it has the reduced form (35).

Therefore, use of (35) in (29), or (30), yields the following reduced form of the constitutive equation for a hyperelastic solid:

$$\mathbf{T}_{\mathrm{R}} = 2\mathbf{F}\frac{\partial\tilde{\Sigma}(\mathbf{C})}{\partial\mathbf{C}} \quad \text{or} \quad \mathbf{T} = 2J^{-1}\mathbf{F}\frac{\partial\tilde{\Sigma}(\mathbf{C})}{\partial\mathbf{C}}\mathbf{F}^{T}. \tag{36}$$

The rule (36) provides the most general form of the constitutive equation that respects the principle of frame indifference. However, it does not yet reflect any inherent symmetries that the material may possess in its response to loading. The effect of material symmetry on the form of the constitutive equation (36) will be described next.

6. Material Symmetry Transformations

Let us imagine that a hyperelastic body is subjected to two tests. In the first experiment, the body is deformed at $\mathbf{X}$ in κ_{R} by a deformation $\mathbf{F}$ relative to ψ so that its stored energy (35) is given by

$$\Sigma_1 \equiv \Sigma(\mathbf{F}) = \tilde{\Sigma}(\mathbf{C}). \tag{37}$$

Now let us return to the same reference configuration κ_{R} to do a second experiment on the same body at the place $\mathbf{X}$. This time, we first subject the body to a specified rigid rotation $\mathbf{Q}$ in frame ψ and afterwards deform it at $\mathbf{X}$ by the same deformation $\mathbf{F}$ used earlier. Thus, the deformation in this experiment is applied in the same manner as before, but now the body has a different orientation in frame ψ so that

$$\hat{\mathbf{F}} = \mathbf{FQ}. \tag{38}$$

In this case, (9a) yields

$$\hat{\mathbf{C}} = \hat{\mathbf{F}}^T\hat{\mathbf{F}} = \mathbf{Q}^T\mathbf{F}^T\mathbf{FQ} = \mathbf{Q}^T\mathbf{CQ},$$

and the stored energy (35) in our second experiment is

$$\Sigma_2 \equiv \Sigma(\hat{\mathbf{F}}) = \tilde{\Sigma}(\mathbf{Q}^T\mathbf{CQ}). \tag{39}$$

In general, of course, we do not expect that the test results (37) and (39) will be the same, i.e., in general it will not happen that $\Sigma_1 = \Sigma_2$. On the other hand, if it does occur that the strain energy is the same for each $\mathbf{F}$ in both experiments, so that the material response at $\mathbf{X}$ is the same before and after the given rigid rotation **Q**, then by (37) and (39) we shall have

$$\tilde{\Sigma}(\mathbf{C}) = \tilde{\Sigma}(\mathbf{Q}^T\mathbf{C}\mathbf{Q}). \tag{40}$$

The special rotation **Q** for which (40) holds is called a material symmetry transformation. It is easy to show that the set of all material symmetry transformations at **X** form a subgroup g_X of the group $\mathscr{G}$ of proper orthogonal tensors. The group g_X is appropriately named the symmetry group at **X** (Gurtin, Ref. 14, Section 21; Ref. 15). The term isotropy group also has been used for this purpose (Truesdell and Noll, Ref. 11, Sections 31, 85; Truesdell, Ref. 10, Chapter 6).

It is important to notice that the symmetry group depends on the reference configuration from which the basal experiments were executed. Since the symmetry group will change with the reference state, and hence with the deformation of the material, it is important to fix the reference configuration relative to which the material symmetry holds. It seems natural, therefore, that the undistorted state of vanishing deformation for which $\mathbf{F} = \mathbf{1}$ should be chosen as the reference state that identifies the inherent symmetries of the material. We shall assume henceforward (at least initially) that the symmetry group is expressed relative to the undistorted state of the material.

If a material has no symmetries whatsoever in its undistorted state at **X**, then $g_X = \{\mathbf{1}\}$; and the material is called triclinic at **X**. A material that has a reflective axis of symmetry in its undistorted state at **X** is characterized by the symmetry group $g_X = \{\mathbf{Q}\colon \mathbf{QH} = \pm\mathbf{H}$ for a fixed direction **H** in $\kappa_R\}$. Thus, this material is said to be transversely isotropic at **X**. Finally, if $g_X = \mathscr{G}$, then every direction at **X** in the undistorted state is an axis of material symmetry. A material having this property is called isotropic at **X**. When the material symmetry group is the same at each particle throughout the body in its undistorted state, the material is identified briefly through its local symmetry group name. For example, a hyperelastic material which is isotropic at every material point in a global undistorted state is called an isotropic hyperelastic material. This most important class of materials will be examined next.

7. Isotropic Hyperelastic Materials

For an isotropic hyperelastic material, the rule (40) must hold for all proper orthogonal tensors **Q**. Therefore, we may take $\mathbf{Q} = \mathbf{R}^T$ in the polar decomposition of **F** and recall (10b) to deduce for every isotropic material the necessary condition

$$\tilde{\Sigma}(\mathbf{C}) = \tilde{\Sigma}(\mathbf{B}). \tag{41}$$

The rule (41) shows that for a given deformation the strain energy function has the same values whether **C** or **B** is used as the independent variable. But for an arbitrary deformation, **C** and **B** generally are distinct tensors. On the other hand, though their principal directions generally are different, their principal values always are the same. Hence, it is the principal invariants $I_k(\mathbf{C})$ of **C** and $I_k(\mathbf{B})$ of **B** that are the same for every deformation **F**. Therefore, (41) suggests that the strain energy must be an isotropic scalar-valued function of these principal invariants alone,

$$\tilde{\Sigma}(\mathbf{C}) = \tilde{\Sigma}(\mathbf{B}) = \Sigma(I_1, I_2, I_3), \tag{42}$$

wherein, specifically,

$$I_1 = \operatorname{tr}\mathbf{B}, \qquad I_2 = \tfrac{1}{2}\,[I_1^2 - \operatorname{tr}(\mathbf{B}^2)], \qquad I_3 = \det \mathbf{B}. \tag{43}$$

On replacing C by $\mathbf{Q}^T\mathbf{C}\mathbf{Q}$ in (42) for all orthogonal **Q**, we shall recover (40). Hence, a hyperelastic material is isotropic if and only if its strain energy function has the representation (42) relative to an undistorted state. The foregoing heuristic argument leading to (42) can be made precise by use of the representation theorem on isotropic scalar-valued functions of a symmetric tensor applied to the rule (40). See Truesdell and Noll (Ref. 11, Section 10) and Truesdell (Ref. 10, Chapter 11).

Bearing in mind (41), we see from (36) that the constitutive equation for an isotropic hyperelastic material may be written as

$$\mathbf{T}_R = 2\mathbf{F}\,\frac{\partial\tilde{\Sigma}}{\partial\mathbf{C}} = 2\,\frac{\partial\tilde{\Sigma}}{\partial\mathbf{B}}\,\mathbf{F} \quad \text{or} \quad \mathbf{T} = 2J^{-1}\,\frac{\partial\tilde{\Sigma}}{\partial\mathbf{B}}\,\mathbf{B}. \tag{44}$$

The last of (44) is expressed entirely in terms of the symmetric tensor **B**, whereas the first involves the local rigid body rotation **R**. Consequently, the last of (44) usually is considered more desirable. Indeed, use of (42) in the last of (44) reveals the following two useful forms of the general constitutive equation for an isotropic hyperelastic solid,

$$\mathbf{T} = \alpha_0\mathbf{1} + \alpha_1\mathbf{B} + \alpha_2\mathbf{B}^2, \tag{45}$$

or, by use of the Cayley–Hamilton theorem,

$$\mathbf{T} = \beta_0\mathbf{1} + \beta_1\mathbf{B} + \beta_{-1}\mathbf{B}^{-1}. \tag{46}$$

The scalar coefficients

$$\alpha_\Lambda = \alpha_\Lambda(I_1, I_2, I_3), \qquad \beta_\Gamma = \beta_\Gamma(I_1, I_2, I_3), \tag{47}$$

where $\Lambda = 0, 1, 2$, and $\Gamma = 0, 1, -1$ are called the material or elastic response functions. These are given in terms of the strain energy function by

$$\beta_0 = \alpha_0 - I_2\alpha_2 = \frac{2}{\sqrt{I_3}}\left[I_2\frac{\partial\Sigma}{\partial I_2} + I_3\frac{\partial\Sigma}{\partial I_3}\right], \tag{48a}$$

$$\beta_1 = \alpha_1 + I_1\alpha_2 = \frac{2}{\sqrt{I_3}}\frac{\partial\Sigma}{\partial I_1}, \tag{48b}$$

$$\beta_{-1} = I_3\alpha_2 = -2\sqrt{I_3}\frac{\partial\Sigma}{\partial I_2}. \tag{48c}$$

It follows from (45) or (46) that in an undistorted state κ_R on which $\mathbf{B} = \mathbf{1}$, the stress need not vanish. Rather, the stress on an undistorted state of an isotropic material is at most a hydrostatic stress $\mathbf{T}_0$ given by

$$T_0 = (\hat{\beta}_0 + \hat{\beta}_1 + \hat{\beta}_{-1})\mathbf{1}, \tag{49}$$

where $\hat{\beta}_\Gamma = \beta_\Gamma(3, 3, 1)$ are the values of the material functions (47) in κ_R. An undistorted reference configuration on which the stress $\mathbf{T}_0 = \mathbf{0}$ is called a stress-free or natural state of the material. Thus, in the natural state, the response functions must satisfy

$$\hat{\beta}_0 + \hat{\beta}_1 + \hat{\beta}_{-1} = 0. \tag{50}$$

Many problems may be investigated without further specification of the material functions, which is most desirable. When this is impossible, special models having experimental foundation generally are used. The determination of the response functions for specific kinds of materials is a principal problem in experimental mechanics. Some special varieties of isotropic materials will be described below.

8. Blatz–Ko Constitutive Equation

It is sometimes useful to replace the principal invariants I_k by another set of independent invariants of $\mathbf{B}$. An example is provided by the invariants J_k defined by

$$J_1 \equiv I_1 = \mathrm{tr}\mathbf{B}, \qquad J_2 \equiv I_2/I_3 = \mathrm{tr}\mathbf{B}^{-1}, \qquad J_3 \equiv I_3^{1/2} = \det\mathbf{F}. \tag{51}$$

In this case, the strain energy function (42) may be rewritten as another function,

$$\Sigma = W(J_1, J_2, J_3). \tag{52}$$

Introducing (51) and (52) into (48) and retaining the same notation for the material functions β_k, we find that

$$\beta_0 = \frac{\partial W}{\partial J_3}, \qquad \beta_1 = \frac{2}{J_3}\frac{\partial W}{\partial J_1}, \qquad \beta_{-1} = -\frac{2}{J_3}\frac{\partial W}{\partial J_2}. \tag{53}$$

These are the elastic response functions in the constitutive equation (46) for the strain energy (52).

Let us consider a special class of materials whose response functions in (53) depend on J_3 alone. Then bearing in mind the assumed functional dependence $\beta_\Gamma = \beta_\Gamma(J_3)$, it may be seen that (53) will hold when and only when $2\partial W/\partial J_1 = \alpha$ and $2\partial W/\partial J_2 = \beta$ are constants. Thus, writing $\partial W/\partial J_3 \equiv W_3(J_3)$, we obtain from (53) the following response functions for this special material:

$$\beta_0 = W_3(J_3), \qquad \beta_1 = \frac{\alpha}{J_3}, \qquad \beta_{-1} = -\frac{\beta}{J_3}. \tag{54}$$

It can be shown that $\beta_1(1) - \beta_{-1}(1) = \alpha + \beta = \mu_0$ is the usual constant shear modulus in the natural state of an isotropic material (Truesdell and Noll, Ref. 11, Section 50). We shall return to this later. Thus, on introducing

$$\alpha \equiv \mu_0 f, \qquad \beta \equiv \mu_0 (1 - f), \tag{55}$$

where f is another constant, and substituting (54) into (46), we reach the general form of the constitutive equation for this special class of isotropic hyperelastic materials,

$$\mathbf{T} = W_3(J_3)\,\mathbf{1} + \frac{\mu_0 f}{J_3}\mathbf{B} - \frac{\mu_0(1-f)}{J_3}\mathbf{B}^{-1}. \tag{56}$$

The stress (56) will vanish in the undistorted state when and only when the response coefficients satisfy

$$W_3(1) + \mu_0(2f - 1) = 0. \tag{57}$$

Equation (56), derived by Beatty and Stalnaker (Ref. 16), was first introduced in an altogether different way by Blatz and Ko (Ref. 2). Therefore, a material described by the constitutive equation (56) is named a Blatz–Ko material.

Thus, as shown by Beatty and Stalnaker, the Blatz–Ko material is the unique hyperelastic material whose elastic response functions depend on J_3 alone.

8.1. Some Results Related to the Blatz–Ko Experiments. Experiments by Blatz and Ko (Ref. 2) on a certain foamed, polyurethane rubber revealed the specific response functions

$$\beta_0 = \mu_0, \tag{58a}$$

$$0 < \beta_1 << 1, \tag{58b}$$

$$\beta_{-1} = -\mu_0/J_3, \tag{58c}$$

where β_1 was considered negligible so that $f = 0$, very nearly, and $W_3 = \mu_0$ (= 32 psi). Thus, (57) holds and (56) reduces to the following constitutive equation for the Blatz–Ko foamed, polyurethane elastomer:

$$\mathbf{T} = \mu_0[\mathbf{1} - J_3^{-1}\,\mathbf{B}^{-1}]. \tag{59}$$

It is easy to show that a cylinder of this material subjected on its ends to a simple tensile (or compressive) loading with principal stress components $T_{33} = T$, $T_{11} = T_{22} = 0$, produces a corresponding extensional (or compressive) deformation with principal stretches $\lambda_3 = \lambda$, $\lambda_1 = \lambda_2$, provided that $\mu_0 > 0$. Hence, $J_3 = \lambda_1^2\lambda$ and (59) yields the two equations

$$T = \mu_0(1 - \lambda_1^{-2}\lambda^{-3}), \tag{60a}$$

$$\lambda_1(\lambda) = \lambda_1^{-1/4}. \tag{60b}$$

It follows from (17) and (18) that the traction-free condition $\mathbf{t}_n = \mathbf{0}$ on the lateral surface and the equilibrium equations without body force, namely, div $\mathbf{T} = \mathbf{0}$, are identically satisfied.

We notice that (60b) is independent of the single material response parameter μ_0 in (59)—it is an example of a universal relation in simple tension valid for every Blatz–Ko foamed rubber material (59). The universal relation (60b) determines uniquely the lateral contraction (or expansion) λ_1 as a function of the axial extension (or compression) λ, and in consequence the extension is called simple (Beatty and Stalnaker, Ref. 16). Hence, the equations (60) yield the following stress–stretch function $T(\lambda)$ and Poisson function $\nu(\lambda)$ for a simple tension (or compression) of the foamed, polyurethane elastomer (59):

$$T(\lambda) = \mu_0(1 - \lambda^{-5/2}), \tag{61a}$$

$$\nu(\lambda) = \frac{1 - \lambda^{-1/4}}{\lambda - 1}. \tag{61b}$$

The stress–stretch function (61a) is a monotone increasing function with a tangent modulus

$$E(\lambda) \equiv \frac{dT(\lambda)}{d\lambda} = \frac{5\mu_0}{2}\lambda^{-7/2}. \tag{62}$$

We thus find by (61b) and (62) that the foamed rubber (59) has a Young's modulus $E_0 \equiv \text{limit}_{\lambda\to 1} E(\lambda) = 5\mu_0/2$ and a Poisson ratio $\nu_0 \equiv \text{limit}_{\lambda\to 1}\nu(\lambda) = 1/4$, which is precisely the experimental value found by Blatz and Ko (Ref. 2). The Cauchy stress–stretch curve has a horizontal, asymptotic limit value for the stress as the stretch becomes indefinitely large, namely, $T(\lambda)|_{\lambda\to\infty} = 2E_0/5$. The Poisson function is monotone decreasing to zero as $\lambda \to \infty$, and it grows indefinitely as $\lambda \to 0$. Plots of the functions (61a) and (61b) are shown in Figs. 3 and 4, respectively.

The engineering stress $T_R(\lambda)$ in the simple extension may be found with the aid of (20). The result is shown in Fig. 3. The engineering stress behaves in tension quite differently from the Cauchy stress; it increases to a maximum value at the universal stretch $\lambda = 6^{2/5} = 2.048$ and then decreases monotonically to zero as $\lambda \to \infty$. But the two stresses are virtually indistinguishable in compression.

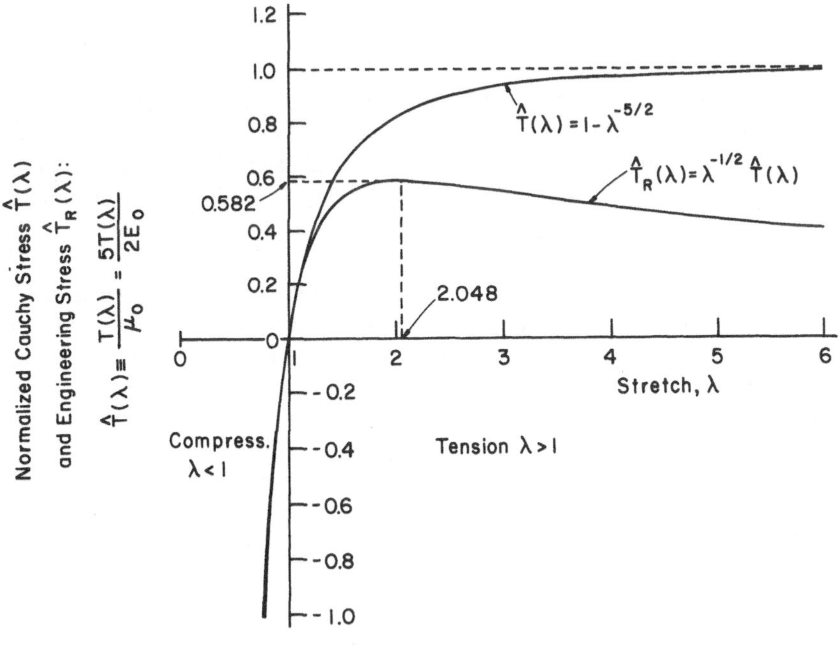

Fig. 3. Normalized Cauchy and engineering stress–stretch graphs for the Blatz–Ko foamed, polyurethane elastomer in a simple tension or compression.

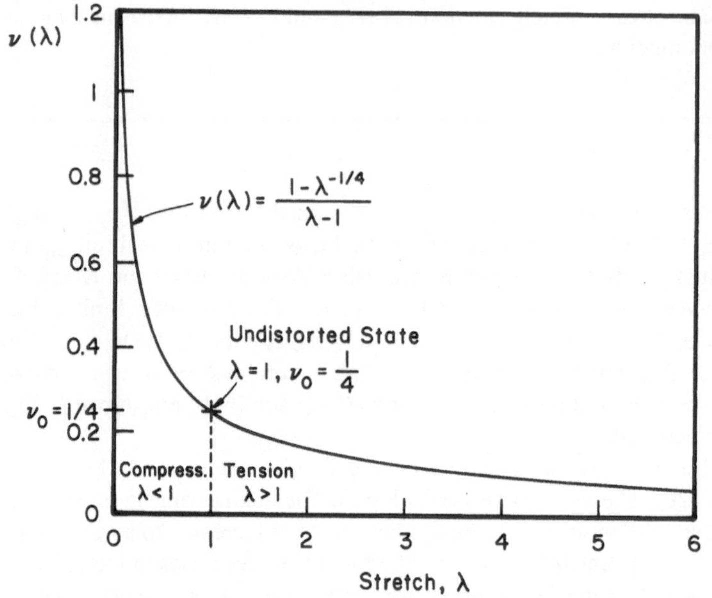

Fig. 4. Graph of the Poisson function for a Blatz–Ko foamed, polyurethane elastomer in a simple tension or compression.

Notice too that the normalized stress–stretch curves in Fig. 3 are the same for every Blatz–Ko material (59). Thus, if the stress–stretch data obtained in a simple tension of a given isotropic hyperelastic material fail to fall on the universal graphs in Figs. 3 and 4, then that particular material cannot be a candidate for inclusion in the special class of Blatz–Ko materials (59). Of course, it remains possible that the material may be a member of the larger class of Blatz–Ko materials (56).

Indeed, other tests by Blatz and Ko (Ref. 2) for a certain solid, polyurethane elastomer yielded the specific empirical constants $f = 1$ and $\alpha = \mu_0$ (= 34 psi). Thus, the response functions (54) for this material class are described by

$$\beta_0 = W_3(J_3), \qquad \beta_1 = \frac{\mu_0}{J_3}, \qquad \beta_{-1} = 0. \tag{63}$$

We thus derive from (56) the Blatz–Ko constitutive equation for the solid, polyurethane rubber,

$$\mathbf{T} = W_3(J_3)\,\mathbf{1} + \frac{\mu_0}{J_3}\mathbf{B}. \tag{64}$$

The natural state condition (57) becomes $W_3(1) = -\mu_0$; otherwise, the first response coefficient in (63) remains undetermined in this theoretical construction. We shall return to this later.

8.2. Poisson Function and Blatz–Ko Volume Control Relation. Blatz and Ko (Ref. 2) made no connection of their data with the Poisson function (61b); rather, as pointed out by Beatty and Stalnaker (Ref. 16), they used a clever ad hoc rule to determine ν_0. Their rule also enabled them to eliminate from (56) the undetermined material function $W_3(J_3)$ and thus achieve consistency of the theory with their experimental data. We shall now show by a different approach how this was accomplished.

For the simple tensile loading described before, but now applied to our general Blatz–Ko material (56), we obtain

$$T = W_3(J_3) + \frac{\mu_0 f}{J_3}\lambda^2 - \frac{\mu_0(1-f)}{J_3}\lambda^{-2}, \tag{65}$$

$$W_3(J_3) = \mu_0\left[-\frac{f}{\lambda} + \frac{(1-f)\lambda}{J_3^2}\right], \tag{66}$$

wherein $J_3 = \lambda_1^2\lambda$ was used to replace λ_1. To obtain these equations, it was necessary to recall Batra's Theorem (Refs. 17 and 18). He showed that a simple tension will always produce a corresponding equibiaxial stretch in every isotropic material provided that the following empirical inequalities hold:

$$\beta_1 > 0, \qquad \beta_{-1} \leq 0. \tag{67}$$

See Truesdell and Noll, Ref. 11, Section 51, and Beatty and Stalnaker, Ref. 16. The general Blatz–Ko material functions (54) adjusted as shown in (55) will satisfy (67) for all $J_3 > 0$ if and only if

$$\mu_0 > 0, \qquad 0 < f \leq 1. \tag{68}$$

These restrictions were not mentioned explicitly by Blatz and Ko (Ref. 2); yet (68) are essential in the biaxial deformation problems they studied. The conditions (68) have been assumed in our formulation of (65) and 66).

If (66) yields a unique relation $\lambda_1 = \lambda_1(\lambda)$ or $J_3 = J_3(\lambda)$, then, in the sense of Beatty and Stalnaker (Ref. 16), the extension will be simple and (65) will determine a unique stress–stretch relation $T(\lambda)$. However, at this point, since $W_3(J_3)$ is unknown we can do no more than place (66) into (65) to obtain the stress as a function $T(\lambda, \lambda_1)$ of the two unrelated stretches. We recall from classical elasticity theory that in a simple tension the lateral contractive stretch is related to the longitudinal stretch by Poisson's ratio, which is identified as a material response constant. Hence, with the usual inequalities imposed on the moduli of the classical theory, every simple tension of an isotropic material will produce a simple extension. A similar approach is used here.

If we had J_3 as a function of λ in (66), then the extension would be simple and (65) and (66) would yield $T(\lambda)$ alone. We observe from (60b), for example, that for the foamed rubber material $J_3 = \lambda^{1/2}$. Thus, let us suppose more generally that for every simple tension the following volume control relation holds:

$$J_3 = \lambda^n. \tag{69}$$

This special rule may be interpreted as a lateral contraction–expansion relation. Indeed, (69) shows that the transverse stretch is now given uniquely by

$$\lambda_1(\lambda) = \lambda^p, \tag{70}$$

where $p \equiv (n - 1)/2$. Thus, the equibiaxial deformation in a simple tension is a simple extension. The common empirical experience that elongation of an isotropic elastic bar in simple tension always is accompanied by lateral contraction, whereas a simple compression induces transverse expansion, implies that

$$n < 1, \qquad p < 0 \tag{71}$$

must hold in (69) and (70).

The Poisson function that derives from (70) is given by

$$\nu(\lambda) = \frac{1 - \lambda^p}{\lambda - 1}. \tag{72}$$

Hence, the value of Poisson's ratio in a simple tension or compression for every material that respects (69) is

$$\nu_0 = \lim_{\lambda \to 1} \nu(\lambda) = - p. \tag{73}$$

That is, we must have $p = -\nu_0$ in (70); and the unique value of n for which (69) holds in a simple tension is $n = 1 - 2\nu_0$. We thus derive the ad hoc volume control, or lateral stretch condition assumed by Blatz–Ko for simple tension, namely,

$$\lambda = J_3^{1/(1 - 2\nu_0)}, \tag{74a}$$

or

$$\lambda_1(\lambda) = \lambda^{-\nu_0}. \tag{74b}$$

Indeed, we see in (60b), for example, that the universal value $\nu_0 = 1/4$ derived earlier for every Blatz–Ko foamed rubber material satisfies the general rule (74b).

8.3. Reduced Form of the General Blatz–Ko Equation. Blatz and Ko (Ref. 2) used (74a) in (66) to express W_3 in terms of J_3 alone. Then with $W(3, 3, 1) = 0$ in the natural state, they integrated the result to deduce the strain energy function for the class of materials studied in their experiments. We shall use (69) to conclude the strain energy function for a Blatz–Ko material that automatically respects the volume control condition (74a), in the simple tension problem,

$$\begin{aligned} W(J_1, J_2, J_3) = {} & \frac{\mu_0 f}{2}\left[(J_1 - 3) - \frac{2}{q}(J_3^q - 1)\right] \\ & + \frac{\mu_0(1-f)}{2}\left[(J_2 - 3) - \frac{2}{q}(J_3^{-q} - 1)\right], \end{aligned} \tag{75}$$

where

$$q \equiv \frac{n-1}{n} = \frac{-2\nu_0}{1-2\nu_0}. \tag{76}$$

This is the strain energy for the Blatz–Ko constitutive equation referenced frequently in the literature dealing with applications for compressible, homogeneous, and isotropic hyperelastic materials. The response functions (54) for this model are given by

$$\beta_0 = \mu_0\left[-fJ_3^{q-1} + (1-f)J_3^{-(q+1)}\right], \tag{77a}$$

$$\beta_1 = \frac{\mu_0 f}{J_3}, \qquad \beta_{-1} = -\frac{\mu_0(1-f)}{J_3}. \tag{77b}$$

And our more general equation (56) for the Blatz–Ko material (75) transforms to the mnemonic constitutive formula

$$\begin{aligned} \mathbf{T} &= \beta_1(\mathbf{B} - J_3^q\mathbf{1}) + \beta_{-1}(\mathbf{B}^{-1} - J_3^{-q}\mathbf{1}) \\ &= \frac{\mu_0 f}{J_3}(\mathbf{B} - J_3^q\mathbf{1}) - \frac{\mu_0(1-f)}{J_3}(\mathbf{B}^{-1} - J_3^{-q}\mathbf{1}). \end{aligned} \tag{78}$$

The stress (78) vanishes in the undistorted state. We note that the foregoing results require that (68) be respected.

Two special models, $f = 1$ and $f = 0$, very nearly are found in applications of (78):

$$\text{Case 1, } f = 0\text{: } \mathbf{T} = \frac{\mu_0}{J_3}(J_3^{-q}\mathbf{1} - \mathbf{B}^{-1}), \tag{79}$$

$$\text{Case 2, } f = 1\text{: } \mathbf{T} = \frac{\mu_0}{J_3}(-J_3^{q}\mathbf{1} + \mathbf{B}). \tag{80}$$

Case 1 characterizes the class of foamed, polyurethane elastomers and case 2 describes the class of solid, polyurethane rubbers studied in the Blatz–Ko experiments. However, exclusive of the values of f, (79) and (80) contain no other specific experimental conditions. For case 1, the Blatz–Ko empirical relation (58a) for the foamed elastomer together with (77a) imply that $q = -1$. Hence, by (76), $\nu_0 = 1/4$, and (79) delivers our previous result (59). We recall, however, that $f = 0$ does not satisfy the empirical inequalities (68).

Use of the empirical Poisson restrictions (71) in (73) and (76) shows that $q \neq 0$ and $\nu_0 > 0$. Equation (69) shows that $n = 0$, hence $\nu_0 = 1/2$, if and only if the simple tension is isochoric. But it does not follow that $J_3 = 1$ need hold in every deformation; hence, $n = 0$, or $\nu_0 = 1/2$, does not imply that the material need be incompressible. Suppose that the simple tension (65) and (66) for our general Blatz–Ko material (56) is isochoric and recall (68). Then use of $J_3 = 1$ in (66) violates (57) for existence of a natural state unless $\lambda = \lambda_1 = 1$. Hence, an isochoric simple tension is impossible in any Blatz–Ko material having a natural state; and, consequently, there are no incompressible materials in the Blatz–Ko class. Thus, accepting the existence of a natural state, we may conclude that as n decreases from 1^- to 0^+, the familiar Poisson's ratio ν_0 for the class of Blatz–Ko materials (75) or (78) increases for 0^+ to $(1/2)^-$. That is,

$$0 < n < 1, \tag{81a}$$

$$0 < \nu_0 < 1/2, \tag{81b}$$

must hold in equations (75) through (78), where $n = 1 - 2\nu_0$. The incompressible limit estimate of an almost incompressible Blatz–Ko material will be discussed later.

9. Incompressible Materials

The volume control relation (69), or any of its variants in (70) and (74), is imposed only for the simple tension or compression of a Blatz–Ko material.

Therefore, it does not constitute a general internal material constraint, because it does not restrict all deformations of the material. An internal material constraint is a kinematical relation defined by a scalar-valued function $\gamma(\mathbf{F})$ that restricts the possible deformations to those for which $\gamma(\mathbf{F}) = 0$ holds. However, application of the principle of material frame indifference (Truesdell and Noll, Ref. 11, Sections 19, 30) reveals that $\gamma(\mathbf{F})$ must have the reduced form $\gamma(\mathbf{U}) = \eta(\mathbf{C}) = \gamma(\mathbf{R}^T\mathbf{B}\mathbf{R}) = 0$, wherein we recall (9a) and 10b). It follows that if η is a function of the principal invariants of $\mathbf{C}$, then $\eta(\mathbf{C}) = \eta(\mathbf{B}) = 0$.

Incompressibility, inextensibility, and rigidity are important examples of internal material constraints. An incompressible material may suffer only isochoric deformations. Thus, in accordance with (3) and (12b) every deformation of an incompressible material is subject to the internal material constraint

$$\gamma(\mathbf{F}) = J - 1 = \det \mathbf{F} - 1 = 0. \tag{82}$$

This transforms to the frame-indifferent relation

$$\eta(\mathbf{C}) = \eta(\mathbf{B}) = I_3 - 1 = \det \mathbf{B} - 1 = 0. \tag{83}$$

The equations (82) and (83) must hold for all deformations of an incompressible body.

It can be shown from (9a) and (13) that $\dot{\mathbf{C}} = 2\mathbf{F}^T\mathbf{D}\mathbf{F}$. Hence, the material time derivative of (83) yields the following equivalent constraint equation in terms of the stretching tensor $\mathbf{D}$, the symmetric part of $\mathbf{L}$ in (13):

$$\tfrac{1}{2}\,\dot{\eta}(\mathbf{C}) = tr\mathbf{D} \equiv \mathbf{1} \cdot \mathbf{D} = 0. \tag{84}$$

9.1. Constraint Reaction Stress. It is intuitively clear that no amount of all-around stress can deform an incompressible body. The Cauchy stress on an incompressible material, therefore, is determined by $\mathbf{F}$ only to within an arbitrary hydrostatic stress $-p\mathbf{1}$. This intuitive idea may be readily established without regard for any further special constitutive properties of the material.

We recall that the stress working is defined by $\mathrm{tr}\mathbf{TD}$ and require that the symmetric, constraint reaction stress $\mathbf{N}$ be workless in any motion that respects the internal material constraint. That is,

$$\mathrm{tr}(\mathbf{ND}) \equiv \mathbf{N} \cdot \mathbf{D} = 0, \tag{85}$$

for all $\mathbf{D}$ that satisfy the constraint $\dot{\eta}(\mathbf{C}) = 0$.

In particular, for an incompressible material, (85) must hold for all symmetric tensors $\mathbf{D}$ for which (84) holds. It thus follows that the workless, constraint reaction stress $\mathbf{N}$ on an incompressible material is proportional to the identity tensor $\mathbf{1}$. We thus write

$$\mathbf{N} = -p\mathbf{l}, \tag{86}$$

where $p = p(\mathbf{x}, t)$ is an undetermined scalar function of $\mathbf{x}$ and t in κ. Thus, the total Cauchy stress $\mathbf{T}$ on an incompressible, elastic material, which need not be hyperelastic, is determined by $\mathbf{F}$ only to within the arbitrary stress (86),

$$\mathbf{T} = -p\mathbf{1} + \mathbf{T}_E(\mathbf{F}). \tag{87}$$

The extra stress $\mathbf{T}_E(\mathbf{F})$ reflects the elastic constitutive response of the material, and hence it must respect the principle of frame indifference and any material symmetry transformations that characterize the material structure. Though only hyperelastic materials are considered here, it can be shown (Truesdell and Noll, Ref. 11, Section 47) for a general isotropic elastic material that (87) may be cast in a form similar to (45) or (46). The constitutive equation for an incompressible, isotropic hyperelastic solid will be derived next.

9.2. Isotropic Hyperelastic Materials. The foregoing discussion does not depend on the existence of an elastic potential for the stress working. If the incompressible material is hyperelastic and isotropic, the invariant strain energy function (42) must respect (83); hence, we have

$$\Sigma = \Sigma(\mathrm{I}_1, \mathrm{I}_2). \tag{88}$$

Moreover, use of (20), (13), (19), and (82) in (26) yields the following form of the differential equation of energy balance for an incompressible material:

$$\dot{\Sigma} = \mathrm{tr}(\mathbf{TD}) \equiv \mathbf{T} \cdot \mathbf{D}. \tag{89}$$

In view of the constraint (84), however, the strain energy determines the stress only to within an arbitrary workless stress (86) that contributes nothing to (89). Thus, with the aid of (43), (13), and (9b), the material time derivative of (88) yields

$$\dot{\Sigma} = \left[2\frac{\partial\Sigma}{\partial I_1}\mathbf{B} + 2\frac{\partial\Sigma}{\partial I_2}(I_1\mathbf{B} - \mathbf{B}^2)\right] \cdot \mathbf{D},$$

and (89) may be rewritten explicitly as

$$\left[\mathbf{T} - 2\frac{\partial \Sigma}{\partial I_1}\mathbf{B} - 2\frac{\partial \Sigma}{\partial I_2}(I_1\mathbf{B} - \mathbf{B}^2)\right] \cdot \mathbf{D} = 0.$$

This energy equation must hold for all symmetric tensors $\mathbf{D}$ for which (84) holds. Therefore, the constitutive equation for an incompressible, isotropic hyperelastic material is given by

$$\mathbf{T} = -p\mathbf{1} + \alpha_1\mathbf{B} + \alpha_2\mathbf{B}^2, \tag{90}$$

or, by use of the Cayley–Hamilton theorem,

$$\mathbf{T} = -p\mathbf{1} + \beta_1\mathbf{B} + \beta_{-1}\mathbf{B}^{-1}. \tag{91}$$

The undetermined parameter p differs in (90) and (91), and the material response coefficients

$$\alpha_\Lambda = \alpha_\Lambda(I_1, I_2), \qquad \beta_\Gamma = \beta_\Gamma(I_1, I_2), \tag{92}$$

with $\Lambda = 1, 2$ and $\Gamma = 1, -1$, are defined by

$$\beta_1 = \alpha_1 + I_1\alpha_2 = 2\frac{\partial \Sigma}{\partial I_1}, \qquad \beta_{-1} = \alpha_2 = -2\frac{\partial \Sigma}{\partial I_2}. \tag{93}$$

From (91), the stress on an undistorted state κ_R of an incompressible material is an arbitrary hydrostatic stress $\mathbf{T}_0$,

$$\mathbf{T}_0 = (-p + \hat{\beta}_1 + \hat{\beta}_{-1})\mathbf{1}, \tag{94}$$

where $\hat{\beta}_\Gamma = \beta_\Gamma(3, 3)$ are the values of the material functions (93) in κ_R. This coincides with our earlier intuitive observation.

10. Rivlin–Saunders Strain Energy Function

Natural rubber, synthetic elastomers, and biological tissue are important examples of real materials that have been modeled as incompressible, isotropic hyperelastic materials. Several special kinds of constitutive equations supported by experiments of various degrees of completeness have been proposed for study. But the thorough, primary experiments by Rivlin and Saunders (Ref. 1) are the most widely recognized among workers in finite elasticity. The Rivlin–Saunders empirical strain energy function for natural gum rubber was determined from a variety of tests designed on the basis of problem solutions obtained by Rivlin (Refs. 19–23). The general constitutive equation (91) was used to characterize several deformations that included simple tension, compression, pure shear, pure shear superimposed on a simple extension, and pure torsion. Then Rivlin and Saunders applied these theoretical results to determine from experiments the form of the response functions (92) for natural gum rubber. I know of no comparable body of analytically based experimental research in finite elasticity that has since equaled this accomplishment for other kinds of incompressible and nonlinearly elastic materials. Moreover, the classical Rivlin–Saunders strain energy function supports two important special analytical models that have been applied in countless cases to demonstrate the application of general work, or to obtain specific analytical results in problems where the solution based on the fully general constitutive equation (91) for an arbitrary strain energy function extends beyond the reach of "exact" analysis. These widely used ideal models, known as the neo-Hookean material and the Mooney–Rivlin material, will be introduced later. We shall focus now on the easy derivation of the Rivlin–Saunders equation from (91) and (93).

The compatibility relation that derives from (93) is

$$\frac{\partial \beta_1}{\partial I_2} + \frac{\partial \beta_{-1}}{\partial I_1} = 0, \tag{95}$$

and the following special results may be read from this equation:

$$\text{(i)} \quad \beta_\Gamma = \beta_\Gamma(I_1) \iff \beta_{-1} \equiv -\beta, \text{ constant}, \tag{96}$$

$$\text{(ii)} \quad \beta_\Gamma = \beta_\Gamma(I_2) \iff \beta_1 \equiv \alpha, \text{ constant}. \tag{97}$$

Use of (97) in (93) and integration of the result with $\Sigma(3, 3) = 0$ yields the Rivlin–Saunders strain energy function,

$$\Sigma(I_1, I_2) = \frac{\alpha}{2}(I_1 - 3) + g(I_2 - 3). \tag{98}$$

Herein $g(I_2 - 3)$ is an unspecified function of I_2 for which $g(0) = 0$. Thus, the Rivlin–Saunders strain energy function obtained for natural gum rubber is the unique strain energy for an incompressible, homogeneous, and isotropic hyperelastic material whose response functions may depend on I_2 alone. The case (96) will be considered further on.

10.1. Mooney–Rivlin and Neo-Hookean Materials. Two important special constitutive models of rubberlike materials that are supported by the

Rivlin–Saunders strain energy function and by independent experiments by Mooney (see Treloar, Ref. 24) are the Mooney–Rivlin and neo-Hookean models. The Mooney–Rivlin material for which $\beta_1 = \alpha$ and $\beta_{-1} = -\beta$ are constants in (93) is the most general theoretical model for which both response functions may be constant. In this case, (98) reduces to the strain energy function for the Mooney–Rivlin material,

$$\Sigma(I_1, I_2) = \frac{\alpha}{2}(I_1 - 3) + \frac{\beta}{2}(I_2 - 3). \tag{99}$$

The neo-Hookean material is a particular kind of Mooney–Rivlin material for which $\beta_{-1} = -\beta = 0$. This model was first derived from the statistical mechanics of a molecular chain network characteristic of the amorphous structure of rubberlike materials; it is the simplest model of rubberlike elastic response (Treloar, Ref. 24).

The similarity of these materials with the Blatz–Ko constitutive equation (56) is evident. On introducing (55) and recalling that $\beta_1 - \beta_{-1} = \alpha + \beta = \mu_0$ identifies the shear modulus in κ_R, we obtain from (91) the constitutive equation for the Mooney–Rivlin material,

$$\mathbf{T} = -p\mathbf{1} + \mu_0 f\mathbf{B} - \mu_0(1 - f)\mathbf{B}^{-1}. \tag{100}$$

The case $\beta = 0$, i.e., $f = 1$, reduces (100) to the constitutive equation for the neo-Hookean material,

$$\mathbf{T} = -p\mathbf{1} + \mu_0\mathbf{B}. \tag{101}$$

The empirical inequalities (67) translate to the restrictions (68). Thus, comparison of (100) and (101) with (56) and (64) suggests that the Mooney–Rivlin and neo-Hookean models for incompressible rubberlike materials correspond to limit models of Blatz–Ko compressible materials for which $J_3 \to 1$ and $W_3(J_3) \to -p$, an undetermined hydrostatic stress. But this limit reduction has never been established.

The following intuitive argument suggests that this might be plausible. Suppose that the Blatz–Ko material described by (77) and based on the volume control relation (69) for simple tension is almost incompressible so that $J_3 = 1 + \epsilon$, where the change in volume per unit volume $\epsilon << 1$ for every deformation. Then, to the first order in ϵ, the response functions (77) may be approximated by

$$\beta_0 = \mu_0(1 - 2f) - \mu_0 q\epsilon, \qquad \beta_1 = \mu_0 f(1 - \epsilon), \quad \beta_{-1} = -\mu_0(1 - f)(1 - \epsilon).$$

In the limit $\epsilon \to 0$, we have $J_3 \to 1$ for all deformations; and hence, at the same time, (69) requires that $n \to 0$. That is, by (76), $q \to -\infty$, or $\nu_0 \to 1/2$; and $J_3 \to 1$ for all deformations apparently describes the virtual incompressibility limit of the material. Thus, adopting this view, if (68) holds and μ_0 is finite, in the limit as $\epsilon \to 0$ and $q \to -\infty$, the response functions may be estimated by

$$\beta_0 = -p \equiv \mu_0(1 - 2f) + p_0, \qquad \beta_1 = \mu_0 f, \qquad \beta_{-1} = -\mu_0(1 - f),$$

in which p_0, and hence p, is an indeterminate scalar. Therefore, in this sense, the Mooney–Rivlin material (100) may be considered as the incompressible limit estimate of the Blatz–Ko material (78). Of course, then the neo-Hookean material (101) may be identified as the incompressible limit estimate of the Blatz–Ko material for which $f = 1$. I recall no analytical model which may be considered an incompressible limit estimate of the Blatz–Ko case $f = 0$. In general, the stress on an undistorted state of an incompressible material does not vanish, so the incompressible limit case is not a genuine Blatz–Ko material.

10.2. Constitutive Equation for Biological Tissue. A formula similar to (98) with α replaced by $-\beta$ and I_1 interchanged with I_2 may be obtained from (96). We have

$$\Sigma(I_1, I_2) = h(I_1 - 3) + \frac{\beta}{2}(I_2 - 3). \tag{102}$$

Of course, $h(0) = 0$. This model includes, for example, a variety derived from statistical mechanics by Ishihara, Hashitsume, and Tatibana (Ref. 25); but we shall not pause to discuss it here. However, the special case for which $\beta = 0$ and

$$\Sigma = h(I_1 - 3) = \frac{\mu_0}{2\gamma}\left[e^{\gamma(I_1 - 3)} - 1\right], \tag{103}$$

where μ_0 is the shear modulus and γ is another constant, has been used often in the biomechanics literature to describe the nonlinearly elastic response of biological tissue. Thus, with the use of (103) in (93), (91) delivers a constitutive equation for hyperelastic biological tissue,

$$\mathbf{T} = -p\mathbf{1} + \mu_0 \mathbf{B}\, e^{\gamma(I_1 - 3)}. \tag{104}$$

This equation reduces to the neo-Hookean material (101) when $\gamma = 0$. The empirial inequalities (67) hold if and only if $\mu_0 > 0$, but they impose no restriction on γ. However, in order that the strain energy (103) shall increase with the deformation from the natural state, $\gamma > 0$ is necessary, the equality holding for the neo-Hookean case.

The experimental foundation for (104) was first investigated by Fung (Ref. 3) in a study of simple extension of certain tissue. The reader is cautioned that one often finds (104) expressed in terms of the second invariant $I_2(\mathbf{B}^{-1})$ of $\mathbf{B}^{-1}$. For an incompressible material, however, $I_2(\mathbf{B}^{-1}) = I_1(\mathbf{B})$. See Demiray (Ref. 26) for an example application.

11. Inflation Response of a Balloon

One may find in the literature other constitutive relations for incompressible elastomers; but none has shared the wide application found for the Mooney–Rivlin and neo-Hookean models. In view of their simplicity, (100) and (101) often are introduced for mathematical convenience, or to illustrate the content of results derived more generally for every incompressible, isotropic, hyperelastic material. It is important to mention too that these models exhibit fairly decent agreement with experiments, though only for moderately large deformations of materials for which the effects of hysteresis are slight and other inelastic effects, such as crystallization, stress softening, and permanent set, are negligible. In a simple tension, for example, satisfactory agreement up to an axial stretch of about 2 to 2.5 is common. And in mathematical applications, I do not know of a single case in which either model fails to provide a satisfactory qualitative picture of physical phenomena for a reasonable range of finite deformation.

An easy example that illustrates this point concerns the inflation of a toy balloon. Everyone knows that to blow up a balloon considerable effort must be exerted initially; but as the balloon grows larger, at some point the inflation task becomes noticeably easier. The inflation pressure required eventually increases again, until, finally, the balloon bursts. The maximum pressure effect is predicted by the simple neo-Hookean model; and our other models can account for the overall phenomenon, except for the bursting pressure limit. The problem also may be solved for an arbitrary isotropic hyperelastic material.

The solution of the inflation problem for an arbitrary strain energy function is demonstrated by Green and Zerna (Ref. 27) as a thin shell limit obtained from the solution for the inflation of a thick-walled, incompressible spherical shell. The same result for a thin spherical shell subsequently was derived by Green and Adkins (Ref. 28) from a general theory of incompressible, hyperelastic membranes, and the inflation phenomenon described above was illustrated for the Mooney–Rivlin material. The solution of the balloon inflation problem will be presented next by use of a simple energy method.

Let us begin with the mechanical energy principle (25); and write

$\Delta\Sigma \equiv \Sigma|_{t_2} - \Sigma|_{t_1}$. Then integration of (25) between any two equilibrium states at the times $t = t_1$ and $t = t_2$, say, shows that the work done by the surface tractions acting over a continuous simple path $\mathcal{G}$ between two equilibrium states without body force is balanced by the change in the total strain energy,

$$\int_{\mathcal{B}} \Delta\Sigma dV = \int_{\mathcal{G}} \int_{\partial\mathcal{B}} \mathbf{t}_n da \cdot d\mathbf{x}. \tag{105}$$

We shall model the balloon as an isotropic, spherical membrane with undeformed radius r_0 and thickness $t_0 << r_0$ at t_1; and we shall suppose that the spherical shape is preserved as the inflation pressure $\hat{p}(r)$, the excess of the internal gas pressure over the external atmospheric pressure, deforms the thin shell uniformly to a radius r and thickness t at t_2. The uniform, isotropic stretch of the membrane is described by $\lambda = r/r_0$; the transverse, normal stretch is given by $\lambda_3 = t/t_0$; and, of course, $\hat{p}(r_0) = 0$. Thus, for any isotropic hyperelastic material, the work–energy principle (105) shows that the work done by the inflation pressure to increase the volume of the sphere is balanced by the elastic energy stored in the membrane material. We shall assume that we are able to write the stored energy as a function $\Sigma = \Sigma(\lambda)$ with $\Sigma(1) = 0$. Then noting that $\mathbf{t}_n \cdot d\mathbf{x} = \hat{p}(r)dr$ is uniform over the membrane surface $\partial\mathcal{B}$ whose total area is $4\pi r^2$ and that $\Sigma = \Sigma(\lambda)$ is uniform over the membrane volume $\mathcal{B}$, we have

$$\int_{\mathcal{B}} \Delta\Sigma dV = 4\pi r_0^2 t_0 \Sigma(\lambda),$$

$$\int_{\mathcal{G}} \int_{\partial\mathcal{B}} \mathbf{t}_n da \cdot d\mathbf{x} = 4\pi \int_{r_0}^{r} r^2 \hat{p}(r) dr = 4\pi r_0^3 \int_{1}^{\lambda} \lambda^2 p(\lambda) d\lambda,$$

wherein $\hat{p}(r) = p(\lambda)$. Thus, the substitution of these integrals into (105) followed by differentiation with respect to λ delivers the general formula for the inflation pressure for an isotropic, hyperelastic spherical membrane,

$$p(\lambda) = \frac{t_0}{r_0 \lambda^2} \frac{d\Sigma(\lambda)}{d\lambda}. \tag{106}$$

The condition obtained from (106) for existence of an extremum pressure p^* at a membrane stretch λ^* is given by

$$\frac{dp}{d\lambda} = \frac{t_0}{r_0 \lambda^3} \left[\lambda \frac{d^2\Sigma}{d\lambda^2} - 2\frac{d\Sigma}{d\lambda} \right] = 0. \tag{107}$$

A similar derivation of (106) was first given by Pipkin (Ref. 29).

In particular, for an incompressible material, $\lambda_3 = \lambda^{-2}$; so (88) yields $\Sigma(I_1, I_2) = \Sigma(\lambda)$. Differentiation of this function and use of (43) and (93) delivers easily from (106) the general rule for the inflation pressure for an incompressible, isotropic hyperelastic spherical membrane,

$$p(\lambda) = \frac{2t_0}{\lambda r_0}\left[1 - \frac{1}{\lambda^6}\right](\beta_1 - \lambda^2\beta_{-1}), \tag{108}$$

wherein $\beta_\Gamma = \beta_\Gamma(\lambda)$. This is the relation obtained differently by Green and Zerna (Ref. 27) and by Green and Adkins (Ref. 28) for an arbitrary strain energy function.

11.1. Neo-Hookean Balloon. Now let us consider a neo-Hookean membrane described by (99) with $\beta_1 = \alpha = \mu_0$, $\beta_{-1} = -\beta = 0$. Then we obtain by (108) the inflation pressure for a neo-Hookean balloon,

$$p(\lambda) = \frac{2\mu_0 t_0}{r_0\lambda}\left(1 - \frac{1}{\lambda^6}\right). \tag{109}$$

Since $p(1) = 0$ in the natural state and $p(\lambda) \to 0$ as $\lambda \to \infty$, then $p(\lambda)$ must attain a maximum value at some intermediate stretch λ^* for which the radius is r^*. The maximum value p^* of the inflation pressure found from (109) is

$$p^* = \frac{12\mu_0 t_0}{r_0 7^{7/6}} = 1.239\,\frac{\mu_0 t_0}{r_0} \tag{110}$$

at the stretch

$$\lambda^* = \frac{r^*}{r_0} = \sqrt[6]{7} = 1.383. \tag{111}$$

Of course, the maximum pressure depends on the ratio t_0/r_0 and the material stiffness $\mu_0 = E_0/3$, where E_0 denotes Young's modulus. Notice, however, that the stretch at which the maximum inflation pressure occurs is the same for every neo-Hookean material. Thus, (111) is a universal solution for this model.

The data obtained from a typical balloon inflation experiment are plotted in Fig. 5. The inflation and deflation curves over one cycle, followed by inflation to failure, are shown. The material response was not studied to determine whether it may be described as neo-Hookean, or anything else. Nevertheless, the estimated maximum pressure occurred in this test at the stretch $\lambda^* = 1.43$, which is only 3.4% greater than the theoretical value (111). There is a notable retracing of a similar curve in the deflation phase, which includes a relative maximum pressure at a stretch larger than λ^*. The smaller maximum pressure attained in the second inflation is consonant with our experience in prestretching a balloon prior to its primary inflation. This stress softening effect induced by preconditioning was first observed in tests by Bouasse and Carrière (Ref. 30) for natural rubber vulcanizates. As a consequence of an extensive experimental study of filler-loaded rubber vulcanizates by Mullins (Ref. 31), however, this prestretch effect is now widely known as the Mullins effect. A thorough list of references on the Mullins effect may be found in the review article by Harwood, Mullins, and Payne (Ref. 32). Of course, hyperelasticity theory is unable to describe these

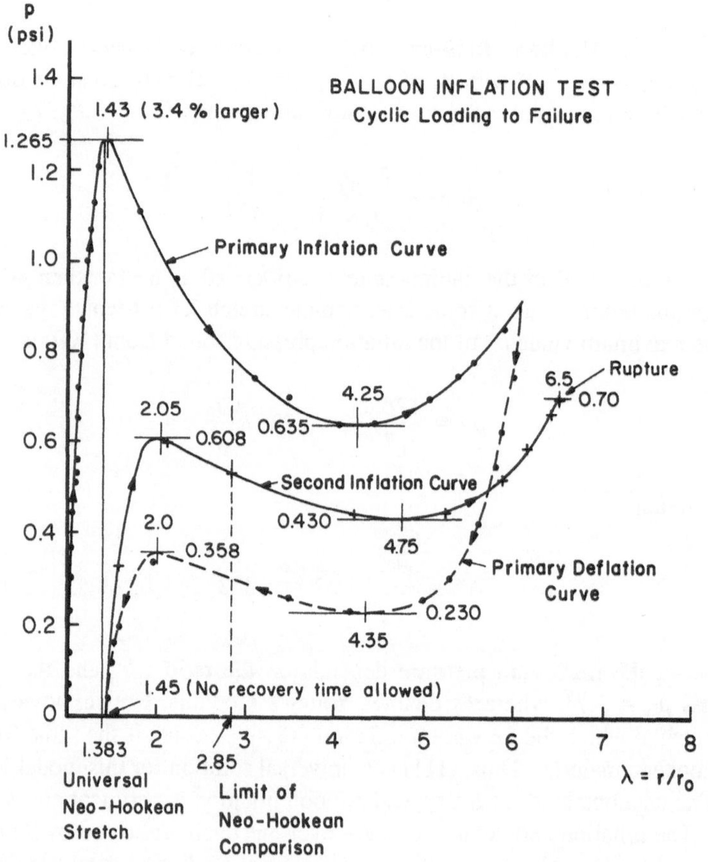

Fig. 5. Typical balloon inflation test showing the inflation pressure as a function of the circumferential stretch.

inelastic effects. A simple constitutive equation that exhibits the essence of the Mullins effect in the balloon inflation and in uniaxial experiments has been studied recently by Johnson and Beatty (Refs. 33–35). Many additional papers on both the experimental and theoretical characterization of the Mullins effect, mostly applied to uniaxial tests, may be found in these works.

It is commonly known that preconditioning of a rubber material induces softening that generally renders test results more consistent and repeatable. The addition of reinforcing filler to gum rubber vulcanizates produces an increased stiffness of the material which may be largely destroyed by deformation. Therefore, the early papers by Mullins (Ref. 31), Mullins and Tobin (Ref. 36), and others concluded that the softening effect is due to rubber filler bond breakage. At the time, Bueche (Ref. 37), for example, attributed the Mullins effect to separation of the rubber molecules from the surfaces of the filler particles. We recall, however, that the primary report by Bouasse and Carrière (Ref. 30) on stress softening involved unfilled natural rubber vulcanizates. Subsequent results by Harwood, Mullins, and Payne (Ref. 38) showed that softening occurs entirely in the rubber phase and is not the primary result of breakdown of the stiffening mechanism introduced by filler. They thus concluded that the extent of softening in filled and unfilled vulcanizates is comparable. More recently, Govindjee and Simo (Refs. 39, 40) applied Bueche's (Ref. 37) earlier ideas to develop a continuum constitutive equation that demonstrates computationally considerable agreement with the results of uniaxial extension experiments by Bueche (Ref. 41). Some kinematical peculiarities in the Govindjee–Simo model are noted by Johnson and Beatty (Ref. 33). Morever, the complexity of these theoretical and mainly computational results in describing the Mullins effect in a simple extension underscores the considerable difficulty one may anticipate in its application to the study of more complex deformations. Besides these few results, the theoretical foundation for the Mullins effect in relation to the mechanical response of elastomers remains largely undeveloped.

Finally, we note that the increased stiffness effect at the larger inflation stretches arises from the ultimate extension of chains in the molecular chain structure. The James–Guth (Ref. 42) statistical mechanical theory of rubber elasticity can account for this ultimate stiffening effect and its influence in the balloon inflation problem; but we shall not explore this here.

11.2. Mooney–Rivlin and Biological Membranes. Other than the special constitutive model introduced by Johnson and Beatty (Ref. 35), I know of no theoretical study of the inelastic effects observed in the balloon experiment. The neo-Hookean model clearly fails to describe completely the primary inflation response, but it does give a reasonably good qualitative picture of the maximum pressure behavior which occurs within the deformation range usually considered satisfactory for this model. Of course, other models may yield sharper results.

The Mooney–Rivlin and tissue materials, for example, can capture more of the primary inflation phenomenon. The inflation pressure (108) for the Mooney–Rivlin material (99) is given by

$$p(\lambda) = \frac{2\alpha t_0}{r_0\lambda}\left[1 - \frac{1}{\lambda^6}\right]\left[1 + \gamma\lambda^2\right], \tag{112}$$

where $\gamma \equiv \beta/\alpha$. The condition (107) for the extreme points translates to

$$\gamma\lambda^8 - \lambda^6 + 5\gamma\lambda^2 + 7 = 0. \tag{113}$$

For the biological tissue (103), (108) yields

$$p(\lambda) = \frac{2\mu_0 t_0}{r_0\lambda}\left[1 - \frac{1}{\lambda^6}\right] e^{\gamma(2\lambda^2+\lambda^{-4}-3)}, \tag{114}$$

and (107) becomes

$$4\gamma\lambda^{12} - \lambda^{10} - 8\gamma\lambda^6 + 7\lambda^4 + 4\gamma = 0. \tag{115}$$

We recall for both cases that $\gamma \geq 0$, and notice that both reduce to the neo-Hookean pressure (109) and stretch (111) when $\gamma = 0$. Otherwise, (113) and (115) may have at most two positive real roots λ_k^* which will depend on the material parameter γ. Hence, the solution(s) $\lambda_k^* = \lambda_k^*(\gamma)$, if any exist, cannot be universal for either case. For the models described here, a universal stretch is a property peculiar to the neo-Hookean model.

When (113) and (115) have two roots λ_k^*, these roots correspond to points of maximum and minimum pressures p_k^* determined by (112) and (114), respectively. However, existence of an extreme point depends on the value of γ. For $\gamma > 0$, it may be seen that $dp/d\lambda > 0$ at $\lambda = 1$ and at ∞. Thus, if a maximum pressure occurs, it must be followed by a minimum. The pressure will attain a maximum value only if the slope $dp/d\lambda < 0$ for some $\lambda > 1$. This will happen for the Mooney–Rivlin model if

$$F(\lambda) \equiv \frac{\lambda^6 - 7}{\lambda^2(\lambda^6 + 5)} > \gamma. \tag{116}$$

The largest value of $F(\lambda)$ occurs for $\lambda = 1.841$, and (116) determines the greatest value $\gamma = 0.214$ for which the pressure may be stationary. Therefore, we predict that if $\gamma < 0.214$, the pressure will rise to a maximum, fall to a minimum, and rise again indefinitely with increasing stretch. The same behavior occurs for the biological tissue provided that $\gamma < 0.067$. This is demonstrated in Figs. 6 and 7 which show the theoretical normalized pressure $\bar{p}(\lambda)$ versus the stretch λ for various values of the parameter γ, where

$$\bar{p}(\lambda) \equiv \frac{r_0 p(\lambda)}{2\alpha t_0} = \frac{k(\lambda)}{\lambda}\left[1 - \frac{1}{\lambda^m}\right], \tag{117}$$

and in the present discussion the exponent $m = 6$. Also,

$$k(\lambda) \equiv \begin{cases} 1, & \text{neo-Hookean case (109)}, & (118a) \\ 1 + \gamma\lambda^2, & \text{Mooney–Rivlin case (112)}, & (118b) \\ e^{\gamma(2\lambda^2 + \lambda^{-4} - 3)}, & \text{biotissue case (114)}. & (118c) \end{cases}$$

In the first and last of (118), we use $\alpha = \mu_0$ in (117).

Figure 6 shows that the neo-Hookean model provides the lower bound inflation pressure for all Mooney–Rivlin materials; therefore, the neo-Hookean

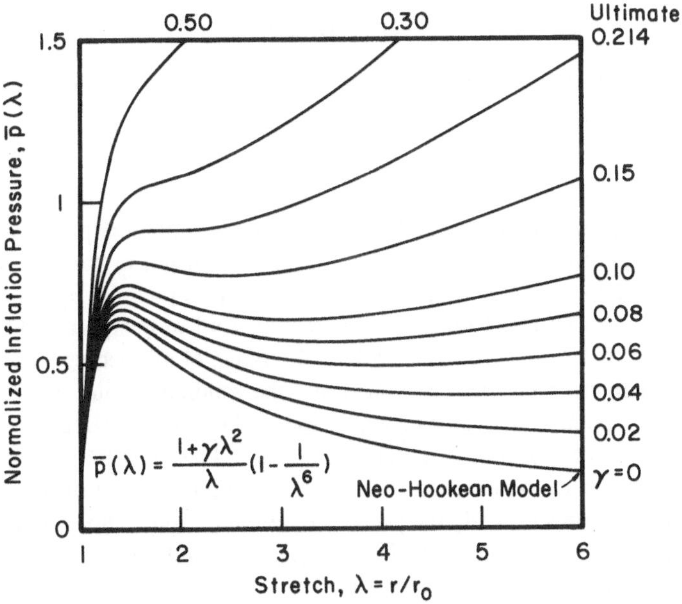

Fig. 6. Normalized inflation pressure as a function of the circumferential stretch of a Mooney–Rivlin balloon for various values of the material parameter γ. The ultimate value for which the pressure may be stationary is $\gamma = 0.214$.

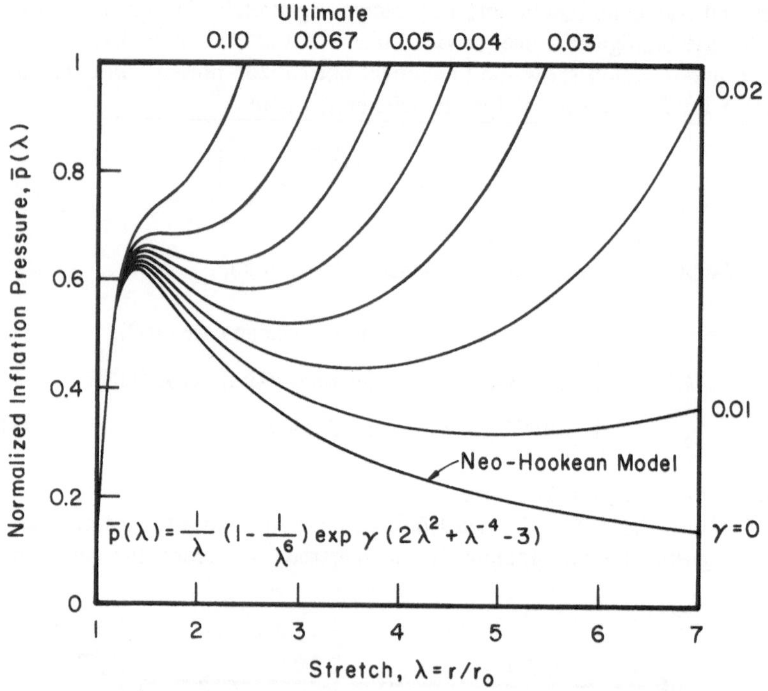

Fig. 7. Normalized inflation pressure as a function of the circumferential stretch of a spherical biomaterial membrane for various values of the material constant γ. The ultimate value for which the pressure may be stationary is $\gamma = 0.067$.

model predicts a safe estimate for the maximum inflation pressure of any Mooney–Rivlin balloon. The extrema in Fig. 6 are listed in Table 1. The data show that both the extreme stretch λ^*_{max} and normalized pressure $\bar{p}^*_{max}$ increase as γ increases from its neo-Hookean value $\gamma = 0$. And in every case the experimental value of 1.43 shown in Fig. 5 is in the range of all theoretical values for λ^*_{max} computed from (113). Notice also that the minimum point moves toward the maximum point as the value of γ approaches the ultimate value $\gamma = 0.214$ at which the pressure may have a stationary value. This may be seen in Fig. 6. Plainly, the shape of the inflation curve is controlled by γ. Use of $\gamma = 0.055$ in (112) and (113) models nicely the primary inflation response shown in Fig. 5.

Figure 7 shows that the inflation response for the biological material is in all respects similar to the Mooney–Rivlin case, including the neo-Hookean lower bound inflation curve. But the minimum effect can occur only for very soft tissue for which $\gamma < 0.067$. The maxima move upward and toward the right as γ is increased. The minima occur at stretch values smaller than those for the

Table 1. Theoretical extrema computed from (113) and (117) for inflation of a Mooney–Rivlin balloon.

γ	λ^*_{max}	$\bar{p}^*_{max}$	λ^*_{min}	$\bar{p}^*_{min}$
0	1.383	0.6197	∞	0
0.02	1.399	0.6437	7.071	0.2828
0.04	1.416	0.6682	4.998	0.4000
0.055	1.429	0.6873	4.250[a]	0.4700
0.06	1.434	0.6933	4.077	0.4898
0.08	1.454	0.7190	3.524	0.5654
0.10	1.476	0.7453	3.143	0.6318
0.15	1.547	0.8144	2.522	0.7718
0.20	1.682	0.8898	2.069	0.8857
0.214	1.842	0.9131	1.842	0.9131
0.30	—	—	—	—
0.50	—	—	—	—

[a] The theoretical value of λ^*_{min} was chosen to match the experimental data in Fig. 5.

Mooney–Rivlin material; but with increasing stretch, the pressure curves rise more rapidly than before. Notice again that the minimum moves toward the maximum as the ultimate value $\gamma = 0.067$ at which the pressure may be stationary is approached.

Finally, it may be noted that as the stretch grows indefinitely large, the slope of the normalized pressure-stretch curve for the Mooney–Rivlin model approaches the constant value γ. Hence, this model does not predict an indefinitely large, bursting pressure at a finite value of the stretch. Nonetheless, the Mooney–Rivlin model does provide a satisfactory qualitative (and probably conservative) description of the overall primary balloon inflation phenomenon provided that $\gamma < 0.214$. It should be noted also that the test balloon was not perfectly spherical, and incompressibility of the material, assumed throughout the analysis, was not confirmed in the experiment. We next illustrate how the compressibility of the material may affect the inflation response of a balloon.

11.3. Inflation of a Blatz–Ko Balloon. The effect of compressibility on the inflation of a spherical membrane may be illustrated for a Blatz–Ko balloon under conditions of isotropic plane stress. It can be shown in this case that if the inequalities (68) hold for the Blatz–Ko material (78), then $J_3^q = \lambda_3^2$, that is,

$$\lambda_3 = \lambda^{2q/(2-q)}, \tag{119}$$

in which q is defined in (76). Consequently, we are able to write an explicit relation for the stored energy (75) in the form $W(J_1, J_2, J_3) = \Sigma(\lambda)$, and hence (106) may be applied. We thereby obtain the normalized inflation pressure for a Blatz–Ko balloon,

$$\bar{p}(\lambda) \equiv \frac{r_0 p(\lambda)}{2\alpha t_0} = \frac{1}{\lambda}\left[1 - \frac{1}{\lambda^m}\right](1 + \gamma\lambda^{m-4}). \tag{120}$$

Herein (55) has been introduced, $\gamma \equiv \beta/\alpha = (1 - f)/f$, and

$$m \equiv -\frac{2(3q-2)}{2-q} = \frac{2(1+\nu_0)}{1-\nu_0}. \tag{121}$$

The stationary points (λ^*, p^*) are found by use of the relation

$$(m-5)\gamma\lambda^{2m-4} - \lambda^m + 5\gamma\lambda^{m-4} + (m+1) = 0. \tag{122}$$

Notice that the normalized inflation pressure (120) is given by our earlier formula (117) in which

$$k(\lambda) \equiv 1 + \gamma\lambda^{m-4}, \qquad \text{Blatz–Ko case (120).} \tag{123}$$

The effect of compressibility on the inflation pressure will be identified in terms of the Poisson exponent m defined in (121). For $0 < \nu_0 < \frac{1}{2}$, (121) shows that $2 < m < 6$. We shall consider all cases within this range.

11.3.1. Case $f = 1$ ($\gamma = 0$). To start with, we see that for $f = 1$ (i.e., $\gamma = 0$) in (55), the extremum condition (122) yields the solution

$$\lambda^* = (1 + m)^{1/m}. \tag{124}$$

Also, (120) shows that $\bar{p}(\lambda) \to 0$ as $\lambda \to \infty$. Thus, with (124), the normalized pressure has the absolute maximum value

$$\bar{p}^* = \bar{p}(\lambda^*) = m(m+1)^{-(m+1)/m}. \tag{125}$$

We see that both the maximum pressure and its stretch depend on m alone. Figure 8 shows that λ^* decreases while $\bar{p}^*$ increases with m. That is, the inflated membrane radius at the maximum pressure becomes smaller as ν_0 increases. Otherwise, the inflation behavior is similar to that of the neo-Hookean model. In fact, (124) and (125) reduce for $m = 6$ to our earlier neo-Hookean relations (110) and (111). The neo-Hookean model yields an upper bound solution for the inflation response of any Blatz–Ko balloon for which $\gamma = 0$.

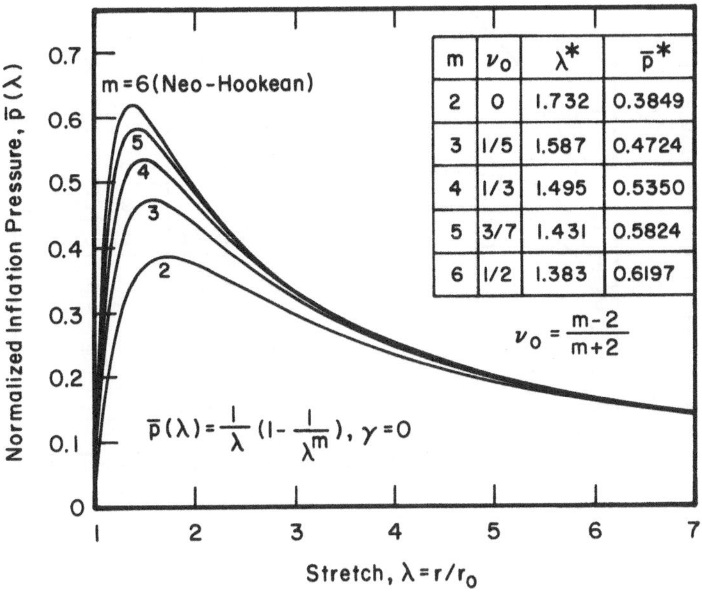

Fig. 8. Inflation response of a Blatz–Ko balloon having $\gamma = 0$ for various values of the Poisson exponent m. The neo-Hookean model gives the upper bound solution for the inflation pressure.

11.3.2. Case $f = 0$. The normalized inflation pressure for a foamed polyurethane-type material for which $f = 0$, very nearly is given by (120),

$$\bar{p}(\lambda) \equiv \lambda^{m-5}\left[1 - \frac{1}{\lambda^m}\right]. \tag{126}$$

Thus, the normalized pressure is again controlled by the Poisson exponent for the material. The pressure has an absolute maximum only for $2 < m < 5$ at the stretch

$$\lambda^* = \left[\frac{5}{5 - m}\right]^{1/m}. \tag{127}$$

Otherwise, for $m = 5$, $\bar{p} \to 1$ as $\lambda \to \infty$; and for $m > 5$, $\bar{p} \to \infty$ with λ. For this model, as shown in Fig. 9, the stretch at the stationary points increases with the Poisson exponent; that is, the inflated balloon radius at the maximum pressure becomes larger for larger values of ν_0. Notice that the inflation response in this case is directly opposite from the inflation response of a Blatz–Ko balloon in the material class for which $f = 1$. It is helpful to note that when $f = 1$, we have

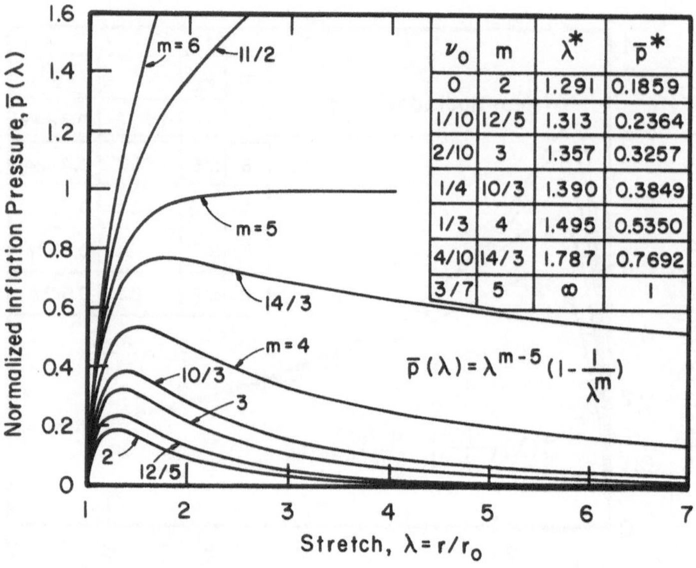

Fig. 9. Inflation behavior of a Blatz–Ko spherical membrane having $f = 0$. This model includes the foamed, polyurethane elastomer for which $\nu_0 = 1/4$ ($m = 10/3$).

$\alpha = \mu_0$, $\beta = 0$, whereas when $f = 0$, $\alpha = 0$ and $\beta = \mu_0$. In either case, $\bar{p}(\lambda) = r_0 p(\lambda)/2\mu_0 t_0$ in (120) and (126).

11.3.3. General Case $0 < f < 1$ ($\gamma > 0$). The case $m = 4$, i.e., $\nu_0 = \frac{1}{3}$, is exceptional. Equation (122) yields for all $\gamma \geq 0$ the invariant stretch

$$\lambda^* = \sqrt[4]{5} = 1.495. \tag{128}$$

But (120) shows that $\bar{p} \to 0$ as $\lambda \to \infty$, and hence for the fixed stretch (128), the pressure has a maximum that increases with γ, namely,

$$\bar{p}^* = \frac{4}{5^{5/4}}(1 + \gamma) = 0.535(1 + \gamma). \tag{129}$$

Notice from the tables in Figs. 8 and 9 that for $m > 4$ the stationary point for the case $f = 0$ is the same as that for $f = 1$; both occur at the invariant stretch (128).

Otherwise, for all $m \in (2, 5)$, i.e., $0 < \nu_0 < 3/7 = 0.428$, the stretch λ^* is found from (122), the maximum pressure is obtained from (120), and it is seen that $p \to 0$ as $\lambda \to \infty$. The behavior is similar for the case $m = 5$, but $\bar{p} \to \gamma$ from above as $\lambda \to \infty$ for each choice of $\gamma \geq 0$.

Finally, for all $m \in (5, 6)$, i.e., for all $\nu_0 \in (3/7, 1/2)$, it is seen from (120) that $\bar{p} \rightarrow \infty$ as $\lambda \rightarrow \infty$. Thus, the pressure will exhibit a minimum provided that, for some $\lambda > 1$,

$$H(\lambda) \equiv \frac{\lambda^m - (m+1)}{\lambda^{m-4}[(m-5)\lambda^m + 5]} > \gamma. \tag{130}$$

The function $H(\lambda)$ has a maximum value H_m for each $m \in (5, 6]$, and therefore the pressure will rise to a maximum, drop to a minimum, and grow again indefinitely provided that $0 < \gamma < H_m$. It was shown earlier that for $\gamma = 0$, $\bar{p} \rightarrow 0$ as $\lambda \rightarrow \infty$. There are no stationary values of $\bar{p}$ for $\gamma > H_m$.

The ultimate value γ_{max} of γ clearly depends on the value of the Poisson exponent $m \in (5, 6)$. It can be shown that γ_{max} decreases from 1.04 to 0.214 as m varies from 5.25 to 6. The results are typical of all others we have seen. The inflated balloon radius at the maximum pressure becomes greater as γ is increased; and the minimum pressure marches toward the maximum as γ nears its ultimate value γ_{max}. The response is similar to that shown earlier for the Mooney–Rivlin and biomaterial membranes.

11.4. Concluding Remarks on the Balloon Inflation Example. Except for a few interesting special quirks, we have seen that the overall physical response for the class of Blatz–Ko materials, though somewhat more difficult to sort out, is not greatly different from the inflation behavior found for the incompressible materials studied earlier. The inflation response is similar to that for both the Mooney–Rivlin and biological materials, but larger values of the material constant γ for which the pressure may have stationary values occur for the Blatz–Ko model. However, the inflation pressure for the Blatz–Ko materials will rise to a maximum, fall to a minimum, and then grow again indefinitely only for $5 < m < 6$; and the ultimate value γ_{max} of γ decreases with increasing m.

We have observed that for the case $\gamma = 0$ the neo-Hookean material for which $m = 6$ provides an upper bound solution for the inflation pressure. More generally, let us consider two balloons having identical geometry and sharing the same material constant $\gamma \geq 0$, one known to be a Mooney–Rivlin material, the other being any member of the Blatz–Ko class for which $m \in (2, 6)$. Then the difference $\Delta\bar{p} \equiv \bar{p}_{MR} - \bar{p}_{BK}$ between the normalized pressure (117) for the Mooney–Rivlin material (118b) and the Blatz–Ko model (123) is given by

$$\Delta\bar{p} = \frac{(1 + \gamma\lambda^{m+2})}{\lambda^{m+7}}\left[\lambda^6 - \lambda^m\right]. \tag{131}$$

It is evident that $\Delta\bar{p} \geq 0$ for $\lambda > 1$ and for all $m \leq 6$, the equality holding for

$m = 6$. Therefore, for an assigned value of $\gamma \geq 0$, the Mooney–Rivlin model provides an upper bound solution for the normalized inflation pressure for every Blatz–Ko balloon.

The effect of the material compressibility in the inflation problem is thus characterized by the Poisson exponent m, that is, by the Poisson ratio of the material. The effect for the same value of γ, in this instance $\gamma = 0.15$, as m is varied over [2, 6] may be seen in Fig. 10. The example is interesting because, in comparison with earlier results for the cases $f = 1$ and $f = 0$, it shows a compound response: the inflated balloon radius at the maximum pressure first decreases as m increases from 2 to about 5, and then it increases again as m increases to 6 at the upper bound inflation solution for the Mooney–Rivlin model with $\gamma = 0.15$. The actual transition value of m for which the turnaround occurs will depend on γ and must be found numerically.

The case $m = 6$ begs further discussion. When $\nu_0 \to \frac{1}{2}$, (121) shows that

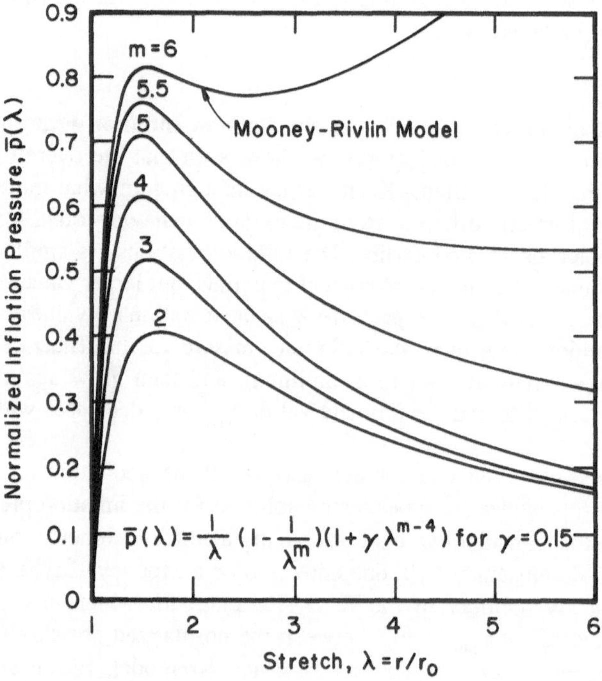

Fig. 10. Inflation response of a Blatz–Ko spherical membrane with $\gamma = 0.15$ plotted for several values of the Poisson exponent m. The radius at the maximum pressure first decreases as m increases to about 5, and afterwards it increases to the Mooney–Rivlin upper bound solution for the inflation pressure.

$m \rightarrow 6$, and (119) gives the incompressibility condition in an equibiaxial deformation. Moreover, when $m = 6$, (120) and (122) for the Blatz–Ko material reduce to (112) and (113) for the Mooney–Rivlin material. Therefore, this example supports our earlier intuitive argument that the Mooney–Rivlin model may be identified as the incompressible limit estimate of an almost incompressible Blatz–Ko material for which both $\nu_0 \rightarrow \frac{1}{2}$ and $J_3 \rightarrow 1$ for all deformations. In general, however, $\nu_0 \rightarrow \frac{1}{2}$ does not imply that and $J_3 = 1$ need hold for all deformations; and it is not clear how the incompressible limit may be effected, if at all, for the class of Blatz–Ko materials (56) or for any other class of compressible, isotropic elastic solids. Indeed, we shall see in the next section that $\nu_o = \frac{1}{2}$ does not generally characterize an incompressible material in finite elasticity.

12. Remarks on Other Kinds of Internal Constraints

It is well known that in classical, linear elasticity theory an incompressible material is characterized by the Poisson ratio $\nu_o = \frac{1}{2}$. In the nonlinear theory, however, the Poisson function $\nu(\lambda)$ for a simple extension of an incompressible material decreases monotonically with increasing stretch in accordance with

$$\nu(\lambda) = \frac{1}{\lambda + \lambda^{1/2}} \tag{132}$$

(Beatty and Stalnaker, Ref. 16). Thus, in the natural state of every incompressible material, the Poisson ratio $\nu_0 = \text{limit}_{\lambda \rightarrow 1} \nu(\lambda) = \frac{1}{2}$. But a natural state Poisson ratio $\nu_0 = \frac{1}{2}$ certainly does not imply that every finite deformation of the material need be isochoric.

Beatty and Stalnaker (Ref. 16) have shown that a Bell constrained material (Bell, Refs. 43, 44) for which $\text{tr}\mathbf{V} = \text{tr}\mathbf{B}^{1/2} = 3$ must hold for all deformations has the constant-valued Poisson function $\nu(\lambda) = \nu_0 = \frac{1}{2}$ in every equibiaxial deformation. But it can be shown also that a Bell constrained material can support no finite isochoric deformations at all. Hence, every Bell constrained material has a Poisson ratio $\nu_0 = \frac{1}{2}$, but none is incompressible (Beatty and Hayes, Refs. 45–47). This is not an isolated case.

Another example is provided by the response of an elastic crystal whose deformation is internally constrained by $\text{tr}\mathbf{B} = 3$ for all deformations (Ericksen, Ref. 48). This constraint restricts every equibiaxial deformation $\lambda_1 = \lambda_2$ and $\lambda_3 = \lambda$ such that $0 < \lambda < \sqrt{3}$ and $0 < \lambda_1 = \sqrt{3/2}$. Whether or not a simple tension of the crystal may be able to produce this deformation will depend on the constitutive equation for the material. Lacking this, we may say that the apparent Poisson function for Ericksen's internally constrained elastic crystal, without further constraints, is given by

$$\nu(\lambda) = \frac{1 - \sqrt{(3 - \lambda^2)/2}}{\lambda - 1}, \tag{133}$$

in every equibiaxial deformation. It follows from (133) that in the natural state $\nu_0 = \frac{1}{2}$, but it can be easily shown that the crystal can support no finite isochoric deformations whatever. In fact, every material constrained by tr$\mathbf{B} = 3$ has a Poisson ratio $\nu_0 = \frac{1}{2}$, but none may be incompressible. Moreover, contrary to one's intuition, Fig. 11 shows that (133) is a monotone increasing function of the normal stretch λ. This does not mean, however, that the cross section swells in extension and contracts in compression. Indeed, the constraint shows that λ_1 decreases as λ increases along the ellipse $\lambda_1^2 = \frac{1}{2}(3 - \lambda^2)$. Notice, on the other hand, that the corresponding monotone decreasing transverse contraction ratio $\alpha(\lambda) \equiv \lambda_1(\lambda)/\lambda$ shown in Fig. 12 supports our intuitive expectation that the cross section ought to contract in simple extension and bulge in compression.

The Bell constraint has an easy geometrical interpretation. We recall that the principal stretches λ_k are the three proper values of the stretch tensors $\mathbf{U}$ and

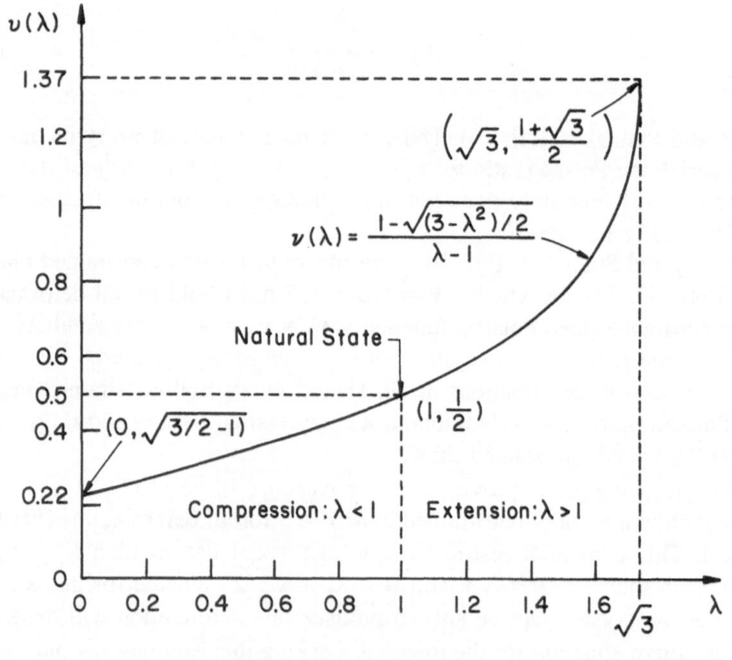

Fig. 11. The apparent Poisson function for Ericksen's constrained elastic crystal.

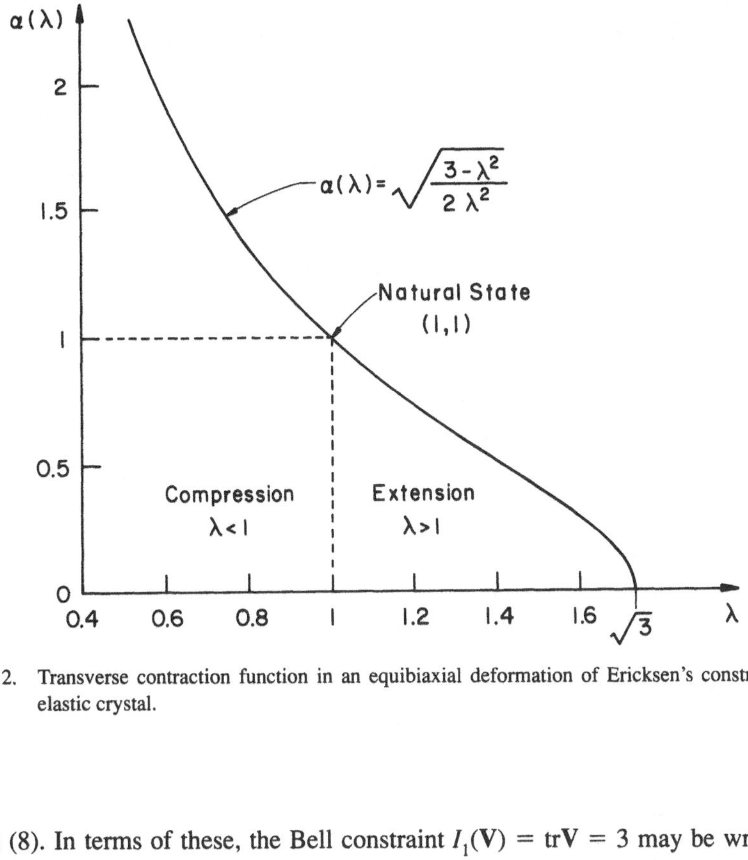

Fig. 12. Transverse contraction function in an equibiaxial deformation of Ericksen's constrained elastic crystal.

V in (8). In terms of these, the Bell constraint $I_1(\mathbf{V}) = \text{tr}\mathbf{V} = 3$ may be written as

$$I_1(\mathbf{V}) = \lambda_1 + \lambda_2 + \lambda_3 = 3. \tag{134}$$

A unit cube of a Bell constrained material subjected to a pure homogeneous deformation $x_k = \lambda_k X_k$ (no sum) becomes a rectangular parallelepiped with edges λ_1, λ_2, λ_3. The deformed volume of this rectangular solid is then $\lambda_1\lambda_2\lambda_3$; and in consequence of the Bell constraint (134), it is seen that the sum of the lengths of its edges is $4(\lambda_1 + \lambda_2 + \lambda_3) = 12$, which is also the length of the perimeter of the undeformed cube. Thus, the geometrical interpretation of the Bell constraint is that the length of the perimeter of a unit cube of a Bell constrained material is unchanged when the cube is homogeneously deformed into a rectangular solid (Beatty and Hayes, Ref. 49).

The internal constraint (134) was discovered by Bell (Refs. 43, 44) in a great body of experiments conducted in the context of finite strain plasticity of annealed metals, whereas the theory of Bell constrained materials developed by Beatty and Hayes (Ref. 45) focuses on nonlinear elastic deformations of isotropic solids characterized by the Bell constraint (134). They show that the constitutive equation for a hyperelastic material constrained by (134) for all deformations has the form

$$\mathbf{T} = p\mathbf{V} + \omega_0\mathbf{1} + \omega_2\mathbf{V}^2, \tag{135}$$

in which $p(\mathbf{x})$ is an arbitrary scalar (Lagrange multiplier) due to the workless constraint and the elastic response functions $\omega_\Gamma = \omega_\Gamma(I_2, I_3)$, $\Gamma = 0, 2$, are defined by

$$\omega_0 \equiv \frac{\partial \Sigma}{\partial I_3}, \qquad \omega_2 \equiv -\frac{1}{I_3}\frac{\partial \Sigma}{\partial I_2}. \tag{136}$$

Here $\Sigma = \Sigma(I_2, I_3)$ and $I_k = I_k(\mathbf{V})$. With (135) in hand, Beatty and Hayes (Ref. 45) derive several remarkable and unanticipated results consonant with the experimental results reported by Bell. They derive precisely Bell's parabolic law for uniaxial loading, precisely Bell's parabolic law for pure shear, and precisely Bell's parabolic law relating the deviatoric stress intensity to the corresponding strain intensity reported in his experiments.

Other kinds of interesting and unusual effects have been reported for materials internally constrained by inextensible fibers. Unusual shear and folding effects that occur in the bending and buckling of bars and plates have been described.

Rogers and Pipkin (Ref. 50) studied the deflection of a transversely loaded fiber-reinforced cantilever beam. They found that plane sections normal to the line of centroids remain plane and parallel to one another, but the beam deforms by shear rather than by flexure, certainly an atypical effect. The shear deflection is independent of the length of the beam and the distance along the beam at which the load is applied. The slope of the beam is discontinuous, the portion between the load and the free end remaining horizontal with zero shear stress.

A similar shear displacement effect was predicted by Kao and Pipkin (Ref. 51) for the finite, plane strain buckling of a fiber-reinforced thick slab under compressive dead loads. They showed that the critical buckling load parallel to the direction of the fibers is proportional to the thickness of the slab but independent of its length!

For a composite material structure in which discrete fibers are uniformly spaced throughout an isotropic, linearly elastic rubber matrix, Schaffers (Ref. 52) has analyzed a shear-induced, in-plane fiber-buckling phenomenon observed in rubber beams reinforced with layers of fibers parallel to the beam length. When the beam is bent sufficiently in the usual sense, the fibers on the compression side of the beam are unable to accommodate their configuration in bending with the matrix, so they buckle laterally in almost sinusoidal patterns. This deformation is communicated through the thickness and may be observed to a lesser extent even on the tension side of the beam. The effect seems to be practically independent of the method of bending. This response is quite different from the aforementioned deflection behavior of the cantilever beam and of the buckled slab reinforced by inextensible fibers. I know of no published experimental studies of either phenomenon; but Charrier (Ref. 53) has also remarked that the collapse of nylon cords and the buckling of steel wires have been observed in certain reinforced transparent rubbers.

Numerous additional references and other examples of inextensible material response may be found in the survey articles by Spencer (Ref. 53) and Pipkin (Refs. 55, 56). It should be mentioned that problems involving constrained materials have been identified with meaningful applications to tires, textiles, inflatable structures, and pressure vessels, to name a few.

13. Boundary-Value Problems and Nonuniqueness in Elastostatics

The condition of equilibrium of a body $\mathscr{B}$ is specified by $\mathbf{a}(\mathbf{X}, t) = 0$ for all particles $\mathbf{X} \in \mathscr{B}$ and for all times t. The general boundary value problem of elastostatics associated with (21) and (22) consists of finding a motion $\mathbf{x}(\mathbf{X})$ that satisfies the equilibrium equation

$$\text{Div}\, \mathbf{T}_{\mathrm{R}}(\nabla \mathbf{x}) + \mathbf{b}_{\mathrm{R}} = \mathbf{0}, \tag{137}$$

everywhere in $\mathscr{B}$, and the boundary conditions of surface traction and place,

$$\mathbf{t}_N = \mathbf{T}_{\mathrm{R}}(\nabla \mathbf{x})\mathbf{N} \text{ prescribed on } \mathscr{S}_{\mathrm{t}}, \tag{138}$$

$$\mathbf{x}(\mathbf{X}) \text{ prescribed on } \mathscr{S}_{\mathrm{X}}, \tag{139}$$

where $\mathscr{S}_{\mathrm{t}}$ and $\mathscr{S}_{\mathrm{X}}$ are disjoint parts of $\partial\mathscr{B}$ in the reference configuration κ_{R}: $\partial\mathscr{B} = \mathscr{S}_{\mathrm{t}} \cup \mathscr{S}_{\mathrm{X}}$. Of course, (23) also must be respected. These equations are expressed in terms of the engineering stress because the deformed body geometry generally is unknown *a priori*; otherwise, (17), (18), and (19) may be used.

For a hyperelastic material defined by (29), it can be shown that the system of equations (137) are the Euler–Lagrange equations and boundary conditions of the variational equation

$$\delta E(\mathbf{F}; \mathbf{X}) = \int_{\partial\mathscr{B}} \mathbf{t}_N \cdot \delta\mathbf{x}\, dA + \int_{\mathscr{B}} \mathbf{b}_R \cdot \delta\mathbf{x}\, dV - \delta \int_{\mathscr{B}} \Sigma(\mathbf{F}; \mathbf{X}) dV = 0, \tag{140}$$

where the integrations are over $\mathcal{B}$ in κ_R. This equation is postulated for arbitrary differentiable variations $\delta\mathbf{x}(\mathbf{X})$ that respect (139). For incompressible materials, the variational principle (140) must be modified to account for the incompressiblity constraint (82) by use of a Lagrange multiplier—in this case, an undetermined hydrostatic pressure. Other constraints are handled similarly (Green and Adkins, Ref. 28). In the dynamical formulation, initial conditions are appended and the energy density is adjusted to include the kinetic energy density. Details of the variational method and some applications to stability and uniqueness questions are given by Ericksen (Ref. 57), Ericksen and Toupin (Ref. 58), Green and Adkins (Ref. 28), and Pearson (Ref. 59). Extension of (140) to include effects of surface and body couples may be found in the article by Toupin (Ref. 60).

Various kinds of traction conditions may be specified in (138). Dead loading is an important and simple example of a traction boundary condition where $\mathbf{t}_N$ is a constant vector $\boldsymbol{\tau}(\mathbf{X})$ at each material point $\mathbf{X}$ in κ_R. That is, $\mathbf{T}_R(\nabla\mathbf{x})\mathbf{N} = \boldsymbol{\tau}(\mathbf{X})$ is a function of only the material point $\mathbf{X}$ in κ_R. In a pure dead load problem, the body force $\mathbf{b}_R$ also is assumed to vary only with $\mathbf{X}$; in fact, it usually is assumed constant, often zero, over κ_R. Dead loads have been used frequently in uniqueness and stability studies. See Bryan (Ref. 61) Pearson (Ref. 59), Hill (Ref. 62), and Beatty (Refs. 63, 64) for examples. Other types of loading will vary with the deforming configuration.

Hydrostatic loading is an example of a traction condition that depends on the motion $\mathbf{x}(\mathbf{X})$. In this case, the traction at each point $\mathbf{x}$ on $\partial\mathcal{B}$ in κ is an all-around stress $p(\mathbf{x})$, either a pressure or a tension, normal to the surface at $\mathbf{x}$. In terms of the Cauchy stress, $\mathbf{T}(\nabla\mathbf{x})\mathbf{n}(\mathbf{x}) = p(\mathbf{x})\mathbf{n}(\mathbf{x})$. The transformation of such expressions in terms of the engineering stress, or conversely, is straightforward. The stability of hyperelastic bodies under hydrostatic loading has been studied by Hill (Ref. 62), Green and Adkins (Ref. 28), and Beatty (Ref. 65). The results show that an equilibrium configuration is stable for hydrostatic loading provided that the response functions of shear and bulk compression are positive. A general theory of configuration-dependent loading in finite strain has been investigated by Sewell (Ref. 66), and Batra (Ref. 67) has reported on the effects of loading devices that characterize the nonlocal interaction between a body and its action environment. We may perceive from these few remarks that uniqueness and stability are related issues that depend on the nature of the loading and of the devices that produce it.

13.1. Simple Examples Describing Nonuniqueness. Some heuristic examples will be presented to demonstrate that uniqueness of the solution to a boundary value problem in finite elasticity generally is not to be expected. We begin with the traction problem.

(i) Consider a thin tube or hemispherical shell in equilibrium with null body force and surface tractions. Then $\mathbf{F} = \mathbf{1}$, i.e., $\mathbf{x} = \mathbf{X}$ is a trivial solution. And any rigid motion is another. But a nontrivial solution for which $\mathbf{F} \neq \mathbf{1}$ also is obtained for the same null tractions and body forces when the tube or hemispherical shell is turned inside out. An everted toy balloon provides a realistic model to illustrate the nonunique, traction-free, everted state of virtually an entire spherical shell.

(ii) A cylinder with traction-free lateral surfaces and subject to equal and oppositely directed end loads may buckle from its straight shape into a distorted shape under the same sufficiently large dead loads.

(iii) Suppose a body is subjected to a uniform hydrostatic loading. Now rotate it rigidly through any angle whatever while maintaining the same hydrostatic loading.

Nonuniqueness also may be expected in the displacement problem. We may imagine a thick-walled and highly elastic tube constrained between and bonded everywhere to the boundaries of infinitely long and rigid, concentric cylinders. Of course, $\mathbf{F} = \mathbf{1}$ is a trivial solution for null boundary displacements. Now imagine that the outer casing is rotated through 360° while the inside shaft is held fixed. The boundary is the same as before, but the interior is now severely deformed. Similar examples have been described by Truesdell and Noll (Ref. 11, Section 44) and by Gurtin (Ref. 15, Chapter 10; Ref. 68).

In the mixed problem, nonuniqueness is exhibited by a cantilever beam fixed at one end, subjected to compressive dead loading at the other, and having traction-free lateral surfaces. The beam may either buckle or remain straight when the end load is sufficiently large. Other examples have been described by Gurtin (Refs. 15, 68).

An especially interesting and unusual example of nonuniqueness in the traction problem for a neo-Hookean material will be illustrated next.

13.2. Rivlin's Cube. Substitution of (101) into (20) and use of (82) and (9b) yields the constitutive equation for the neo-Hookean material in terms of the engineering stress,

$$\mathbf{T}_R = -p\mathbf{F}^{-T} + \beta_1\mathbf{F}, \tag{141}$$

where $\beta_1 = \mu_0 > 0$ is a constant and p is an unknown pressure. Some interesting and unexpected results obtained by Rivlin (Refs. 19, 69, 70) concern the homogeneous deformation of a homogeneous unit cube of neo-Hookean material loaded uniformly by three identical pairs of equal and oppositely directed forces acting normally to its faces. The dead load boundary condition on the face with the unit coordinate normal $\mathbf{N}_k$ is

$$\mathbf{T}_R \mathbf{N}_k = \tau \mathbf{N}_k. \tag{142}$$

Rivlin studied homogeneous solutions of the form

$$\mathbf{T}_R = \tau \mathbf{1}, \text{ with } \mathbf{F} = \sum_{k=1}^{3} \lambda_k \mathbf{N}_k \otimes \mathbf{N}_k, \tag{143}$$

and $\lambda_k > 0$, constant. Thus, the equilibrium equations (137) with zero body force are identically satisfied. The incompressibility condition (82) requires

$$\lambda_1 \lambda_2 \lambda_3 = 1 \tag{144}$$

to hold for all $\lambda_k > 0$. The constitutive equation (141) relates the applied forces to the homogeneous deformation (143); we have

$$\tau = -\frac{p}{\lambda_k} + \beta_1 \lambda_k. \tag{145}$$

Equations (144) and (145) provide four simultaneous equations for λ_k and p. But the solution of this system is not unique. Elimination of p from (145) yields

$$(\lambda_k - \lambda_\ell)\left[\frac{\tau}{\beta_1} - (\lambda_k + \lambda_\ell)\right] = 0, \tag{146}$$

for $k, \ell = 1, 2, 3$. Thus, if the applied forces τ are specified, (146) and (144) reveal that the equilibrium configuration of pure homogeneous deformation may not be uniquely determined.

For a uniform tension $\tau > 0$ on all of the faces, Rivlin (Ref. 69) found that seven possible equilibrium states exist. The trivial state

$$\text{(i)} \quad \lambda_1 = \lambda_2 = \lambda_3 = 1 \tag{147}$$

is always a solution. But there are two further nontrivial cases for which

$$\text{(ii)} \quad \lambda_1 = \lambda_2, \qquad 0 < \lambda_3 < \frac{\tau}{3\beta_1}, \tag{148}$$

$$\text{(iii)}\quad \lambda_1 = \lambda_2, \qquad \frac{\tau}{3\beta_1} < \lambda_3 < \frac{\tau}{\beta_1}. \tag{149}$$

And there are four other states obtained from these two by cyclic permutation of the λ's.

The stability of each of these states was investigated with respect to superimposed, arbitrary, infinitesimal deformations $\mathbf{u}(\mathbf{x})$ with gradient $\mathbf{H} \equiv \nabla\mathbf{u}$. The homogeneous equilibrium states corresponding to a specified dead loading condition are those for which the energy functional

$$E \equiv \int_{\mathcal{B}} [\Sigma(\mathbf{F}) - \operatorname{tr}(\mathbf{T}_R\mathbf{F}) - p(J - 1)]dV \tag{150}$$

has stationary values with respect to arbitrary infinitesimal deformations $\mathbf{H}$ compatible with the incompressibility constraint (82) and the assigned motion $\mathbf{x} = \mathbf{x}(\mathbf{X})$. The Lagrange multiplier is denoted by the undetermined pressure $p = p(\mathbf{X})$, and $\mathbf{T}_R$ is the engineering stress associated with the state of homogeneous deformation whose stability is in question. The same formulation may be used for compressible materials by removing the constraint and putting $p = 0$ in (150) to obtain

$$E \equiv \int_{\mathcal{B}} [\Sigma(\mathbf{F}) - \operatorname{tr}(\mathbf{T}_R\mathbf{F})]dV. \tag{151}$$

In either case, the equilibrium states are given by

$$\delta E = 0 \tag{152}$$

for all allowable $\mathbf{H}$. An equilibrium state so obtained is called stable if the corresponding stationary value of E is a minimum. That is, if

$$\delta^2 E > 0 \tag{153}$$

for all allowable infinitesimal deformations $\mathbf{H}$. Neutral stability, the case when $\delta^2 E = 0$ for some $\mathbf{H}$ but otherwise is always positive, is neglected.

With the aid of this energy criterion of stability Rivlin (Ref. 69) found that the state (ii) and its two permuted relatives are stable, while the state (iii) and its two relatives are unstable. The state (i) is always a solution, but it may be stable or unstable according as $\tau/\beta < 2$ or > 2, respectively. Hence, for sufficiently large tensile loads there are seven different solutions of homogeneous deformation! Three are inherently unstable, and three are always stable. And there may be even more solutions for a larger class of deformations under the same loads.

For compressive loads $\tau < 0$, equations (144) and (146) yield only the trivial solution $\lambda_1 = \lambda_2 = \lambda_3 = 1$. Beatty (Ref. 64) showed that this undeformed state is unstable when $\tau < 0$, and pointed out that this arises essentially from instability with respect to rigid body rotations. Other stable equilibrium states of homogeneous deformation maintained by three pair of equal and opposite forces, of which at least two pair are distinct, exist. The stable state which actually may be attained will depend on the order in which the forces are applied. Additional details are provided by Rivlin (Refs. 19, 69). See also Gurtin (Ref. 15).

Some further remarks concerning stability and uniqueness will be presented later. We now turn to some fundamental general elastostatic solutions obtained by inverse methods.

14. Universal Inverse Solutions

Suppose we are given a certain rubberlike material, and we are assigned the task of characterizing its elasticity by an appropriate constitutive equation that will enable us to predict its response to specified loading and displacement boundary conditions. Although many materials, such as an ordinary sheet of paper, have oriented structures and may exhibit time-dependent behavior, let us assume as a first approximation that our rubberlike material may be modeled as either a compressible or an incompressible, homogeneous, and isotropic hyperelastic material. Then our task is reduced to the study of some series of boundary value problems that may be helpful in the design of an independent body of experiments by which the forms of the elastic response functions β_Γ in (46) or (91) may be determined. Of course, the independent kinematic constraint of incompressibility (82) will allow us to decide if, in fact, our material may be incompressible or not. Homogeneous deformations that are not automatically isochoric clearly will suffice. Beatty and Stalnaker (Ref. 16) have shown that a plot of data for $\ln\lambda_1^{-1}$ versus $\ln\lambda$ obtained in a simple tension test, for example, might be used for this purpose. For an incompressible material, these kinematical data must map a straight line of slope $\frac{1}{2}$.

But what experiments should be designed to determine the elastic response functions β_Γ for either a compressible or an incompressible, isotropic hyperelastic material? And what ultimate independent experiment may be designed to test the final form of the mathematical model used to characterize the given material? It is clear, in particular, that the experimenter must know *a priori* the class of deformations that actually may be produced in every compressible or incompressible, homogeneous and isotropic, hyperelastic material by the application of surface loading alone. Also, the surface loads needed to effect them must be known in order to select the kinds of loading devices that may be used.

On the mathematical side, it is equally clear that the system consisting of the equilibrium equations (137), the boundary conditions (138) and (139), and the constitutive equation (46) or (91) for an isotropic hyperelastic solid comprise a formidable system of nonlinear, partial differential equations. The complexity of this system and its potential for generating nonunique solutions for even the simplest of boundary value problems, as illustrated in the example of Rivlin's cube, overwhelm our ability to solve them generally. Consequently, instead of seeking general solutions for specified boundary data, we are essentially forced by mathematical difficulties to adopt a different, inverse strategy.

A suitable class of smooth deformations of physical interest and characterized by a number of parameters is chosen for study. Each assigned internal, kinematic constraint, such as incompressibility, is used to find restrictions on the deformation parameters. Then the constitutive equation is used to determine the stress distribution that will satisfy the differential equations of equilibrium without the introduction of peculiar body forces. Finally, the surface loading necessary to maintain the deformation in this equilibrium configuration is determined. This so-called inverse or semi-inverse method was used by Rivlin (Refs. 19–23) to construct by special examples a collection of exact solutions to a number of traction boundary value problems that yielded significant results of physical interest to both analysts and experimenters. This work marked the birth in 1948 of the modern theory of finite elasticity. It is, in fact, a thorough, but incomplete response to just what the experimenter ordered in our initial task assignment. A different and more general approach to the investigation of inverse solutions was introduced by Ericksen (Ref. 71). We shall see that Ericksen's results provide the kinds of tools requested by our experimenter.

14.1. Ericksen's Problem. A deformation that can be produced in a material by the application of surface tractions alone[2] is called a controllable deformation. A controllable deformation that can be effected in every homogeneous, isotropic hyperelastic material is called a universal deformation. The problem of determining all such universal deformations for the two important classes of compressible and incompressible, homogeneous, and isotropic hyperelastic materials was initiated by Ericksen (Refs. 71, 72), and is now widely known as Ericksen's problem.

14.1.1. Solution for Compressible Materials. Ericksen (Ref. 72) proved that homogeneous deformations are the only controllable deformations possible

[2]Because the arbitrary pressure function in (91) may be adjusted to remove the body force potential in application of (137), it is easily seen that a deformation which is possible in an incompressible material with zero body force is possible in the same material acted on by any conservative body force whatever (Truesdell and Noll, Ref. 11, Section 56).

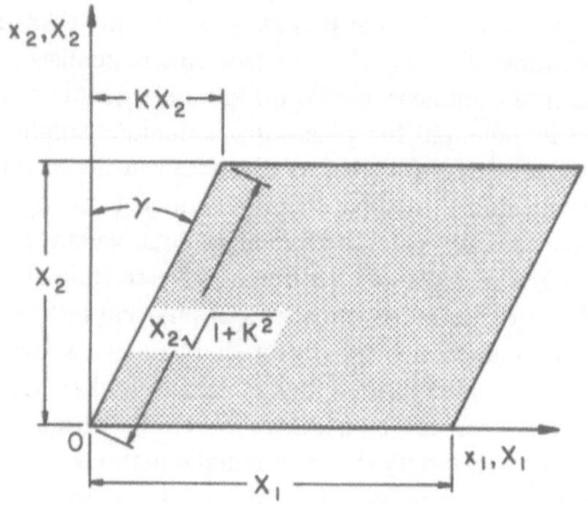

Fig. 13. Simple shear of a block. The amount of shear is $K = \tan\gamma$.

in every compressible, homogeneous, and isotropic hyperelastic material.[3] These are described as

Family 0: Homogeneous deformations.

$$\mathbf{x} = \mathbf{F}\mathbf{X} + \mathbf{c}, \tag{154}$$

wherein **F** is a constant tensor and **c** is a constant vector.

The most interesting example of a homogeneous deformation is a simple shear defined by

$$x_1 = X_1 + KX_2, \qquad x_2 = X_2, \qquad x_3 = X_3, \tag{155}$$

in a common rectangular Cartesian frame in which x_k denote the coordinates of a material point in κ whose coordinates are X_k in κ_R. The geometry is illustrated in Fig. 13. The simple shear is an isochoric deformation that is possible in every compressible, homogeneous, and isotropic hyperelastic material. The constitutive equation (46) shows that the shear stress required to effect a shear of amount $K = \tan\gamma$ is given by

$$T_{12} = K\mu(K^2), \tag{156}$$

[3]A somewhat easier and direct proof of Ericksen's theorem on controllable deformations in compressible materials was found by Shield (Ref. 73). See also Truesdell and Noll (Ref. 11, Section 91).

wherein the generalized shear response function is defined by

$$\mu(K^2) \equiv \beta_1(K^2) - \beta_{-1}(K^2). \tag{157}$$

It is seen that the shear stress is an odd function of the amount of shear. Notice that the shear stress is in the direction of the shear if and only if $\mu(K^2) > 0$. We see that the empirical inequalities (67) support this physically essential requirement. However, shear stress alone does not suffice to produce a simple shear.

It also follows from (46) that additional normal stresses must be supplied on all pairs of the plane faces. In fact, the response functions $\beta_1(K^2)$ and $\beta_{-1}(K^2)$ are determined by the normal stress differences; we have

$$T_{11} - T_{33} = \beta_1 K^2, \tag{158a}$$

$$T_{22} - T_{33} = \beta_{-1} K^2, \tag{158b}$$

where

$$T_{33} = \beta_0 + \beta_1 + \beta_{-1} \equiv \tau(K^2)K^2. \tag{159}$$

The last relation determines $\tau(K^2)$, hence $\beta_0(K^2)$. Since the response functions are even functions of K, the normal stresses are unchanged when the shear is reversed. If these are not furnished, the block will tend to contract or to expand. Such normal stress effects are typical of problems in finite elasticity. And there is more.

The most striking feature of the simple shear problem is that the results (158) and (159) are not determined by the shear stress. On the contrary, the shear stress is determined by the normal stress difference

$$KT_{12} = T_{11} - T_{22}, \tag{160}$$

and it is determined in the same way for every homogeneous, isotropic hyperelastic material regardless of the form of the response functions. The formula (160) is an example of a universal relation in finite elasticity theory. If a material cannot satisfy the rule (160) in a properly designed simple shear experiment, at least in principle,[4] then that material cannot be modeled as an isotropic hyperelastic material, whatever may be its response functions. We shall have more to say about universal relations later. We continue with Ericksen's problem for incompressible materials.

[4]In practice, however, it is not likely that a global simple shear deformation may be produced in any real material.

14.1.2. Solution for Incompressible Materials. In addition to the Family 0 of homogeneous deformations restricted to satisfy the incompressibility condition (82), Ericksen (Ref. 71) found for incompressible materials four other families of nonhomogeneous, controllable deformations. These are described below.

Family 1: Bending, stretching, and shearing of a rectangular block,

$$r = \sqrt{2AX}, \qquad \theta = BY, \qquad z = \frac{Z}{AB} - BCY. \tag{161}$$

Family 2: Straightening, stretching, and shearing of a sector of a hollow cylinder,

$$x = \frac{1}{2}AB^2R^2, \qquad y = \frac{\Theta}{AB}, \qquad z = \frac{Z}{B} + \frac{C\Theta}{AB}. \tag{162}$$

Family 3: Inflation, bending, torsion, extension, and shearing of an annular wedge,

$$r = \sqrt{AR^2 + B}, \qquad \theta = C\Theta + DZ, \qquad z = E\Theta + FZ, \tag{163}$$

with $A(CF - DE) = 1$.

Family 4: Inflation or eversion of a sector of a spherical shell,

$$r = \left[\pm R^3 + A\right]^{1/3}, \qquad \theta = \pm\Theta, \qquad \phi = \Phi. \tag{164}$$

In these relations (x, y, z), (r, θ, z), and (r, θ, ϕ) are, respectively, the usual rectangular, cylindrical, and spherical coordinates of the material point in κ, and (X, Y, Z), (R, Θ, Z), and (R, Θ, Φ) have the same meaning in κ_R. The parameters A, B, C, D, E, F are constants. Of course, $AB \neq 0$ in (161) and (162). The deformation families are difficult to visualize, except in stages. The deformation Family 1 is illustrated in Figs. 14 and 15, for example. All of these problems are summarized in Section 57 of the comprehensive treatise by Truesdell and Noll (Ref. 11); Ericksen's theorems are outlined in Section 91. See also Wang and Truesdell (Ref. 12).

14.2. Elusive Conclusion of Ericksen's Problem. Completeness of the foregoing solutions was established by Ericksen except for two cases:

(i) Two equal principal stretches with at least one nonconstant principal strain invariant.

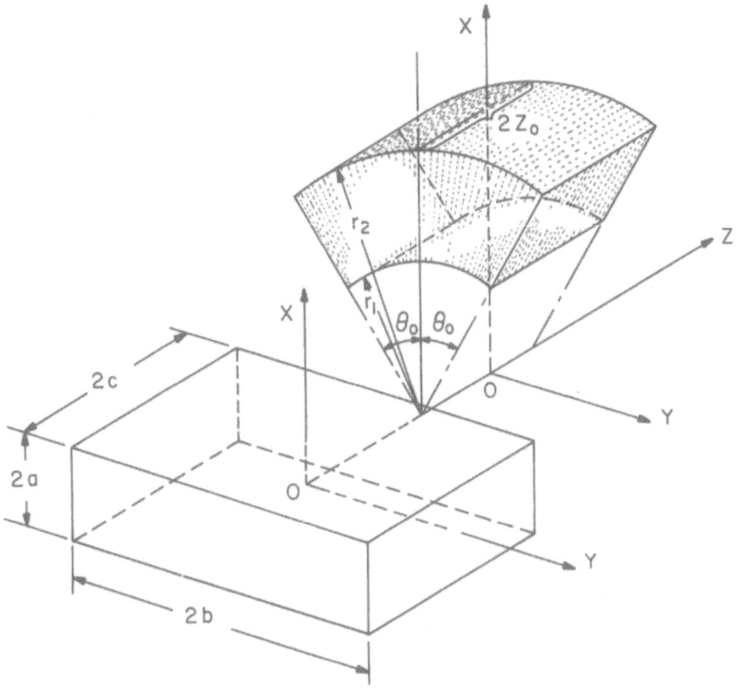

Fig. 14. Bending of a block with stretch: $r = (2AX)^{1/2}$, $\theta = BY$, $z = Z/(AB)$.

(ii) Constant principal strain invariants I_1 and I_2.

These unresolved aspects of Ericksen's problem subsequently led others to search for its conclusion.

Marris and Shiau (Ref. 74) showed that there were no further solutions in case (i). And with this result in hand, Martin and Carlson (Ref. 75) confirmed that case (i) contains none of the known universal deformation families. But the conclusion of case (ii) has proved more elusive. Ericksen had conjectured earlier that deformations in case (ii) must necessarily be homogeneous deformations. Later, however, Fosdick (Ref. 76) noticed that the counterexample

$$r = A^{1/2}R\,, \qquad \theta = C\Theta, \qquad z = FZ, \qquad ACF = 1, \tag{165}$$

a special case among the examples characterized by Ericksen as Family 3, has constant strain invariants and is not homogeneous if $C \neq 1$. Following Fosdick's lead, Klingbeil and Shield (Ref. 77) and Singh and Pipkin (Refs. 78, 79) discovered independently a distinct additional family of controllable, nonhomo-

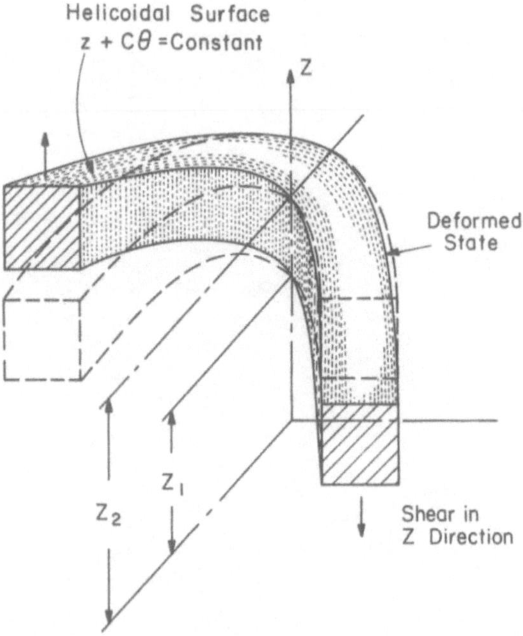

Fig. 15. Bending, stretching, and shearing of a rectangular block: $r = (2AX)^{1/2}$, $\theta = BY$, $z = \lambda Z - BCY$, $\lambda \equiv 1/(AB)$.

geneous deformations characterized by constant strain invariants. This is identified as

Family 5: Inflation, bending, extension, and azimuthal shearing of an annular wedge,

$$r = A^{1/2}R, \qquad \theta = D\log(BR) + C\Theta, \qquad z = FZ, \qquad ACF = 1. \quad (166)$$

This is essentially the form obtained by Singh and Pipkin.[5] It is seen that Fosdick's example (165) is included in (166) when $D = 0$; and the special case $AC = 1$, $C^2 + D^2 = 1$ yields the result of Klingbeil and Shield (Ref. 77).

Completeness of the solution family (166) has not been established. Therefore, the question of existence of other families having constant principal invari-

[5]Holden (Ref. 80) has used complex variables to derive (166); and Huilgol (Ref. 81) has extended Family 5 to transversely isotropic materials.

ants remains open. But several important advances toward its solution have been made.

Fosdick and Schuler (Ref. 82) showed that for plane deformations with uniform transverse stretch, other than homogeneous deformations, Family 5 represents the complete class of constant invariant, controllable plane[6] deformations that can be produced in every incompressible, isotropic hyperelastic material. Hence, any other solution in this class must be three dimensional in character. Subsequently, Fosdick (Ref. 83) showed further that there are no additional new solutions for the class of radially symmetric deformations.

Further attempts to characterize the terminal link in this chain of studies of Family 5 have produced no new solutions. Müller (Refs. 84, 85) studied certain generalized plane, cylindrical, and spherical deformations. Kafadar (Ref. 86) proved that a solution cannot be described in a holonomic coordinate system in which one of the proper vectors $\mathbf{e}_k$ of $\mathbf{B}$ is normal to a coordinate surface. And Marris (Ref. 88) showed that no additional solutions exist for which any two of the abnormalities $\pi_{kk} \equiv \mathbf{e}_k \cdot \text{curl}\, \mathbf{e}_k$, $k = 1, 2, 3$ (no sum), are constant, hence he concluded that any new solution can have at most one principal vector $\mathbf{e}_k$ determining a vector field of constant, nonzero abnormality. There is no doubt that if any new solution with constant principal invariants exists, it will be complicated.

Finally, it should be pointed out that in addition to homogeneous deformations, certain controllable, nonhomogeneous deformations may be produced in special kinds of compressible, homogeneous, and isotropic hyperelastic materials, that is, in materials having specific response functions β_Γ. Some examples are provided by Currie and Hayes (Ref. 89), Holden (Ref. 90), and Parry (Ref. 91). Of course, Ericksen's solutions are different in that they constitute universal solutions; they apply to every homogeneous, isotropic, hyperelastic material regardless of the form of the response functions β_Γ. Ericksen (Refs. 57, 92) has introduced new techniques to simplify and improve semi-inverse methods by application of certain generalized coordinates, and he demonstrates these in analysis of tension, torsion, and bending of prisms, circular beams, and helical springs.

[6] Knowles (Ref. 87) examined Ericksen's problem for a finite antiplane shear deformation defined on a cylindrical body by the axial displacement vector field $\mathbf{u} = u(x_1, x_2)\mathbf{k}$ in Cartesian coordinates x_k. He showed that for homogeneous, isotropic, and compressible hyperelastic materials the only universal solution is a simple shear. For incompressible materials, he proved that there is one and only one further universal antiplane deformation. This deformation carries a typical cross section of the cylinder into a portion of a right helicoid, as illustrated in Fig. 15. These results, however, as emphasized by Knowles, are special cases of the universal deformations obtained more generally by Ericksen (Refs. 71, 72).

Ericksen's universal solutions are precisely the kinds of practical mathematical results sought by experimenters to guide their design of tests and loading devices for practical evaluation of material response, and they provide the tools that may serve ultimately to test hyperelasticity theory itself. The classical experiments by Rivlin and Saunders (Ref. 1) were designed in just this way; and together with other early basal experiments by Mooney, Treloar, and other co-workers, many of which are described in the treatise by Treloar (Ref. 24), the Rivlin–Saunders experiments provided essential nourishment from which the seeds of finite elasticity have grown to a meaningful physical theory.

An example of a controllable, nonhomogeneous deformation that can be produced in a particular kind of compressible material will be investigated next. The noncontrollable aspect of a more general class of materials also will be discussed.

15. Nonuniversal Inverse Solutions: Example

We have learned that only homogeneous deformations are possible in every unconstrained, isotropic hyperelastic material. Of course, universal deformations are not the only types of deformation possible in an isotropic material. But nonuniversal, controllable deformations can be discussed only with respect to specific constitutive equations. Hence, Ericksen's theorem does not preclude the possibility of controllable, nonhomogeneous deformations in special kinds of compressible, isotropic materials. This will be demonstrated for a Blatz–Ko cylinder of radius a subjected to a pure torsion described by (163) with $B = E = 0$, $A = C = F = 1$. This is an isochoric deformation; hence, $J_3 = \det \mathbf{F} = 1$. In the physical tensor basis $\mathbf{e}_{ij} = \mathbf{e}_i \otimes \mathbf{e}_j$ of the current cylindrical reference system (r, θ, z), we obtain by (9b)

$$\mathbf{B}^{-1} = \mathbf{1} + K^2\mathbf{e}_{33} - K(\mathbf{e}_{23} + \mathbf{e}_{32}), \tag{167}$$

wherein $K \equiv Dr$ corresponds locally to the amount of shear in a simple shear of amount K, as described before. Of course, D is the angle of twist per unit length of the cylinder; and, for brevity, we shall call it the twist. The invariants (51) for the pure torsion problem are the same as those for a simple shear of amount K, namely, $J_1 = J_2 = I_1 = I_2 = 3 + K^2$.

Let us consider the special Blatz–Ko material (79). Then with (167), the stress distribution is described by

$$\mathbf{T} = \mu_0 K(\mathbf{e}_{23} + \mathbf{e}_{32} - K\mathbf{e}_{33}). \tag{168}$$

That is, the only nonzero physical stress components are the axial and shear stress components given by

$$T_{33} = -\mu_0 K^2, \tag{169a}$$

$$T_{23} = \mu_0 K. \tag{169b}$$

Thus, for every Blatz–Ko material of the kind (79), (169a) and (169b) yield the following universal relation for the pure torsion of a cylinder:

$$T_{33} = -KT_{23}. \tag{170}$$

Equation (169b) shows that the shear stress is in the direction of the shear if and only if $\mu_0 > 0$. This supports the empirical inequalities (68) only for f very nearly zero, the common assumption for this model. Equation (169a) indicates that a compressive axial stress arises in response to the twist; and if appropriate compressive end loads are not supplied to maintain the cylinder at its undeformed length, the twist will induce elongation and the pure torsion condition will fail.

It can be shown that the equilibrium equations with zero body force are identically satisfied. Therefore, a pure torsion is controllable in every homogeneous Blatz–Ko material in the class (79).

The null traction condition on the lateral surface is satisfied. The stress distribution (169) over the ends of the cylinder requires the application of a compressive end thrust of magnitude N and equal and oppositely directed axial torques of magnitude M. These are given by

$$N = GD^2, \qquad M = GD, \tag{171}$$

in which the torsional stiffness $G \equiv \mu_0 I_0$ and $I_0 \equiv \pi a^4/2$, as usual. Thus, unlike the classical theory, compressive end loads proportional to the square of the twist must be supplied to prevent elongation of the cylinder. The torque is proportional to the twist, the constant of proportionality being the torsional stiffness, precisely as in the linear theory of torsion. And further, we derive by (171) the following special universal rule relating the compressive thrust and the applied torque to the angle of twist per unit length of the cylinder:

$$N = DM. \tag{172}$$

Thus, for the special class of Blatz–Ko materials (79), the compression must increase with the torque in the same way for every cylinder in the class, regardless of its torsional stiffness. This phenomenon raises the natural question of the potential instability of the bar under sufficiently large twist. But we shall leave this question for another place.

It can be shown that for the more general Blatz–Ko material (56) for which (57) holds, the equilibrium equations are satisfied with zero body force if and only if $f = 0$, which is the case studied above. However, if we admit an artificial body force by rotating the cylinder rigidly about its axis with a constant angular speed ω while subjecting it to a pure torsion in the spinning reference frame, the equations of motion can be satisfied provided that

$$\omega = Dc_T f^{1/2}. \tag{173}$$

Herein $c_T \equiv (\mu_0/\rho_0)^{1/2}$ is the classical speed of transverse waves in κ_R. The case $\omega = 0$ holds only for the model $f = 0$. The other special case (80), for which $f = 1$, requires $\omega = Dc_T$. The result (173) is independent of the stress, but it varies with the amount of twist D.

The deformation in this case is not controllable; it requires the special and variable body force $\mathbf{b} = \rho_0 r\omega^2 \mathbf{e}_r$ per unit volume to maintain the pure torsion. The lateral surface of the cylinder must be free. It turns out that the end thrust and torque required for the general Blatz–Ko material are given by

$$N = G(1 - f)D^2, \qquad M = GD, \tag{174}$$

and these yield the rule

$$N = D(1 - f)M \tag{175}$$

relating the thrust and the torque in the rotating frame. We thus reach the interesting result that a Blatz–Ko material with $f = 1$ requires no end thrust whatever to produce the pure torsion provided that the spin (173) can be maintained at each value of the twist D. Hence, no amount of twist can induce elongation or contraction of a properly rotated Blatz–Ko cylinder belonging to the class (80). More generally, we find that if the empirical inequalities (68) hold, compressive end loads must be furnished to prevent elongation of the cylinder when $f \neq 1$. For every Blatz–Ko cylinder under a pure torsion, the end thrust is proportional to the square of the twist and the torque is proportional to the twist, both being dependent on the torsional stiffness. Their ratio N/M, however, is independent of the torsional stiffness. Equation (175) is not a special universal relation because it depends on the material parameter f.

We observe that the universal relation (170) after adjustment of the index labels, is really a special case of the universal formula (160) found for simple shear. We shall learn in the next section that the universal relations for simple shear and torsion are members of a larger class of universal relations for both compressible and incompressible, isotropic materials.

16. Class of Universal Relations

Universal relations are equations that hold for every material in a specified class, regardless of the response functions. Their importance is underscored by the fact that if a universal relation cannot be satisfied by data obtained in a suitably designed laboratory test of a given material, then that material cannot be a candidate for inclusion in the class. A general class of universal relations for isotropic elastic materials characterized by the constitutive equations (46) and (91) will be described.

Beatty (Ref. 93) recognized that a class of universal relations for isotropic elastic materials is characterized by the tensor equation

$$\mathbf{TB} = \mathbf{BT}. \tag{176}$$

Of course, a similar relation holds for $\mathbf{B}^{-1}$. This simple rule is an immediate consequence of (46) or (91). It obviously holds for every compressible or incompressible, isotropic elastic material, whatever may be the form of the response functions β_Γ. Moreover, (176) stands independently of the equations of motion, either static or dynamic; it states that the tensor $\mathbf{TB}$ is symmetric. Therefore, the general universal relation (176) yields the following three scalar equations expressed in terms of the physical components T_{ij} and B_{ij} of $\mathbf{T}$ and $\mathbf{B}$, respectively:

$$B_{12}(T_{11} - T_{22}) = (B_{11} - B_{22})T_{12} + B_{13}T_{32} - T_{13}B_{32}, \tag{177a}$$

$$B_{23}(T_{22} - T_{33}) = (B_{22} - B_{33})T_{23} + B_{21}T_{13} - T_{21}B_{13}, \tag{177b}$$

$$B_{31}(T_{33} - T_{11}) = (B_{33} - B_{11})T_{31} + B_{32}T_{21} - T_{32}B_{21}. \tag{177c}$$

It is well known that the rule (176) is necessary and sufficient for coincidence of the principal directions of the stress $\mathbf{T}$ and the deformation $\mathbf{B}$ in an isotropic material. The result that the proper directions of $\mathbf{T}$ coincide with those of $\mathbf{B}$ is a trivial consequence of (46), or (91), alone; but the converse, particularly when there is a multiplicity of principal values, requires more. The relations (177) have been applied by Batra (Ref. 17) to show, conversely, that the principal vectors of $\mathbf{B}$ coincide with those of $\mathbf{T}$. If the principal values of $\mathbf{T}$ are distinct, so are those of $\mathbf{B}$ and nothing more is required. Hence, in this case, the theorem expresses a universal condition. However, if two or three of the principal values of $\mathbf{T}$ are equal, $\mathbf{T}$ and $\mathbf{B}$ will certainly have identical principal directions, but it turns out that $\mathbf{B}$ also will have corresponding equal principal values provided the empirical inequalities (67) hold for all deformations of the material. This result also expresses a property valid for every compressible and incompressible, isotropic elastic material whose response functions, in the sense of (67), are reasonably well-behaved but otherwise unspecified; it is virtually universal.

Furthermore, the equations (177) are the generators of many universal relations for isotropic elasticity theory, including the well-known universal rule for a simple shear (Beatty, Ref. 93). To see this, it suffices to consider a class of problems for which the deformation tensor (9b) has the representation

$$\mathbf{B} = B_{11}\mathbf{e}_{11} + B_{22}\mathbf{e}_{22} + B_{33}\mathbf{e}_{33} + B_{12}(\mathbf{e}_{12} + \mathbf{e}_{21}) \tag{178}$$

in the physical basis $\mathbf{e}_{ij}$. Clearly, $\mathbf{e}_3$ is a principal vector for $\mathbf{B}$. It follows for an isotropic material (46) or (91) that $\mathbf{e}_3$ also is a proper vector for $\mathbf{T}$; and hence $\mathbf{T}$ has the same physical component form as (178). Conversely, if $\mathbf{T}$ has the form (178), Batra's theorem (Ref. 17) shows that $\mathbf{B}$ does too. Hence, in every problem for which either $\mathbf{B}$ or $\mathbf{T}$ has the representation (178), the system (177) reduces to the single universal rule

$$\frac{T_{11} - T_{22}}{T_{12}} = \frac{B_{11} - B_{22}}{B_{12}}. \tag{179}$$

This rule shows clearly that the orthogonal principal directions for $\mathbf{T}$ and $\mathbf{B}$ in the plane normal to $\mathbf{e}_3$ coincide, one direction ϕ being given by the familiar formula $\tan 2\phi = 2B_{12}/(B_{11} - B_{22})$. But there is more.

In particular, for the simple shear (155), (9b) becomes

$$\mathbf{B} = (1 + K^2)\mathbf{e}_{11} + \mathbf{e}_{22} + \mathbf{e}_{33} + K(\mathbf{e}_{12} + \mathbf{e}_{21}). \tag{180}$$

Hence, use of (180) in (179) yields easily the familiar rule (160) relating the shear stress to the normal stress difference.

This method, based only on the use of (176), provides universal relations possible for both compressible and incompressible, homogeneous, and isotropic elastic materials. Several additional examples of universal relations generated by (177) including those for the universal deformation families, have been derived in this way (Beatty, Ref. 93). In particular, for the Family 3 defined by (163), by a suitable numbering of the physical components, (179) provides the universal relation

$$T_{11} - T_{22} = T_{12}\frac{r^2(C^2 + D^2R^2) - (E^2 + F^2R^2)}{r(CE + DFR^2)}, \tag{181}$$

which determines the local principal directions in the plane. If the material is incompressible, $A(CF - DE) = 1$ must hold. However, the same universal relation may hold for special compressible materials for which the nonhomogeneous deformation family (163) may be controllable but no longer universal. When $B = E = 0$ and $A = C = F = 1$, (181) yields the universal relation (160) in which $K \equiv Dr$ for a pure torsion. Also, for the special Blatz–Ko material (79) we may recover from (181) our earlier universal formula (170). We merely change the order of the indices from 123 to 231 in (181) and afterwards note that $T_{11} = T_{22} = 0$ in (168). Universal relations for certain controllable, nonuniversal deformations have been derived by Currie and Hayes (Ref. 89) using an entirely different approach. Their results, together with some additional universal formulas, are described by Beatty (Ref. 93). Another method for the study of universal relations has been described by Hayes and Knops (Ref. 94). And Beatty and Hayes (Ref. 45) have discussed a class of constrained, isotropic materials that yields a universal relation of the form $\mathbf{TV} = \mathbf{VT}$.

Other universal relations exist that are not members of the class characterized by a rule of the type (176). Rivlin's universal relation relating the torsional modulus to the tensile force and the stretch in torsion and extension of a rod, a special case in Family 3 above, is an example valid for every incompressible isotropic material (Truesdell and Noll, Ref. 11, Section 57). In addition, we have observed earlier several universal relations valid only for special kinds of isotropic materials. We recall the lateral contraction function (60b) for the Blatz–Ko material (59), the stretch (111) at the maximum inflation pressure for a neo-Hookean balloon or (128) for the special Blatz–Ko balloon, and the thrust–twist relation (172) for the pure torsion of a Blatz–Ko cylinder.

Universal deformations and universal results of various kinds are road signs posted to direct and to warn the experimenter in his exploration of the constitutive properties of real materials. If these signs are ignored, he can only wander aimlessly along an otherwise uncharted labyrinth in the vast realm of materials science. If they are thoughtfully evaluated, he will discover the rich rewards locked in the smaller domain of the science of highly elastic materials. But there is more to bear in mind. To avoid wasteful detours, he also must be aware of potential constraints imposed by constitutive inequalities and the long accepted behavior of the classical moduli of isotropic elasticity theory. Constitutive inequalities will be discussed next.

17. Truesdell's Problem: Restrictions on Constitutive Equations

The general nature of the strain energy function has been the key to the remarkable progress achieved in finite elasticity since its rebirth during World War

II. It is, of course, most desirable, though not always possible, to seek problem solutions for arbitrary material response functions. A significant number of universal solutions of problems of physical importance have been obtained by the inverse method; and the same tools have proved effective in the study of nonuniversal solutions for specific isotropic materials. In either case, except for sufficient smoothness tacitly assumed to ensure the differentiability and integration of response functions, in general, no restrictions have been imposed on the nature of the strain energy function. Certainly, the elastic response cannot be fully arbitrary. Even in the linear theory where the strain energy is a quadratic function, it is not arbitrary; the strain energy must be positive definite to ensure reasonable physical results. The question of what restrictions should be imposed on the strain energy function of hyperelasticity theory to capture in the mathematical model the actual physical behavior of isotropic materials in finite deformation forms the substance of Truesdell's problem.

The problem set by Truesdell (Ref. 95) (see also Truesdell and Noll, Ref. 11) concerns the characterization of the class of functions $\Sigma(I_1, I_2, I_3)$ that may serve as strain energy functions for hyperelastic materials. It seeks to identify restrictions to be imposed on constitutive equations to assure, by analysis, sensible physical behavior that the mathematical model is intended to describe, and to ensure existence of solutions with proper smoothness.

The question, as I perceive it, has been studied in two parts. The first part addresses the issue of viable, sometimes empirically motivated, physical restrictions to be set for all isotropic elastic material response. These restrictions, in my mind, provide tools to aid in the physical interpretation of analytical results derived for every isotropic elastic material, hence also for specific isotropic materials. The second part addresses the analytical structure of the theory itself. Certain mathematical conditions essential to assure proper smoothness and existence of solutions at a more abstract level have been investigated. Understandably, these restrictions often lack interpretation in physical terms, though in fact they may be related in some sense to matters of material stability. In some cases, they are shown to be consistent with the physical restrictions set in the first part. This will become evident in the examples below.

17.1. Empirical Inequalities. I do not know if Truesdell's problem has a definitive solution. However, at the least, to model real material behavior, I believe the response functions β_Γ should be compatible with fairly general empirical descriptions of mechanical response derived from carefully controlled large deformation tests of isotropic materials of special kinds. Truesdell and Noll (Ref. 11, Sections 51–53, 153–171) recognized that experimental data available at the time, though of limited extent, appeared to support the empirical inequalities

$$\text{Compressible:} \quad \beta_0 \leq 0, \qquad \beta_1 > 0, \qquad \beta_{-1} \leq 0, \tag{182a}$$

$$\text{Incompressible:} \quad \beta_1 > 0, \qquad \beta_{-1} \leq 0 \tag{182b}$$

for compressible and incompressible materials. In fact, a variety of tests by Rivlin and Saunders (Ref. 1), Treloar (Ref. 24), and others, on rubberlike materials support (182b). Naturally, this also provides reasonable ground for accepting the same pair of inequalities in (182a); and we shall say more about these later. However, no theoretical foundation for (182) is known.

The empirical inequalities (182) are imposed for all deformations of an isotropic material. Hence, let us consider a material which is isotropic relative to a natural state for which (50) holds. Then the classical shear modulus μ_0 is determined by

$$\mu_0 = \hat{\beta}_1 - \hat{\beta}_{-1}. \tag{183}$$

We note that (182) imply that $\mu_0 > 0$ is a necessary condition for material response typical in shear. This was shown in (157) for the simple shear of a block of isotropic material. Notice that (157) reduces to (183) when $K = 0$.

We have also seen earlier that the empirical inequalities are useful in physical situations. They revealed, for example, that compressive, not tensile end loads are necessary to effect a noncontrollable, pure torsion of a rotating, compressible Blatz–Ko cylinder. More generally, in the pure torsion of an arbitrary incompressible and isotropic, circular cylinder of radius a and having a stress-free lateral surface, it is known (Truesdell and Noll, Ref. 11, Sections 57, 91) that in addition to a torque that produces the twist D about the cylinder axis, a normal axial force

$$N = -\pi D^2 \int_0^a r^3(\beta_1 - 2\beta_{-1})dr \tag{184}$$

also must be applied. If the empirical inequalities (182b) hold, the force (184) is always compressive. And if this end thrust is not supplied, the twisted cylinder will elongate. This result is universal for incompressible materials. Other examples by Batra (Refs. 17, 18) illustrate for isotropic hyperelastic materials the physical principle that the stress **T** and stretch **B** share the same principal axes and corresponding equal principal values when the empirical inequalities hold. Hence, a simple tensile load produces a simple extension, provided the empirical inequalities are satisfied. See also Beatty and Stalnaker (Ref. 16).

Finally, let us observe that the second pair of empirical inequalities (182a) for the Blatz–Ko material (56) are equivalent to (68). These conditions were not imposed by Blatz and Ko (Ref. 2), but (68) are essential in their biaxial deformation problems. Let us recall that the Blatz–Ko material (56) has a natural state if and only if, by (57), $\hat{\beta}_0 \equiv \beta_0(1) = \mu_0(1 - 2f)$. Thus, in view of (68), we must have $-\mu_0 \leq \hat{\beta}_0 < \mu_0$. Now, if the first of (182a) also is imposed for all deformations, then $\hat{\beta}_0 \leq 0$ holds with $\mu_0 > 0$ if and only if $1/2 \leq f$. Therefore, the empirical inequalities hold for the Blatz–Ko material (56) having a natural state if and only if $-\mu_0 \leq \beta_0(1) \leq 0$ and $1/2 \leq f \leq 1$. We see from (58) that the Blatz–Ko foamed, polyurethane rubber model for which $f = 0$, very nearly, and $\hat{\beta}_0 = \mu_0 > 0$ fails to satisfy these criteria (Beatty, Ref. 4, Beatty and Stalnaker, Ref. 16). Of course, one may argue that a foamed rubber ought not to be modeled as a homogeneous, materially uniform, and isotropic hyperelastic continuum. Nevertheless, the Blatz–Ko data seem to show good agreement with this model. We shall say more about this later. It should be emphasized also that their data for the compressible, solid polyurethane rubber support all of the empirical inequalities (182a).

17.2. Other Inequalities. Other analytical restrictions that express the physical behavior of materials have been derived or postulated as expressing stable material response to loading and deformation. The Baker–Ericksen inequalities (Truesdell and Noll, Ref. 11, Section 51), for example, are based on the intuitive mechanical principle that in an isotropic material the greater principal stress t_k should occur always in the direction of the larger principal stretch λ_k; that is, $(t_i - t_j)(\lambda_i - \lambda_j) > 0$ if $\lambda_i \neq \lambda_j$ (no sum). This criterion leads to the following restrictions derived by Baker and Ericksen (Ref. 96) for both compressible and incompressible materials:

$$\mu_{ab} > 0, \text{ if } \lambda_a \neq \lambda_b, \qquad \mu_{aa} \geq 0, \text{ if } \lambda_a = \lambda_b, \tag{185}$$

where

$$\mu_{ab} \equiv \beta_1 - \frac{1}{\lambda_a^2 \lambda_b^2} \beta_{-1}. \tag{186}$$

The constitutive inequalities (185) for the case when the λ_i are distinct were conjectured earlier by Truesdell (Ref. 97). Clearly, the empirical inequalities (182) imply the Baker–Ericksen inequalities (185); but not conversely.

The Baker–Ericksen inequalities have proved useful in the physical interpretation of the content of problem analysis. They demonstrate, for example, that for every incompressible, isotropic material in simple extension, tension produces lengthening, while compression produces shortening. In the absence of instability, this plainly supports natural material response typical of tensile and compressive loading. Further, it can be shown that if the inequalities (185) hold, the shear modulus (157) is a positive function (Truesdell and Noll, Ref. 11, Sections 54 and 55); and Ericksen (Ref. 98) proved that (185) are necessary and sufficient for the speeds of all principal waves in an incompressible and isotropic hyperelastic material to be real (Truesdell and Noll, Ref. 11, Section 78). These conclusions illustrate how (185) have been used to express the physical content of general analytical results.

Other technical questions directed at the qualitative behavior, existence, uniqueness, and stability of inverse solutions also have attracted considerable attention. Whether semi-inverse problems always possess solutions may be questionable; and certain analytical restrictions on the response functions have proved useful in providing rather general answers to such matters. Antman (Ref. 99), for example, has studied a family of semi-inverse problems of the type described as Family 1 in (161), but for general compressible, elastic materials whose elasticities satisfy the following strong ellipticity condition:

$$\frac{\partial T_{Ri\alpha}}{\partial F_{j\beta}} \mu_i \mu_j \nu_\alpha \nu_\beta > 0 \tag{187}$$

for arbitrary nonzero vectors $\boldsymbol{\mu}$, $\boldsymbol{\nu}$. Notice that the former inequalities (182) and (185) are imposed only for isotropic elastic materials, whereas (187) is a condition fit for all elastic materials, hyperelastic or not. In fact, Antman made no use of (29) for hyperelastic materials, yet he was able to show that "reasonable" semi-inverse boundary value problems, i.e., those for which (187) holds, always have solutions, and that a variety of their qualitative features of monotonicity, growth, and uniqueness may be determined without specification of $\Sigma(\mathbf{F})$. Similar results were demonstrated earlier by Antman (Ref. 100) for the one-dimensional problem of flexure, extension, and shear of a circular ring under hydrostatic pressure. The strong ellipticity condition also plays a significant role in Antman's study (Ref. 101) of the existence and regularity of solutions of a class of semi-inverse equilibrium problems of a nonlinearly elastic sector of a tube. Antman's clear examples and others by Ball (Refs. 102, 103) demonstrate the manner in which physically important, but purely mathematical questions of existence and regularity depend on a consistent collection of physical assumptions. They reveal also the comprehensive character of the strong ellipticity inequality. This is further illustrated below.

Knowles and Sternberg (Ref. 104) have pointed out that there may be some difficulties with the empirical basis of the Blatz–Ko model for the foamed polyurethane material for which the average values of the material parameters $\mu_0 = 32$ psi, $\nu_0 = \frac{1}{4}$ and $f = 0$ were used to reduce the general constitutive equation (78) to the special form (59). Use of these average data in the graph shown as Fig. 19 in the Blatz–Ko paper (Ref. 2) reveals a favorable comparison with their strip biaxial tension (plane strain, uniaxial tension) data. But when the same averaged data are used in their Figs. 15 and 23 for the uniaxial tension and the homogeneous biaxial tension (equibiaxial, plane stress)[7] tests, respectively, it may be seen that there are substantial departures from the actual test results for these cases, as mentioned by Knowles and Sternberg (Ref. 104). There is evident scatter in the data for the equibiaxial, plane stress experiment, which, among the three types of tests, yielded the poorest empirical value for f, namely, $f = -0.19$, as compared with the adopted average $f = 0$. Moreover, $f < 0$ stands in contradiction to the empirical inequalities (68), whereas the data for the other Blatz–Ko tests respect them. On the other hand, it may be seen easily with (58) that the Baker–Ericksen inequalities (185) hold for all homogeneous deformations of this material if and only if $\mu_0 > 0$. However, accepting the special Blatz–Ko model, Knowles and Sternberg found that despite the reasonable material response exhibited by (59) for a variety of homogeneous deformations, the strong ellipticity condition[8] (187) holds for this model with $\mu_0 > 0$ if and only if the corresponding principal stretches are restricted to the range

$$2 - \sqrt{3} < \frac{\lambda_i}{\lambda_j} < 2 + \sqrt{3}, \qquad i \neq j. \tag{188}$$

Knowles and Sternberg (Ref. 105) also have derived in terms of the local principal stretches for finite plane equilibrium deformations conditions necessary and sufficient for strong ellipticity for a homogeneous and isotropic, but otherwise arbitrary compressible hyperelastic material. For the special Blatz–Ko foamed rubber material, they recover (188) valid for all plane deformations. One will find that the stretch data in the Blatz–Ko experiments for the foamed polyurethane material satisfy (188).

17.3. Antiplane Shear Problem. We have seen that nonhomogeneous deformations which are not universal may still be produced by surface tractions alone in special kinds of compressible and incompressible, homogeneous, and isotropic hyperelastic materials provided that certain conditions on the material

---

[7]The parenthetical terms are used by Knowles and Sternberg (Ref. 104), the others by Blatz and Ko (Ref. 2).

[8]Knowles and Sternberg (Ref. 104) proved this result for the local ellipticity condition $\det[A_{i\alpha j\beta}\nu_\alpha\nu_\beta] \neq 0$ for all vectors $\boldsymbol{\nu} \neq \mathbf{0}$. However, they show further that for the special Blatz–Ko model their local ellipticity condition holds if and only if strong ellipticity holds. See Section 18.2 below.

parameters may be satisfied. An antiplane shear deformation provides an interesting example for which general restrictions on the strain energy function may be derived which determine whether a given class of materials is actually capable of sustaining antiplane shear deformations. These conditions are discussed below for both non-axisymmetric and axisymmetric antiplane shear deformations. The results are illustrated in an important example.

A general antiplane shear deformation is defined by

$$x_1 = X_1, \qquad x_2 = X_2, \qquad x_3 = X_3 + u(x_1, x_2), \tag{189}$$

in which x_k are the current rectangular Cartesian coordinates of the material point with referential coordinates X_k in the undeformed state, and $u(x_1, x_2)$ denotes the out-of-plane displacement function. For this class of deformations, Knowles (Refs. 106, 107) determined essentially algebraic conditions on the strain energy function in order that the equilibrium equations without body force are satisfied under antiplane shear deformations. If the response functions of a given material cannot be chosen to satisfy Knowles's conditions under antiplane shear deformations, then the material is not capable of sustaining this class of isochoric deformations. Our study of the antiplane shear problem begins with the simpler axisymmetric case.

The telescopic shear of a circular cylindrical tube is an axisymmetric, antiplane shear deformation described by

$$r = \mathrm{R}, \qquad \theta = \Theta, \qquad z = Z + u(R), \tag{190}$$

where (r, θ, z) are the current cylindrical coordinates of a material point initially at (R, Θ, Z) in the same Cartesian reference system. This is a special case of (189) for which $u(x_1, x_2) = u(R)$, where $R^2 = x_1^2 + x_2^2$. Although the axisymmetric, antiplane shear deformation (190) cannot be sustained by surface tractions alone in every compressible or incompressible, isotropic and homogeneous hyperelastic material, it may be controllable in special kinds of compressible and incompressible materials.

Rivlin (Ref. 22) has shown that the telescopic shear problem for an arbitrary incompressible, isotropic and homogeneous, hyperelastic material leads to a nonlinear ordinary differential equation for $u(R)$ whose solution may be obtained only on specification of the strain energy function. In particular, for an incompressible Mooney–Rivlin material the exact displacement function is given by

$$u(R) = A\ln(BR), \tag{191}$$

where A and B are constants determined by assigned boundary conditions. The remaining equilibrium equation then determines the unknown pressure function, and hence the stress components and traction conditions may be found for the class of Mooney–Rivlin materials. Thus, except for the details leading eventually to the determination of $u(R)$, the axisymmetric, antiplane shear problem for any specified incompressible material may be considered solved. Therefore, for an incompressible material, a simple shear and an axisymmetric, antiplane shear deformation are identified as trivial cases among the general class of antiplane shear problems[9] (189) studied subsequently by Adkins (Ref. 108) for cylinders of arbitrary cross section. In the general case (189) the two differential equations to which Adkins (see also Green and Adkins, Ref. 28) reduced the overdetermined system of three equilibrium equations restrict both the form of the displacement function and the form of the strain energy function for which the antiplane displacement may be controllable. For the special Mooney–Rivlin material, Adkins shows that nontrivial states of antiplane shear are always possible and some general results are derived. For a compressible material, however, the situation is less clear and hence only the simple shear case may be considered trivial. In fact, the axisymmetric, antiplane shear deformation (190) may not be possible for a specified class of compressible materials unless certain auxiliary conditions on the strain energy function are satisfied.

Necessary and sufficient conditions on the form of the strain energy function for homogeneous compressible and incompressible, isotropic, hyperelastic materials for which nontrivial states of antiplane shear may be admissible have been derived by Knowles (Refs. 106, 107). To recall these theorems, we first note that the shear stress response function $\tau(\kappa) \equiv [T_{13}^2 + T_{33}^2]^{1/2}$ is given by

$$\tau(\kappa) \equiv \kappa\mu(\kappa^2), \tag{192}$$

in which $\kappa^2 \equiv (\partial u/\partial x_1)^2 + (\partial u/\partial x_2)^2$ and, with (48),

$$\mu(\kappa^2) \equiv \beta_1 - \beta_{-1} = 2(\Sigma_1 + \Sigma_2) \tag{193}$$

is called the shear response function. Here and below the numerical subscripts denote partial derivatives of Σ with respect to the principal invariants I_k. For either the antiplane shear (189) or (190), we have

$$I_1 = I_2 = 3 + \kappa^2, \qquad I_3 = 1. \tag{194}$$

[9]For incompressible materials, both Adkins (Ref. 108) and Knowles (Ref. 106) consider a more general deformation in which the body first undergoes a specified axial stretch λ and is then subjected to the antiplane shear displacement $u(x_1, x_2)$. Here we consider only the case $\lambda = 1$.

For the axisymmetric case (190), $\kappa \equiv \partial u/\partial R$. We note also from (177) that the physical, Cauchy stress components for the axisymmetric, antiplane shear problem satisfy the universal relation $T_{zz} - T_{rr} = \kappa T_{rz}$, which has the same form as (160) for a simple shear. For the non-axisymmetric problem (189), however, the system (177) yields three universal relations,

$$u_2(T_{33} - T_{22}) = \kappa^2 T_{23} + u_1 T_{12}, \tag{195a}$$

$$u_1(T_{33} - T_{11}) = \kappa^2 T_{13} + u_2 T_{12}, \tag{195b}$$

$$u_1 T_{23} = u_2 T_{13}. \tag{195c}$$

Notice that we have made no explicit use of the constitutive equation to derive these rules from (177).

Returning to the axisymmetric case, we note that the last condition in (194) shows that independently of the compressibility of the material, the antiplane shear deformations (189) and (190) are isochoric and are therefore candidate deformations for both compressible and incompressible materials. Thus, cast in the above terms, Knowles proved the following theorems on nonaxisymmetric, antiplane shear deformations:

Incompressible Case (Knowles, Ref. 106): If the strain energy function $\Sigma(I_1, I_2)$ for a given incompressible material is such that the ellipticity condition $d\tau(\kappa)/d\kappa > 0$ holds for all $\kappa \geq 0$, then the associated incompressible hyperelastic material is capable of sustaining nontrivial states of antiplane shear (189) if and only if Σ also satisfies the condition

$$b\Sigma_1 + (b - 1)\Sigma_2 = 0, \tag{196}$$

for some constant b and for $I_1 = I_2 \geq 3$.

Compressible Case (Knowles, Ref. 107): If the strain energy function $\Sigma(I_1, I_2, I_3)$ for a given compressible material is such that $d\tau(\kappa)/d\kappa > 0$ holds for all $\kappa \geq 0$, then the associated compressible hyperelastic material can sustain nontrivial states of antiplane shear (189) if and only if Σ, in addition to satisfying (196), also satisfies the condition

$$\Sigma_{11} + I_1\Sigma_{12} + (I_1 - 1)\Sigma_{22} + \Sigma_{13} + \Sigma_{23} + \tfrac{1}{2}\Sigma_2 = 0, \tag{197}$$

for $I_1 = I_2 \geq 3, I_3 = 1$.

We note, however, that the axisymmetric case is excluded as a trivial case for incompressible materials and may be considered complete, as discussed earlier. We further note that the mathematical structure used to derive the conditions for compressible materials, as emphasized by Knowles (Ref. 107), excludes the nontrivial axisymmetric problem. The latter case was studied recently by Polignone and Horgan (Ref. 109). They derived from the equilibrium equations two necessary conditions on Σ in order that an axisymmetric, antiplane shear deformation (190) may be possible in a specified compressible material. Their conditions on Σ, however, are expressed in terms of two ordinary nonlinear differential equations for $u(R)$.

The finite amplitude vibration of an axisymmetric load supported by a cylindrical shear mounting subjected to a quasistatic deformation consisting of simultaneous axial and gyratory shear is investigated in a recent paper by Beatty and Khan (Ref. 110). The pure axial motion requires solution of the antiplane shear problem (190) for both compressible and incompressible materials. Two general classes of materials are studied there; those having a constant shear response function $\mu(\kappa^2) = \mu_0$ and those for which $\mu(\kappa^2) = \mu_0 + 2\mu_1\kappa^2$ is a quadratic function of the total shear strain $\kappa(R)$, where μ_0 and μ_1 are constants. Otherwise, identification of specific kinds of materials having these response functions was not necessary. Motivated by considerations in this work, Jiang and Beatty (Ref. 111) revisited the axisymmetric, antiplane shear and azimuthal shear problems to identify specific classes of compressible materials for which these separate and simultaneous deformations may be possible. For the axial shear problem, for example, they show that κ cannot be constant and they deduce a single necessary and sufficient algebraic condition that characterizes all homogeneous and isotropic, compressible hyperelastic materials for which axisymmetric, antiplane shear deformations are possible. They prove also that Knowles's conditions (196) and (197) for nontrivial, antiplane shear deformations of a compressible material suffice, but are not necessary for the material to support axisymmetric, antiplane shear deformations. They prove the following result:

A compressible, isotropic and homogeneous, hyperelastic material whose strain energy function $\Sigma(I_1, I_2, I_3)$ satisfies the condition $\mu(\kappa^2) > 0$ for all $\kappa \in (-\infty, \infty)$ is capable of sustaining controllable, nontrivial ($\kappa \neq 0$), axisymmetric, antiplane shear deformations if and only if Σ also satisfies the condition

$$\begin{aligned}(\Sigma_1 + \Sigma_2)[\Sigma_{11} + I_1\Sigma_{12} + (I_1 - 1)\Sigma_{22} + \Sigma_{13} + \Sigma_{23} + \tfrac{1}{2}\Sigma_2] \\ = (I_1 - 3)[\Sigma_1(\Sigma_{12} + \Sigma_{22}) - \Sigma_2(\Sigma_{11} + \Sigma_{21})],\end{aligned} \tag{198}$$

for $I_1 = I_2 \geq 3, I_3 = 1$.

Moreover, Jiang and Beatty (Ref. 111) show that the shear response function (193) in axisymmetric, antiplane shear deformations is constant if and only if the strain energy function satisfies

$$\Sigma_{11} + 2\Sigma_{12} + \Sigma_{22} = 0, \qquad \text{for } I_1 = I_2 \geq 3, I_3 = 1. \tag{199}$$

Hence, use of (194) and (199) in (198) shows that for the subclass of hyperelastic materials having a constant shear response function, the condition (198) necessary and sufficient for the material to sustain axisymmetric, antiplane shear deformations reduces to

$$2(\Sigma_{12} + \Sigma_{22} + \Sigma_{13} + \Sigma_{23}) + \Sigma_2 = 0, \tag{200}$$

for $I_1 = I_2 \geq 3, I_3 = 1$. For every compressible material in this subclass, the antiplane shear displacement function has the form (191). It is evident that for all materials with constant shear response $\mu(\kappa^2) = \mu(0) \equiv \mu_0 > 0$. Moreover, we note that Knowles's ellipticity condition $d\tau(\kappa)/d\kappa = \mu_0 > 0$ is also automatically satisfied. The condition (199) is precisely the relation assumed by Polignone and Horgan (Ref. 109) to obtain from their nonlinear differential equations a more tractable system of equations to explore a special subclass of hyperelastic materials capable of supporting axisymmetric, antiplane shear deformations. They used (199) to show that their equations simplify to our (200); but they did not recognize the substance of (199) as a condition on the shear response function itself.

It is useful to look at an important example set by Jiang and Beatty (Ref. 111). Consider a special class of hyperelastic materials for which the strain energy function Σ is defined by

$$\begin{aligned}\Sigma(I_1, I_2, I_3) &= \alpha(I_1 - 3) + \beta(I_2 - 3) \\ &\quad + \gamma[(I_1 - 3)^2 - (I_2 - 3)^2] + \delta(I_2 - 3)^2(I_3 - 1),\end{aligned} \tag{201}$$

where α, β, γ, and δ are constants. Clearly, $\Sigma(3, 3, 1) \equiv 0$ in the undeformed state and one can show that the stress vanishes in the undeformed state provided that $2\beta = -\alpha$. Hence, by (193),

$$\mu(\kappa^2) = \alpha = \mu_0 > 0, \text{ for all } \kappa, \quad \beta = -\frac{\mu_0}{2}. \tag{202}$$

Thus, our primary condition on the shear response function is satisfied for all κ; and we note in passing that Knowles's ellipticity condition on the shear stress response function also holds for $\kappa \geq 0$. The strain energy function satisfies the condition (200) for

$$\gamma = -\frac{\mu_0}{8}, \qquad \delta = -\frac{\mu_0}{16}. \tag{203}$$

The conditions (202) and (203) are necessary and sufficient for the material (201) to sustain axisymmetric, antiplane shear deformations. Therefore, Jiang and Beatty (Ref. 111) conclude that axisymmetric, antiplane shear deformations are possible in every material for which the strain energy function has the form

$$\Sigma(I_1, I_2, I_3) = \mu_0\{I_1 - 3 - \frac{1}{2}(I_2 - 3) - \frac{1}{8}[(I_1 - 3)^2 - (I_2 - 3)^2] - \frac{1}{16}(I_2 - 3)^2(I_3 - 1)\}. \tag{204}$$

On the other hand, it may be seen that both (196) and (197) are violated for any compressible material having the strain energy function (204). Consequently, Knowles's conditions necessary and sufficient for nontrivial, nonaxisymmetric, antiplane shear deformations are not necessary for a compressible material to sustain axisymmetric, antiplane shear deformations when either of the preliminary conditions on the shear response function or the shear stress response function hold.

This concludes our illustration of restrictions on the strain energy function in order that a specified class of nonhomgeneous deformations may be controllable in a particular class of compressible, isotropic, and homogeneous hyperelastic materials. It should be noted, however, that the nonaxisymmetric, antiplane shear problem for anisotropic compressible materials has been studied by Tsai and Rosakis (Ref. 112) and the corresponding axisymmetric case is mentioned by Jiang and Beatty (Ref. 111). The latter authors also remark on similar necessary and sufficient conditions derived for other problems of azimuthal shear and simultaneous axial and gyratory shear deformations for both isotropic and anisotropic materials. These topics will appear in future papers.

17.4. Concluding Remarks. Dunn (Ref. 113) has studied a specific material model for which neither the strain energy nor the Cauchy stress vanishes in an undistorted state. Dunn's hypothetical material has no natural state, and the hydrostatic stress in every isotropic deformation is a constant. Hence, no further stress is needed to effect an arbitrary uniform dilatation of the material from an undistorted state. This response is certainly unrealistic. He shows, nonetheless, that the empirical, Baker–Ericksen, strong ellipticity, and two further physical constitutive inequalities yield physically reasonable results for a physically unrealistic material model. Thus, some restrictions commonly imposed on constitutive equations neither alone nor collectively are sufficiently restrictive to weed out physically unrealistic material models. It may be noted, on the other hand, that Dunn's model may be eliminated by insisting that relative to an undistorted state of an isotropic compressible material on which the stress may be at most hydrostatic, the volume must be increased by hydrostatic tension and decreased by a pressure. That is, according to Truesdell and Noll (Ref. 11, Section 53), in every uniform expansion or contraction $\mathbf{V} = \lambda\mathbf{1}$, the hydrostatic stress τ must be a strictly increasing function $\tau = \hat{\tau}(\lambda)$ of the dilatational stretch λ,

$$(\tau^* - \tau)(\lambda^* - \lambda) > 0, \tag{205}$$

for $\lambda^* \neq \lambda$ and $\tau^* = \hat{\tau}(\lambda^*)$. Now, for Dunn's model, $\tau^* = \tau$ for all $\mathbf{V} = \lambda\mathbf{1}$, hence it is not supported by the pressure–compression criterion (205). His further example of a material having a natural state and for which $\mathbf{T}(\lambda) = 0$ for all $\mathbf{V} = \lambda\mathbf{1}$ also fails the test of (205).

There is now a large literature on various constitutive inequalities. These have been conveniently summarized by Wang and Truesdell (Ref. 12) and by Truesdell and Noll (Ref. 11, Sections 51–53, 153–171). A variety of physically motivated inequalities on material response functions have been proposed and applied in examples to demonstrate their fundamental utility in the interpretation of physical results. The few examples provided here illustrate that some restrictions, like the empirical and Baker–Ericksen inequalities, commonly are applied *a posteriori*, i.e., at essentially the "terminal stages" of a problem analysis, to interpret the physical content of results otherwise derived without prior restrictions. Others, however, like the strong ellipticity condition, usually are introduced *a priori*, as part of the mathematical structure to be respected in the analysis. These deliver additional restrictive relations which, if violated, may express a breakdown or discontinuity in the nature of a problem solution, for example (Knowles and Sternberg, Ref. 104). It appears, therefore, that restrictions to be imposed on the mechanical response generally fall into two classes: one kind provides an interpretive element of the physical theory, while the other provides an additional structural component to the mathematical framework of the theory. It is interesting that heretofore only the latter type have given rise to controversy in literature directed at Truesdell's problem. The critical remarks by Rivlin (Ref. 114) and Ericksen's views on the solution of Truesdell's problem (Ref. 57, pp. 220–223) are particularly noteworthy.

Thus, while many interconnections and implications of constitutive inequalities have been discovered and studied in a variety of cases (see Wang and Truesdell, Ref. 12), some general, some specific, as demonstrated here in a few examples, the situation, though certainly much clearer than it was when the problem was first proposed in 1956, remains unsettled and is sometimes just plain controversial. To this day, no universal body of constitutive inequalities has been adopted, and no inclusive, universal inequality has been found; but there has advanced considerable progress toward understanding the role of constitutive inequalities in the theory of constitutive equations. Hence, the investigation of Truesdell's problem continues. Its connection with elastic stability theory will be described below.

18. Elastic Stability and Nonuniqueness

The problem of elastic stability of a perfectly flexible and inextensible thin rod known as the elastica was first studied by Euler (1707–1783). He solved the problem for arbitrarily large deflections by using a theory based on the constitutive equation that the bending moment M at any section of the rod is proportional to its curvature κ at that place,

$$M = B\kappa\,, \tag{206}$$

where B is the constant bending stiffness. It is seen at once that a rod subject to pure end moments $M = M_e$ always is bent to a circular shape of radius B/M_e. More generally, however, determination of the exact shape under compressive end loads is a difficult nonlinear problem that Euler solved exactly. Investigation of the solution led to invention of a number of mathematical tools now in common use and often attributed to others. The calculus of variations, solutions by infinite series, Bessel functions, Fresnel integrals, the Cornu spiral, and other topics in differential equations grew from this early problem. And so did the theory of elastic stability.

During the century following Euler's death, a few stability problems were solved; and in 1888 the first unifying energy theory of stability of thin bodies was formulated by Bryan (Ref. 61). Since then, stability theories of varying degrees of generality have been proposed; and in 1955 Pearson (Ref. 59) successfully developed the energy criterion within the framework of finite elasticity theory. But even today, there is no universally accepted stability criterion; and use of various criteria has led to different theories, some exhibiting deficiencies of one kind or another. Beatty (Ref. 63) mentions, for example, that some authors have required the body to be in a homogeneous state of stress prior to instability; and others who have used Euler's method of adjacent equilibrium states, omitted the anisotropy induced by a change of reference configuration. Further details concerning various criteria may be found in the basic article by Knops and Wilkes[10] (Ref. 115). We note, however, that stability theories, including the fundamental dynamical theory of Liapounov, generally use the linearized equations for small motions superimposed on an assigned, possibly finitely deformed, equilibrium configuration, or other motion state. This was Euler's approach too.

18.1. Remarks on Euler's Criterion. Euler used the idea that an equilibrium configuration of a body is unstable for dead loads if there exists another equilibrium configuration situated in the neighborhood of the assigned one. Hence, Euler's method also is known as the method of adjacent equilibrium. The success of this classical concept of elastic stability theory may be measured by the massive body of literature on its applications. There are, nevertheless, serious limitations of Euler's method. One of these concerns cases of instability where an adjacent equilibrium configuration does not exist, and others for which the loading is not conservative.

An example of the first kind is provided by a spherical cap cut from a tennis ball. The cap is in equilibrium without surface loads.[11] However, we recall from studies of the eversion problem for both compressible and incompressible isotropic materials that the equilibrium equations for a spherical shell under zero surface tractions have at least two equilibrium solutions that differ by more than a rigid motion. These equilibrium states generally are not neighboring states, and though each may be stable for small disturbances, one may be transformed to the other by a readily observable instability.

Indeed, if a tennis ball shell of any depth less than a hemisphere is simply supported on a smooth surface and subjected to a sufficiently large normal dead load applied at its crown point, there will occur at a certain critical thrust an instantaneous snap-through instability from an equilibrium state of the deformed cap to a remote and traction-free, everted equilibrium configuration. For shallow shells, the snap-through will occur at a lesser thrust, and hence somewhat closer

[10]This is an excellent survey of the foundations of the theory of elastic stability, particularly from the mathematician's point of view. The text by Leipholz (Ref. 116) is directed toward structural mechanics and may be attractive to engineers interested in a more applied, but theoretically sound development. See also Bolotin (Ref. 117) and Zeigler (Ref. 118).

[11]Antman (Ref. 119) has shown that every homogeneous, compressible, and isotropic spherical cap has an everted image in equilibrium under zero traction on the spherical boundary. However, only the total force, not the traction, over the edge (thickness) boundary is zero. This result explains why the actual everted shell with zero traction exhibits flaring at the edge—it arises due to the discrepancy between the actual zero traction on the flared edge and the zero force required to effect the assumed ideal deformation from a straight radial edge into an everted straight radial edge. The eversion of an incompressible shell is described by Truesdell and Noll (Ref. 11, Sections 57, 95).

to the undeformed configuration. The same thing happens, of course, if the everted shell is loaded similarly. But due to the presence of the initial stress distribution in the everted state, the snap-back to the natural state occurs at a smaller critical thrust than before. If the undeformed shell is shallow, the snap-back to the undeformed state from the everted shape occurs under fairly small load, hence close to the unloaded, everted configuration. It is not clear what effects the construction of the ball and the flared edge may have in this experiment. In any event, the stability problem of the snap-through to eversion and return effect for a spherical shell apparently has never been studied. Our example shows that the ultimate configuration will not be infinitesimally close to the initial equilibrium configuration whose stability under the prescribed dead load is questioned. Hence, Euler's method of adjacent equilibrium states cannot be applied.

In general, the Euler method also cannot be applied to questions of stability of a body subjected to nonconservative loading. It can be shown, for example, that Euler's method gives erroneous results for the critical thrust produced by a compressive follower load applied to the free end of a cantilever beam. In this case, the load has a fixed magnitude but its direction remains tangent to the axis of the beam at the free end, which is free to oscillate. Since Euler's method compares only nearby equilibrium states, it automatically excludes all dynamical motions. Thus, for the nonconservative follower load problem, Euler's method is replaced by the dynamic method of small vibrations about the straight equilibrium configuration. The book by Bolotin (Ref. 117) is devoted entirely to the investigation of these kinds of nonconservative problems of elastic stability. He shows that Euler's method is applicable in all cases for which the external forces acting on a body are conservative,[12] and in this case loss of stability can occur only in the form of static instability. Hence, according to Bolotin, for a conservative system, such as a cantilever beam under compressive dead loads, Euler's method and the dynamic method will give the same correct solution. The same thing sometimes happens for nonconservative loading, but, in general, Euler's method must be considered unreliable in nonconservative problems. In particular, for the beam under follower loads, the Euler method of adjacent equilibrium states predicts a critical thrust that is only about one-eighth of the correct critical value derived by the dynamic method of small vibrations. We shall see later on that Euler's method also is related to questions of uniqueness.

---

[12]Certain smallness assumptions are embedded in Bolotin's argument; and apparently these approximations preclude consideration of the snap-through, dead load problem. Otherwise, the snap-through problem is a clear counterexample which shows that conservative loading does not suffice for the use of Euler's method.

18.2. Other Stability Criteria. Heuristic counterexamples mentioned earlier demonstrate that unqualified uniqueness is neither expected nor desired in finite elasticity theory. Gurtin and Spector (Ref. 120) have shown, however, that uniqueness holds in any convex, stable set of deformations. Their criterion of stability specifies that a deformation **F** is stable if the incremental power required to move a body from **F** is strictly positive; otherwise, it is considered unstable. Spector (Refs. 121, 122) used this criterion to derive additional uniqueness theorems for pure traction and general loading problems.

This is similar to the criterion of infinitesimal stability introduced by Truesdell and Noll (Ref. 11, Sections 68bis, 89). Existence of a stored energy function is not essential for this. The static deformation is called infinitesimally stable if the work done in every further infinitesimal deformation compatible with the boundary data of place and tractions is not less than that needed to produce the same infinitesimal deformation subject to dead loading, i.e., at the same state of stress as in the ground state of strain,

$$\int_{\mathcal{B}} tr[(\mathbf{T_R} - \mathbf{T}_0)\mathbf{H}^T]dV \geq 0. \tag{207}$$

The integration extends over the body in the strained ground state on which the stress is $\mathbf{T}_0$ and where gradients $\mathbf{H} = \nabla\mathbf{u}$ of all superimposed displacements compatible with the boundary data are to be considered. This is equivalent to

$$\int_{\mathcal{B}} A_{i\alpha j\beta} u_{i,\alpha} u_{j,\beta} dV \geq 0, \tag{208}$$

where $A_{i\alpha j\beta} \equiv \partial T_{Ri\alpha}/\partial F_{j\beta}$ in Cartesian coordinates. For a hyperelastic material, $A_{i\alpha j\beta} = \partial^2\Sigma/\partial F_{i\alpha}\partial F_{j\beta}$. If the strict inequality is used in (208), the configuration in which it holds is called infinitesimally superstable for the boundary data considered.

Hadamard proved that a configuration of an elastic body will be infinitesimally stable for boundary conditions of place and traction provided that the inequality

$$A_{i\alpha j\beta}\mu_i\mu_j\nu_\alpha\nu_\beta \geq 0 \tag{209}$$

holds for all vectors $\boldsymbol{\mu}$ and $\boldsymbol{\nu}$ at each material point. This is called Hadamard's inequality. The proof is given by Truesdell and Noll (Ref. 11, Section 68bis). This is only slightly weaker than the condition of strong ellipticity mentioned earlier in (187). Hadamard's inequality provides an algebraic first check for stability; for if (209) fails to hold, the equilibrium configuration cannot be stable. In general, however, Hadamard's inequality, like the strong ellipticity inequality, may be quite difficult to analyze.

It is natural to try to relate stability criteria to some comprehensive *a priori* constitutive inequality. But we know that unqualified uniquenss is undesirable in finite elasticity, so any inequality strong enough to establish it cannot be a suitable candidate. Moreover, no inequality that may guarantee that all solutions are stable is acceptable. Thus, we know to some extent what the comprehensive inequality should not be; but what comprehensive inequality should be imposed for general study remains an open question in finite elasticity theory.

In addition to the strong ellipticity (SE) condition, another general monotonicity condition known as the generalized Coleman–Noll (GCN) condition was proposed by Truesdell and Toupin (Ref. 123). We shall not discuss this inequality here; the reader may consult Truesdell and Noll (Ref. 11, Sections 51 and 52) and Wang and Truesdell (Ref. 12, pp. 210–262) for full details. Rather we mention only that neither of these inequalities implies the other. The SE condition is considered too strong because it excludes fluids, which from the point of view of solid mechanics appears unimportant; and the GCN inequality is believed too weak to yield definite stability results, for example. Another comprehensive monotonicity (M) condition has been introduced by Krawietz (Ref. 124). The M condition implies the GCN and the Hadamard stability condition (209), which we have observed is a weaker form of the strong ellipticity inequality (187). Krawietz (Ref. 124) has shown also that the M condition ensures reasonable behavior in the areas of statics, work theorems, stability, uniqueness, and wave propagation. Thus, the M condition may be imposed as a tentative comprehensive *a priori* inequality for constitutive equations in finite elasticity theory. It is a relatively new candidate to be considered for future study and debate.

Some connections between constitutive inequalities and stability have been found for special cases. It is known for the boundary value problem of place, for example, that for infinitesimal deformations from a homogeneous reference configuration the following implications hold (Truesdell, Ref. 10, Chapter 19):

$$\text{SE} \Rightarrow \text{superstability} \Rightarrow \text{uniqueness} \Rightarrow \text{existence}.$$

However, the general connection between restrictions to be imposed on constitutive equations and the theory of elastic stability has not been established.

18.3. Energy Method, Uniqueness, and Euler's Criterion. The aforementioned criteria of infinitesimal stability have been introduced to circumvent introduction of the strain energy for hyperelasticity; otherwise, in effect, they really are not much different from the energy criterion of stability. This specifies that an equilibrium configuration is stable for boundary conditions of place and traction, if and only if in every virtual displacement satisfying the boundary data of place, the virtual work W done by the loading does not exceed the corresponding increase in the total internal energy U. Otherwise, it is called unstable. For dead load surface tractions and body forces, the criterion $W \leq U$ is equivalent to the requirement that the second variation of the energy functional E in (150) or (151) be nonnegative for all allowable superimposed infinitesimal deformations.

Ericksen and Toupin (Ref. 58) and Hill (Ref. 62) show that Hadamard stability of a stressed state implies uniqueness for the boundary value problem associated with superimposed infinitesimal deformations. Beatty (Ref. 63) showed that this uniqueness prevails when a zero moment constraint is added to exclude undesirable trivial instabilities due to rigid rotations in the traction boundary value problem. In sum, these criteria yield the following uniqueness theorem for small superimposed deformations: If a configuration of an elastic body is infinitesimally superstable for mixed boundary conditions of place and tractions, the mixed boundary value problem for superimposed infinitesimal strain has at most one solution, to within an arbitrary infinitesimal rotation. Thus, in this sense,

$$\text{stability} \Rightarrow \text{uniqueness.}$$

But the converse does not follow. For the same argument with < 0 in (208) also yields uniqueness. However, it does follow that

$$\text{nonuniqueness} \Rightarrow \text{instability.}$$

And this conclusion provides the basis in finite elasticity for the Euler method of stability analysis under dead loads.

Euler's method of determining a nontrivial solution of the boundary value problem of small deformations superimposed on a large deformation is a common method of stability analysis used in applications. One example concerns the indentation of an arbitrary hyperelastic half space bounded by a plane surface, subjected to all-around pressure in planes parallel to the surface, and additionally loaded normal to its otherwise free surface by an axisymmetric rigid punch. This situation may model, for example, a structure at rest on a locally plane surface of the earth. Usmani and Beatty (Ref. 125) showed that if the radial pressure is sufficiently large, the surface will collapse under any infinitesimal, normal indentation load.

The balloon inflation problem studied earlier illustrates the variety of effects that one may find in the solution of the same problem for the various constitutive models commonly used in applications. But there is considerably more to the analysis of this example. It is well known that the maximum pressure is a point of bifurcation from the ideal spherical state to an aspherical configuration. This instability effect may be readily observed from the deformation of a grid drawn on the balloon in its undeformed state (Gent, Ref. 126). A similar effect occurs in an inflated cylindrical membrane. An outline of the stability analysis for the spherical membrane and several additional references to original works, including papers on the cylindrical membrane problem, are provided in the text by Ogden (Ref. 6).

The familiar formation of a neck that appears in the simple tension test of metals and certain polymeric materials is a similar instability phenomenon. The effect in polymers, however, is more striking. The neck forms when the load rises to a maximum; then the load drops rather suddenly to a minimum and rises again slightly to a fairly constant value at which the neck propagates steadily along the length of the specimen. The phenomenon is commonly known as cold drawing. An analysis of this problem for a nonlinear elastic solid is presented by Hutchinson and Neale (Ref. 127). Chater and Hutchinson (Ref. 128) report on a closely similar kind of propagation of instability observed in the inflation of a cylindrical balloon. The initial bulge instability of a long cylindrical balloon occurs at a critical inflation pressure as described earlier for spherical balloons, then the inflation pressure falls to an essentially constant value at which the bulge propagates quasistatically down the length of the balloon.

We have seen that the application of dead loads often leads to nonunique deformation states. In particular, we recall the unusual example of Rivlin's cube for which at least seven homogeneously deformed equilibrium configurations are possible under uniform dead loads applied to each face. Three of the states are always stable, three are always unstable. There are no solutions for which all three principal stretches λ_k are distinct; but only the undeformed state, which may be stable or unstable depending on the intensity of the loads, shares the same symmetry as the loading. Sawyers (Ref. 129) considered the case when only two of the three forces are equal. Interesting new phenomena occur for the Mooney–Rivlin model. Ball and Shaeffer (Ref. 130) have shown that, depending on the ratio β/α of the Mooney–Rivlin constants in (99), new asymmetric solutions exist for which all three λ_k are distinct; and these solutions result in a variety of unusual stability effects, including secondary bifurcation from a deformation branch along which only two stretches are equal onto a branch where none are equal (see also Marsden and Hughes, Ref. 9). Asymmetric, homogeneous equilibrium solutions and their stability for a Mooney–Rivlin cube under equal triaxial dead loads also have been studied both analytically and numerically by Tabaddor (Ref. 131).

Similar asymmetric deformations under symmetric loading may occur in the plane stretching of a thin square sheet. In 1948, Treloar (Ref. 132) observed in experiments that in some instances of sufficiently large loading of a thin rubber sheet by equal tensile forces, the corresponding principal stretches sometimes were unequal. In an effort to model this phenomenon, Kearsley (Ref. 133) showed that both symmetric and asymmetric equilibrium solutions exist, the effect depending on the ratio Σ_2/Σ_1 of the derivatives of the strain energy function $\Sigma(I_1, I_2)$. Only symmetric stretching may occur in a neo-Hookean sheet, a stable result found earlier by Rivlin (Ref. 19); but Kearsley finds that both asymmetric and symmetric deformations may occur in a Mooney–Rivlin sheet under equal forces. His heuristic discussion of the stability suggests that the asymmetric response, when it exists, is stable; the symmetric response is not. The same problem has been investigated by Macsithigh (Ref. 134) using a minimization method. He finds a critical value for the Mooney–Rivlin ratio β/α at which the number of asymmetric solutions changes, but not all are found to be potential minimizers for the energy functional. Tabaddor (Ref. 135) has shown both analytically and numerically that multiple stable and unstable states may exist; but the equilibrium state actually attained depends on the loading path.

Finally, an interesting phenomenon that has not been studied analytically, so far as I know, concerns a sufficiently short and thick-walled cylindrical tube compressed between lubricated rigid, parallel end plates (Beatty, Ref. 136). Experiments by Willis (Ref. 137) have shown that both the inside and outside, initially straight lateral surfaces will bulge outward to form convex,[13] barrel-shaped surfaces at fairly small strains. But at about 15–20% compression, these convex surfaces become straight again near the platen ends while remaining convex in the major central portion of the sample. At a certain critical compressive load intensity, the inside wall reverses its curvature (Payne and Scott, Ref. 138) and collapses to form a cusp-shaped inside surface directed toward the outside boundary in the central plane, while the outside wall continues to bulge as usual. The effect is illustrated in Fig. 16. This instability usually occurs only when the wall thickness is less than the height of the tube. Also the compressive deformation at the cusp formation appears to be independent of the particular rubber used. For sufficiently short specimens, however, the Willis instability effect disappears and both the inside and outside cylinder walls bulge to form a convex barrel shape.

19. Concluding Remarks

Elastic stability theory, possibly the most abused and maligned topic of finite elasticity, should be more systematically and carefully reviewed, not so much with the view to censure its foundations, which may beg further attention,

[13]The actual internal shape of the tube wall during deformation, except for the cusp formation, was not known precisely by Willis. Payne and Scott (Ref. 138), however, have described both surfaces as convex, and this is consistent with the decreasing internal volume measurements described by Willis. I have adopted the same view here and in Fig. 16(b), but the reader is cautioned that this may be inaccurate.

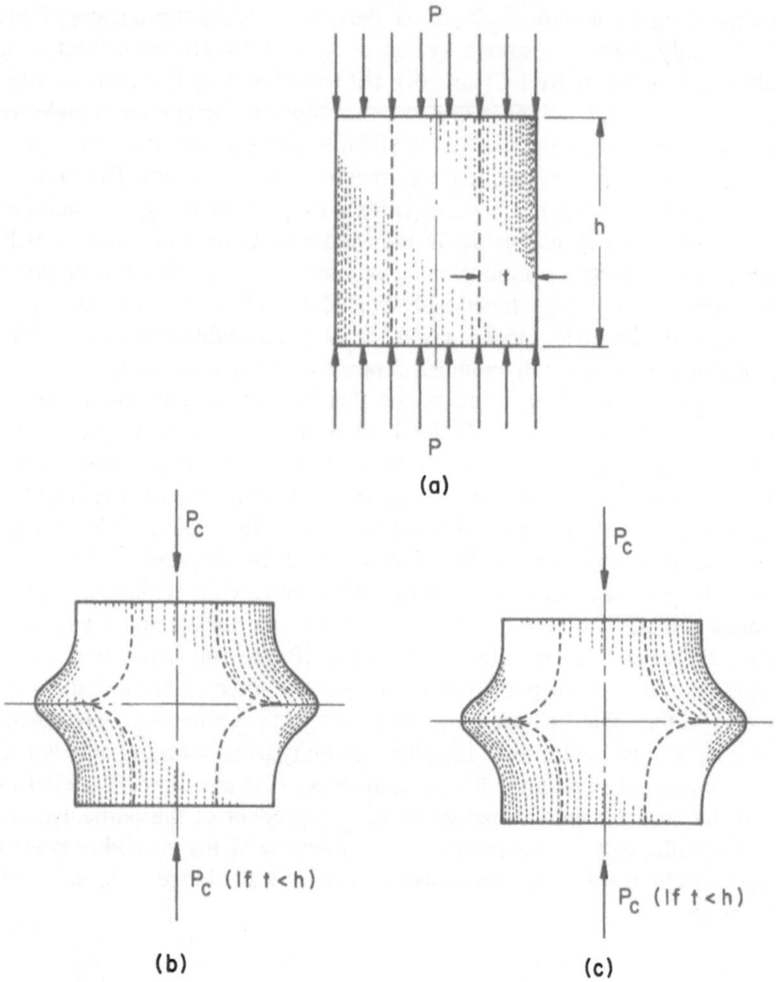

Fig. 16. (a) Willis instability phenomenon of a sufficiently short, thick-walled tube. Initially, both walls bulge as shown in (b). At a critical load P_c, the inside wall collapses to form the cusp geometry shown in (c).

but to remedy various technical defects and, more importantly, by direct appeal to clearly stated criteria and first principles, to demonstrate its methods and its utility in both analytical and technical applications. The foundations and their application to some analytical topics have been studied by Knops and Wilkes (Ref. 115), for example. But there is presently no definitive catalog of known analytical results and solutions for problems of elastic stability in finite strain, either static or dynamic, hence no evaluation of their content or value is possible. Unfortunately, as a consequence of these deficiencies, a serious reader is soon overwhelmed by a superabundant, confusing, and tedious literature which has brought to this important topic the label of an unhappy subject (Truesdell and Noll, Ref. 11, Section 68bis.). Hence, there is much that needs to be done to correct this. For, on the contrary, elastic stability is a fascinating area of finite elasticity theory, which, in addition to describing technical effects of buckling, warping, and wrinkling of structural elements, provides a physically meaningful mechanism for the characterization of uniqueness in analysis, and it may provide the key to further contributions toward the solution of Truesdell's problem. Without doubt, elastic stability theory deserves deeper analysis and more thoughtful review than may be provided in the few remarks assembled here.

There are, of course, other important and exciting topics in finite elasticity that have not been mentioned, and others of equal significance that fall outside the scope and purpose of this survey. Dynamical problems solved by Knowles (Refs. 139, 140), Truesdell (Ref. 141), Wang (Ref. 142), Shahinpoor and Nowinski (Ref. 143), and Rogers and Baker (Ref. 144) are primary examples among many others, some of which have been collected in the text by Eringen and Suhubi (Ref. 145). Exact solutions of the finite amplitude vibrations of rigid bodies supported by hyperelastic springs whose deformation is characterized by results from finite elasticity theory have been found by Beatty (Refs. 146–148), Beatty and Chow (Ref. 149), and Beatty and Khan (Ref. 110). The results of Beatty and Chow (Ref. 150) deduced from nonlinear elasticity theory provide, for example, an accurate prediction of the nearly constant transverse vibrational frequency of a highly stretched rubber cord, an unusual phenomenon observed in experiments by Baker (Ref. 151) and von Lang (Ref. 152) at the turn of the century. Topics on wave propagation in finitely deformed materials have been omitted, but the reference text by Chen (Ref. 153) fills this gap.

Subjects concerning the thermodynamics of materials have been ignored for the purpose of simplicity. This is a difficult area by itself. The book by Ericksen (Ref. 154) provides an excellent introduction to this subject, with many interesting applications, including the Treloar instability in the biaxial stretching of rubber sheets. Another resource is the treatise by Truesdell and Noll (Ref. 11), the standard reference for all matter studied here. Discussion of the elegant existence theorems due to Ball (Refs. 102, 103), a major advance in three-dimensional, nonlinear elastostatics, may be found in the foundations text by Marsden and Hughes (Ref. 9). Variational methods in finite elastostatics are treated by Lee and Shield (Ref. 155).

Additional topics of interest may be found in the survey article by Shield (Ref. 156), the ASME symposium volume edited by Rivlin (Ref. 157), and the proceedings of the Symposium on Finite Elasticity (Carlson and Shield, Ref. 158). There is much work completed for anisotropic materials in finite strain, and none of it has been mentioned here. A thorough description of primary research in this area may be found in the work by Green and Adkins (Ref. 28). A contemporary review of this subject, including work on elastic materials with inextensible constraints, deserves consideration. The constitutive theory of biological materials in finite deformation and its applications to topics in biomechanics, like the simple inflation problem illustrated in an earlier example, also deserve special, and possibly critical, review. I must, however, dismiss the temptation to elaborate further other areas for which general studies may be helpful. We must leave these matters and the discussion of other relevant topics in finite elasticity for another place.

References

1. RIVLIN, R. S., and SAUNDERS, D. W., *Large Elastic Deformations of Isotropic Materials, VII, Experiments on the Deformation of Rubber*, Philosophical Transactions of the Royal Society of London, Vol. A243, pp. 251–288, 1951.
2. BLATZ, P. J., and KO, W. L., *Application of Finite Elasticity to the Deformation of Rubbery Materials*, Transactions of the Society for Rheology, Vol. 6, pp. 223–251, 1962.
3. FUNG, Y. C. B., *Elasticity of Soft Tissues in Simple Elongation*, American Journal of Physiology, Vol. 213, pp. 1532–1544, 1967.
4. BEATTY, M. F., *A Lecture on some Topics in Nonlinear Elasticity and Elastic Stability*, IMA Preprint No. 99, Institute for Mathematics and its Applications, University of Minnesota, Minneapolis, 1984.
5. BEATTY, M. F., *Topics in Finite Elasticity: Hyperelasticity of Rubber, Elastomers, and Biological Tissues-with Examples*, Applied Mechanics Reviews, Vol. 40, pp. 1699–1734, 1987.
6. OGDEN, R. W., *Nonlinear Elastic Deformations*, John Wiley and Sons, New York, New York, 1984.
7. BOWEN, R. M., and WANG, C.-C., *Introduction to Vectors and Tensors*, Plenum Press, New York, New York, 1976.
8. ATKIN, R. J., and FOX, N., *An Introduction to the Theory of Elasticity*, Longman Group, London, England, 1980.
9. MARSDEN, J. E., and HUGHES, T. J. R., *Mathematical Foundations of Elasticity*, Prentice–Hall, Englewood Cliffs, New Jersey, 1993.
10. TRUESDELL, C., *The Elements of Continuum Mechanics*, Springer-Verlag, Berlin, Germany, 1966.
11. TRUESDELL, C., and NOLL, W., *The Nonlinear Field Theories of Mechanics*, Flügge's Handbuch der Physik, Springer-Verlag, Berlin, Germany, Vol. III/3, 1965.
12. WANG, C.-C., and TRUESDELL, C., *Introduction to Rational Elasticity*, Noordhoff International Publishing, Leyden, The Netherlands, 1973.
13. BEATTY, M. F., *Principles of Engineering Mechanics, Vol. 1, Kinematics—The Geometry of Motion*, Plenum Press, New York, New York, 1986.

14. GURTIN, M. E., *The Linear Theory of Elasticity*, Mechanics of Solids II, Flügge's Handbuch der Physik, Springer-Verlag, Berlin, Germany, Vol. VIa/2, pp. 1–295, 1972.
15. GURTIN, M. E., *Topics in Finite Elasticity*, CBMS-NSF Regional Conference Series in Applied Mathematics, SIAM, Philadelphia, Pennsylvania, 1981.
16. BEATTY, M. F., and STALNAKER, D. O., *The Poisson Function of Finite Elasticity*, ASME Journal of Applied Mechanics, Vol. 53, pp. 807–813, 1986.
17. BATRA, R. C., *On the Coincidence of the Principal Axes of Stress and Strain in Isotropic Elastic Bodies*, Letters on Applied Science, Vol. 3, pp. 435–439, 1975.
18. BATRA, R. C., *Deformation Produced by a Simple Tensile Load in an Isotropic Elastic Body*, Journal of Elasticity, Vol. 6, pp. 109–111, 1976.
19. RIVLIN, R. S., *Large Elastic Deformations of Isotropic Materials, II, Some Uniqueness Theorems for Pure Homogeneous Deformations*, Philosophical Transactions of the Royal Society of London, Vol. A240, pp. 491–508, 1948.
20. RIVLIN, R. S., *Large Elastic Deformations of Isotropic Materials, IV, Further Developments of the General Theory*, Philosophical Transactions of the Royal Society of London, Vol. A241, pp. 379–397, 1948.
21. RIVLIN, R. S., *Large Elastic Deformations of Isotropic Materials, V, The Problem of Flexure*, Proceedings of the Royal Society of London, Vol. A195, pp. 463–473. 1949.
22. RIVLIN, R. S., *Large Elastic Deformations of Isotropic Materials, VI, Further Results in the Theory of Torsion, Shear and Flexure*, Philosophical Transactions of the Royal Society of London, Vol. A242, pp. 173–185, 1949.
23. RIVLIN, R. S., *A Note on the Torsion of an Incompressible Highly Elastic Cylinder*, Proceedings of the Cambridge Philosophical Society, Vol. 45, pp. 485–487, 1949.
24. TRELOAR, L. R. G., *The Physics of Rubber Elasticity*, 3rd Edition, Clarendon Press, Oxford, England, 1975.
25. ISHIHARA, N., HASHITSUME, N., and TATIBANA, M., *Statistical Theory of Rubber-like Elasticity IV*, Journal of Chemistry and Physics, Vol. 19, pp. 1508–1520, 1951.
26. DEMIRAY, H., *Stresses in Ventricular Wall*, ASME Journal of Applied Mechanics, Vol. 98, pp. 194–197, 1976.
27. GREEN, A. E., and ZERNA, W., *Theoretical Elasticity*, Clarendon Press, Oxford, England, 1954.
28. GREEN, A. E., and ADKINS, J. E., *Large Elastic Deformations and Non-linear Continuum Mechanics*, Clarendon Press, Oxford, England, 1960.
29. PIPKIN, A. C., *Nonlinear Phenomena in Continua*, Nonlinear Continuum Theories in Mechanics and Physics and their Applications, Edited by R. S. Rivlin, Centro Internazionale Matematico Estivo, II Ciclo-Bressanone, Italy, September 3–11, 1969, pp. 57–58, 1970.
30. BOUASSE, H., and CARRIÈRE, Z., *Courbes de Traction du Caoutchouc Vulcanisé*, Annales de la Faculté des Sciences, Vol. 5, pp. 257–283, 1903.
31. MULLINS, L., *Effect of Stretching on the Properties of Rubber*, Journal of Rubber Research, Vol. 16, pp. 275–289, 1947.
32. HARWOOD, J. A. C., MULLINS, L., and PAYNE, A. R., *Stress Softening in Rubbers—A Review*, Journal of the IRI, Vol. 1, pp. 17–27, 1967.
33. JOHNSON, M. A., and BEATTY, M. F., *The Mullins Effect in Uniaxial Extension and*

its Influence on the Transverse Vibration of a Rubber String, Continuum Mechanics and Thermodynamics, Vol. 5, pp. 83–115, 1993.

34. JOHNSON, M. A., and BEATTY, M. F., *A Constitutive Equation for the Mullins Effect in Stress Controlled Uniaxial Extension Experiments*, Continuum Mechanics and Thermodynamics, Vol. 5, pp. 301–318, 1993.
35. JOHNSON, M., and BEATTY, M. F., *The Mullins Effect in Equibiaxial Extension and its Influence on the Inflation of a Balloon*, International Journal of Engineering Science, Vol. 33, pp. 223–245, 1995.
36. MULLINS, L., and TOBIN, N. R., *Stress Softening in Rubber Vulcanizates, Part I*, Journal of Polymer Science, Vol. 9, pp. 2993–3009, 1965.
37. BUECHE, F., *Molecular Basis for the Mullins Effect*, Journal of Applied Polymer Science, Vol. 4, pp. 108–114, 1960.
38. HARWOOD, J. A. C., MULLINS, L., and PAYNE, A. R., *Stress Softening in Natural Rubber Vulcanizates, Part II: Stress Softening Effects in Pure Gum and Filler Loaded Rubbers*, Journal of Applied Polymer Science, Vol. 9, pp. 3011–3021, 1965.
39. GOVINDJEE, S., and SIMO, J., *A Micro-Mechanically Based Continuum Damage Model for Carbon Black Filled Rubbers Incorporating the Mullins Effect*, Journal of the Mechanics and Physics of Solids, Vol. 39, pp. 87–112, 1991.
40. GOVINDJEE, S., and SIMO, J., *Transition from Micro-Mechanics to Computationally Efficient Phenomenology: Carbon Black Filled Rubbers Incorporating Mullins' Effect*, Journal of the Mechanics and Physics of Solids, Vol. 40, pp. 213–233, 1992.
41. BUECHE, F., *Mullins Effect and Rubber–Filled Interaction*, Journal of Applied Polymer Science, Vol. 5, pp. 271–281, 1961.
42. JAMES, H. M., and GUTH, E., *Theory of the Elastic Properties of Rubber*, Journal of Chemistry and Physics, Vol. 11, pp. 455–481, 1943.
43. BELL, J. F., *Continuum Plasticity at Finite Strain for Stress Paths of Arbitrary Composition and Direction*, Archive for Rational Mechanics and Analysis, Vol. 84, pp. 139–170, 1983.
44. BELL, J. F., *Contemporary Perspectives in Finite Strain Plasticity*, International Journal of Plasticity, Vol. 1, pp. 3–27, 1985.
45. BEATTY, M. F., and HAYES, M. A., *Deformations of an Elastic, Internally Constrained Material, Part 1: Homogeneous Deformations*, Journal of Elasticity, Vol. 29, pp. 1–84, 1992.
46. BEATTY, M. F., and HAYES, M. A., *Deformations of an Elastic, Internally Constrained Material, Part 2: Nonhomogeneous Deformations*, Quarterly Journal of Mechanics and Applied Mathematics, Vol. 45, pp. 663–709, 1992.
47. BEATTY, M. F., and HAYES, M. A., *Deformations of an Elastic, Internally Constrained Material, Part 3: Small Superimposed Deformations and Waves*, Zeitschrift für angewandte Mathematik und Physik, Vol. 46, pp. S72–S106, 1995.
48. ERICKSEN, J. L., *Constitutive Theory for Some Constrained Elastic Crystals*, IMA Preprint No. 132, Institute for Mathematics and its Applications, University of Minnesota, Minneapolis, 1985.
49. BEATTY, M. F., and HAYES, M. A., *On Bell's Constraint in Finite Elasticity*, Advances in Modern Continuum Dynamics, Proceedings of the International Conference at Elba Island, June 6–11, 1991, Edited by G. Ferrarese, Pitagora Editrice, Bologna, Italy, 1993.

50. Rogers, T. G., and Pipkin, A. C., *Small Deflections of Fiber-Reinforced Beams or Slabs*, ASME Journal of Applied Mechanics, Vol. 38, pp. 1047–1048, 1971.
51. Kao, B., and Pipkin, A. C., *Finite Buckling of Fiber-Reinforced Columns*, Acta Mechanica, Vol. 13, pp. 265–280, 1972.
52. Schaffers, W. J., *Buckling in Fiber-Reinforced Elastomers*, Textile Research Journal, Vol. 47, pp. 502–512, 1977.
53. Charrier, J. M., *Large Elastic Deformations of some Cord Reinforced Rubber Shells*, Rubber Chemistry and Technology, Vol. 43, pp. 282–303, 1970.
54. Spencer, A. J. M., *Deformations of Fibre-Reinforced Materials*, Clarendon Press, Oxford, England, 1972.
55. Pipkin, A. C., *Finite Deformations in Materials Reinforced with Inextensible Cords*, Finite Elasticity, Edited by R. S. Rivlin, ASME, AMD Vol. 27, pp. 91–102, 1977.
56. Pipkin, A. C., *Stress Analysis of Fiber Reinforced Materials*, Advances in Applied Mechanics, Vol. 19, pp. 1–51, 1979.
57. Ericksen, J. L., *Special Topics in Elastostatics*, Advances in Applied Mechanics, Vol. 17, pp. 189–244, 1977.
58. Ericksen, J. L., and Toupin, R. A., *Implications of Hadamard's Conditions for Elastic Stability with Respect to Uniqueness Theorems*, Canadian Journal of Mathematics, Vol. 8, pp. 432–436, 1956.
59. Pearson, C. E., *General Theory of Elastic Stability*, Quarterly of Applied Mathematics, Vol. 14, pp. 133–144, 1955.
60. Toupin, R. A., *Theories of Elasticity with Couple Stress*, Archive for Rational Mechanics and Analysis, Vol. 17, pp. 85–112, 1964.
61. Bryan, G. H., *On the Stability of Elastic Systems*, Proceedings of the Cambridge Philosophical Society, Vol. 6, pp. 199–210, 287–292, 1888.
62. Hill, R., *On Uniqueness and Stability in the Theory of Finite Elastic Strain*, Journal of the Mechanics and Physics of Solids, Vol. 5, pp. 229–241, 1957.
63. Beatty, M. F., *Some Static and Dynamic Implications of the General Theory of Elastic Stability*, Archive for Rational Mechanics and Analysis, Vol. 19, pp. 167–188, 1965.
64. Beatty, M. F., *A Theory of Elastic Stability for Incompressible, Hyperelastic Bodies*, International Journal of Solids and Structures, Vol. 3, pp. 23–37, 1967.
65. Beatty, M. F., *Stability of Hyperelastic Bodies Subject to Hydrostatic Loading*, International Journal for Nonlinear Mechanics, Vol. 5, pp. 367–383, 1970.
66. Sewell, M. J., *On Configuration-Dependent Loading*, Archive for Rational Mechanics and Analysis, Vol. 23, pp. 327–351, 1967.
67. Batra, R. C., *On Non-Classical Boundary Conditions*, Archive for Rational Mechanics and Analysis, Vol. 48, pp. 163–191, 1972.
68. Gurtin, M. E., *On Uniqueness in Finite Elasticity*, Proceedings of the IUTAM Symposium on Finite Elasticity, Lehigh University, 1980, Edited by D. E. Carlson and R. T. Shield, Martinus Nijhoff Publishers, The Hague, The Netherlands, 1982.
69. Rivlin, R. S., *Stability of Pure Homogeneous Deformations of an Elastic Cube under Dead Loading*, Quarterly of Applied Mathematics, Vol. 32, pp. 265–272, 1974.
70. Rivlin, R. S., *Some Thoughts on Material Stability*, Proceedings of the IUTAM Symposium on Finite Elasticity, Lehigh University, 1980, Edited by D. E. Carlson and R. T. Shield, Martinus Nijhoff Publishers, The Hague, The Netherlands, pp. 105–122, 1982.

71. ERICKSEN, J. L., *Deformations Possible in every Isotropic, Incompressible, Perfectly Elastic Body*, Zeitschrift für angewandte Mathematik und Physik, Vol. 5, pp. 466–489, 1954.
72. ERICKSEN, J. L., *Deformations Possible in every Compressible, Isotropic, Perfectly Elastic Material*, Journal of Mathematical Physics, Vol. 34, pp. 126–128, 1955.
73. SHIELD, R. T., *Deformations Possible in Every Compressible, Isotropic, Perfectly Elastic Material*, Journal of Elasticity, Vol. 1, pp. 91–92, 1971.
74. MARRIS, A. W., and SHIAU, J. F., *Universal Deformations in Isotropic Incompressible Hyperelastic Materials when the Deformation Tensor has Equal Proper Values*, Archive for Rational Mechanics and Analysis, Vol. 36, pp. 135–160, 1970.
75. MARTIN, S. E., and CARLSON, D. E., *A Note on Ericksen's Problem*, Journal of Elasticity, Vol. 6, pp. 105–108, 1976.
76. FOSDICK, R. L., *Remarks on Compatibility*, Modern Developments in the Mechanics of Continua, Proceedings of the International Conference on Rheology, Edited by S. Eskinazi, Academic Press, New York, New York, pp. 109–127, 1966.
77. KLINGBEIL, W., and SHIELD, R. T., *On a Class of Solutions in Plane Finite Elasticity*, Zeitschrift für angewandte Mathematik und Physik, Vol. 17, pp. 489–501, 1966.
78. SINGH, M., and PIPKIN, A. C., *Note on Ericksen's Problem*, Zeitschrift für angewandte Mathematik und Physik, Vol. 16, pp. 706–709, 1965.
79. SINGH, M., and PIPKIN, A. C., *Controllable States of Elastic Dielectrics*, Archive for Rational Mechanics and Analysis, Vol. 21, pp. 169–210, 1966.
80. HOLDEN, J. T., *Note on a Class of Controllable Deformations of Incompressible Elastic Materials*, Applied Scientific Research, Vol. 24, pp. 98–104, 1971.
81. HUILGOL, R. R., *A Finite Deformation Possible in Transversely Isotropic Materials*, Zeitschrift für angewandte Mathematik und Physik, Vol. 17, pp. 787–789, 1966.
82. FOSDICK, R. L., and SCHULER, K. W., *On Ericksen's Problem for Plane Deformations with Uniform Transverse Stretch*, International Journal of Engineering Science, Vol. 7, pp. 217–233, 1969.
83. FOSDICK, R. L., *Statically Possible Radially Symmetric Deformations in Isotropic Incompressible Elastic Solids*, Zeitschrift für angewandte Mathematik und Physik, Vol. 22, pp. 590–607, 1971.
84. MÜLLER, W. C., *A Characterization of the Five Known Families of Solutions of Ericksen's Problem*, Archive for Applied Mechanics, Vol. 22, pp. 515–522, 1970.
85. MÜLLER, W. C., *Some Further Results on the Ericksen Problem for Deformation with Constant Strain Invariants*, Zeitschrift für angewandte Mathematik und Physik, Vol. 21, pp. 633–636, 1970.
86. KAFADAR, C. B., *On Ericksen's Problem*, Archive for Rational Mechanics and Analysis, Vol. 47, pp. 15–27, 1972.
87. KNOWLES, J. K., *Universal States of Finite Anti-Plane Shear, Ericksen's Problem in Miniature*, American Mathematical Monthly, Vol. 86, pp. 109–113, 1979.
88. MARRIS, A. W., *Universal Deformations in Incompressible Isotropic Elastic Materials*, Journal of Elasticity, Vol. 5, pp. 111–128, 1975.
89. CURRIE, P. K., and HAYES, M., *On Non-Universal Finite Elastic Deformations*, Proceedings of the IUTAM Symposium on Finite Elasticity, Lehigh University, 1980, Edited by D. E. Carlson and R. T. Shield, Martinus Nijhoff Publishers, The Hague, The Netherlands, pp. 143–150, 1982.

90. HOLDEN, J. T., *A Class of Exact Solutions for Finite Plane Strain Deformations of a Particular Elastic Material*, Applied Scientific Research, Vol. 19, pp. 171–181, 1968.
91. PARRY, G. P., *Corollaries of Ericksen's Theorems on the Deformations Possible in every Isotropic Hyperelastic Body*, Archive for Applied Mechanics, Vol. 31, pp. 757–760, 1979.
92. ERICKSEN, J. L., *Semi-Inverse Methods in Finite Elasticity Theory*, Finite Elasticity, Edited by R. S. Rivlin, ASME, AMD Vol. 27, pp. 11–21, 1977.
93. BEATTY, M. F., *A Class of Universal Relations in Isotropic Elasticity Theory*, Journal of Elasticity, Vol. 17, pp. 113–121, 1987.
94. HAYES, M., and KNOPS, R. J., *On Universal Relations in Elasticity Theory*, Zeitschrift für angewandte Mathematik und Physik, Vol. 17, pp. 636–639, 1966.
95. TRUESDELL, C. A., *The Main Unsolved Problem in Finite Elasticity Theory*, Zeitschrift für angewandte Mathematik und Physik, Vol. 36, pp. 97–103, 1956. English translation in *Foundations of Elasticity Theory*, Edited by C. Truesdell, International Science Review Series, Gordon and Breach, New York, New York, 1965.
96. BAKER, M., and ERICKSEN, J. L., *Inequalities Restricting the Form of Stress Deformation Relations for Isotropic Elastic Solids and Reiner–Rivlin Fluids*, Journal of the Washington Academy of Sciences, Vol. 44, pp. 33–35, 1954. See also *Foundations of Elasticity Theory*, Edited by C. Truesdell, International Science Review Series, Gordon and Breach, New York, New York, 1965.
97. TRUESDELL, C. A., *The Mechanical Foundations of Elasticity and Fluid Dynamics*, Journal of Rational Mechanics and Analysis, Vol. 1, pp. 125–300, 1952. *Corrections and Additions*, Ibid., Vol. 2, pp. 593–616, 1953.
98. ERICKSEN, J. L., *On the Propagation of Waves in Isotropic, Incompressible, Perfectly Elastic Materials*, Journal of Rational Mechanics and Analysis, Vol. 2, pp. 329–337, 1953.
99. ANTMAN, S. S., *A Family of Semi-Inverse Problems in Nonlinear Elasticity*, Contemporary Developments in Continuum Mechanics and Partial Differential Equations, Edited by G. M. de LaPenha and L. A. Medeiros, North-Holland Publishing Company, The Netherlands, pp. 1–24, 1978.
100. ANTMAN, S. S., *Monotonicity and Invertibility Conditions in One-Dimensional Nonlinear Elasticity*, Nonlinear Elasticity, Academic Press, New York, New York, pp. 57–92, 1973.
101. ANTMAN, S. S., *Regular and Singular Problems for Large Elastic Deformations of Tubes, Wedges and Cylinders*, Archive for Rational Mechanics and Analysis, Vol. 83, pp. 1–52, 1983.
102. BALL, J. M., *Convexity Conditions and Existence Theorems in Nonlinear Elasticity*, Archive for Rational Mechanics and Analysis, Vol. 63, pp. 337–403, 1977.
103. BALL, J. M., *Constitutive Inequalities and Existence Theorems in Nonlinear Elastostatics*, Nonlinear Analysis and Mechanics I, Edited by R. J. Knops, Pitman, London, England, pp. 187–241, 1977.
104. KNOWLES, J. K., and STERNBERG, E., *On the Ellipticity of the Equations of Nonlinear Elastostatics for a Special Material*, Journal of Elasticity, Vol. 5, pp. 341–361, 1975.

105. KNOWLES, J. K., and STERNBERG, E., *On the Failure of Ellipticity of the Equations of Finite Elastostatic Plane Strain*, Archive for Rational Mechanics and Analysis, Vol. 63, pp. 321–336, 1977.
106. KNOWLES, J. K., *On Finite Anti-Plane Shear for Incompressible Elastic Materials*, Journal of the Australian Mathematical Society, Series B, Vol. 19, pp. 400–415, 1976.
107. KNOWLES, J. K., *A Note on Anti-Plane Shear for Compressible Materials in Finite Elastostatics*, Journal of the Australian Mathematical Society, Series B, Vol. 20, pp. 1–7, 1977.
108. ADKINS, J. E., *Some Generalizations of the Shear Problem for Isotropic Incompressible Materials*, Proceedings of the Cambridge Philosophical Society, Vol. 50, pp. 334–345, 1954.
109. POLIGNONE, D. A., and HORGAN, C. O., *Axisymmetric Finite Anti-Plane Shear of Compressible Nonlinearly Elastic Circular Tubes*, Quarterly of Applied Mathematics, Vol. 50, pp. 323–341, 1992.
110. BEATTY, M. F., and KHAN, R. A., *Finite Amplitude, Free Vibrations of an Axisymmetric Load Supported by a Highly Elastic Tubular Shear Spring*, Journal of Elasticity, Vol. 37, pp. 179–242, 1995.
111. JIANG, Q., and BEATTY, M. F., *On Compressible Materials Capable of Sustaining Axisymmetric Shear Deformations, Part 1: Anti-Plane Shear of Isotropic Hyperelastic Materials*, Journal of Elasticity, Vol. 39, pp. 75–95, 1995.
112. TSAI, H., and ROSAKIS, P., *On Anisotropic Compressible Materials that can Sustain Elastodynamic Anti-Plane Shear*, Journal of Elasticity, Vol. 35, pp. 213–222, 1994.
113. DUNN, J. E., *Certain A Priori Inequalities and a Peculiar Elastic Material*, Quarterly Journal of Mechanics and Applied Mathematics, Vol. 36, pp. 351–363, 1983.
114. RIVLIN, R. S., *Some Restrictions on Constitutive Equations*, Proceedings of the International Symposium on the Foundations of Continuum Thermodynamics, Bassaco, Italy, 1973.
115. KNOPS, R. J., and WILKES, E. W., *Theory of Elastic Stability*, Flügge's Handbuch der Physik, Springer-Verlag, Berlin, Germany, Vol. VIa/3, pp. 125–302, 1973.
116. LEIPHOLZ, H., *Stability Theory*, Academic Press, New York, New York, pp. 176–270, 1970.
117. BOLOTIN, V. V., *Nonconservative Problems of the Theory of Elastic Stability*, Macmillan Co., New York, New York, 1963.
118. ZEIGLER, H., *Principles of Structural Stability*, Blaisdell Publishing Company, Waltham, Massachusetts, 1968.
119. ANTMAN, S. S., *The Eversion of Thick Spherical Shells*, Archive for Rational Mechanics and Analysis, Vol. 70, pp. 113–123, 1979.
120. GURTIN, M. E., and SPECTOR, S. J., *On Stability and Uniqueness in Finite Elasticity*, Archive for Rational Mechanics and Analysis, Vol. 70, pp. 153–165, 1979.
121. SPECTOR, S. J., *On Uniqueness in Finite Elasticity with General Loading*, Journal of Elasticity, Vol. 10, pp. 145–161, 1980.
122. SPECTOR, S. J., *On Uniqueness for the Traction Problem in Finite Elasticity*, Journal of Elasticity, Vol. 12, pp. 367–383, 1982.
123. TRUESDELL, C., and TOUPIN, R. A., *Static Grounds for Inequalities in Finite Strain*

of Elastic Materials, Archive for Rational Mechanics and Analysis, Vol. 12, pp. 1–33, 1963.

124. KRAWIETZ, A., *A Comprehensive Inequality in Finite Elastic Strain*, Archive for Rational Mechanics and Analysis, Vol. 58, pp. 127–149, 1975.
125. USMANI, S. A., and BEATTY, M. F., *On the Surface Instability of a Highly Elastic Half-Space*, Journal of Elasticity, Vol. 4, pp. 249–263, 1974.
126. GENT, A. N., *Rubber Elasticity: Basic Concepts and Behavior*, Chapter 1 in Science and Technology of Rubber, Academic Press, New York, New York, 1978.
127. HUTCHINSON, J. W., and NEALE, K. W., *Neck Propagation*, Journal of Mechanics and Physics of Solids, Vol. 31, pp. 405–426, 1983.
128. CHATER, E., and HUTCHINSON, J. W., *On the Propagation of Bulges and Buckles*, Journal of Applied Mechanics, Vol. 51, pp. 269–277, 1984.
129. SAWYERS, K. N., *Stability of an Elastic Cube under Dead Load: Two Equal Forces*, International Journal of Nonlinear Mechanics, Vol. 11, pp. 11–23, 1976.
130. BALL, J. M., and SCHAEFFER, D., *Bifurcation and Stability of Homogeneous Equilibrium Configurations of an Elastic Body under Dead Load Tractions*, Proceedings of the Cambridge Philosophical Society, Vol. 94, pp. 315–340, 1983.
131. TABADDOR, F., *Rubber Elasticity Models for Finite Element Analysis*, Computers and Structures, Vol. 26, pp. 33–40, 1987.
132. TRELOAR, L. R. G., *Stresses and Birefringence in Rubber Subjected to General Homogeneous Strain*, Proceedings of the Physical Society, Vol. 60, pp. 135–142, 1948.
133. KEARSLEY, E. A., *Asymmetric Stretching of a Symmetrically Loaded Elastic Sheet*, International Journal of Solids and Structures, Vol. 22, pp. 111–119, 1986.
134. MACSITHIGH, G. P., *Energy-Minimal Finite Deformations of a Symmetrically Loaded Elastic Sheet*, Quarterly Journal of Mechanics and Applied Mathematics, Vol. 39, pp. 111–123, 1986.
135. TABADDOR, F., *Elastic Stability of Rubber Products*, Rubber Chemistry and Technology, Vol. 60, pp. 957–965, 1987.
136. BEATTY, M. F., *Elastic Stability of Rubber Bodies in Compression*, Finite Elasticity, Edited by R. S. Rivlin, ASME, AMD Vol. 27, pp. 125–150, 1977.
137. WILLIS, A. H., *Instability in Hollow Rubber Cylinders Subjected to Axial Loads*, VII International Congress of Applied Mechanics, London, England, Vol. 1, pp. 280–296, 1948.
138. PAYNE, A. R., and SCOTT, J. R., *Engineering Design with Rubber*, Interscience Publishers, New York, New York, Chapter 6, 1960.
139. KNOWLES, J. K., *Large Amplitude Oscillations of a Tube of Incompressible Elastic Material*, Quarterly of Applied Mathematics, Vol. 18, pp. 71–77, 1960.
140. KNOWLES, J. K., *On a Class of Oscillations in the Finite Deformation Theory of Elasticity*, ASME Journal of Applied Mechanics, Vol. 29, pp. 283–286, 1962.
141. TRUESDELL, C., *Solutio Generalis et Accurata Problematum Quamplurimorum de moto Corporum Elasticorum Incomprimibilium in Deformationibus Valde Magnis*, Archive for Rational Mechanics and Analysis, Vol. 11, pp. 106–113. 1962. *Adendum*, Ibid, Vol. 12, pp. 427–428, 1963. See also Truesdell and Noll, Ref. 11, Section 61; and Wang and Truesdell, Ref. 12, pp. 315–341.
142. WANG, C.-C., *On the Radial Oscillations of a Spherical Thin Shell in Finite Elasticity Theory*, Quarterly of Applied Mathematics, Vol. 23, pp. 270–274, 1965.

143. SHAHINPOOR, M., and NOWINSKI, J. L., *Exact Solution to the Problem of Forced Large Amplitude Radial Oscillations of a Thin Hyperelastic Tube*, International Journal for Nonlinear Mechanics, Vol. 6, pp. 193–207, 1971.
144. ROGERS, C., and BAKER, J. A., *The Finite Elastodynamics of Hyperelastic Thin Tubes*, International Journal for Nonlinear Mechanics, Vol. 15, pp. 225–233, 1980.
145. ERINGEN, A. C., and SUHUBI, E. S., *Elastodynamics, Vol. 1, Finite Motions*, Academic Press, New York, New York, 1974.
146. BEATTY, M. F., *Finite Amplitude Oscillations of a Simple Rubber Support System*, Archive for Rational Mechanics and Analysis, Vol. 83, pp. 195–219, 1983.
147. BEATTY, M. F., *Finite Amplitude Vibrations of a Body Supported by Simple Shear Springs*, ASME Journal of Applied Mechanics, Vol. 51, pp. 361–366, 1984.
148. BEATTY, M. F., *Finite Amplitude Vibrations of a Neo-Hookean Oscillator*, Quarterly of Applied Mathematics, Vol. 44, pp. 19–34, 1986.
149. BEATTY, M. F., and CHOW, A. C., *Free Vibrations of a Loaded Rubber String*, International Journal for Nonlinear Mechanics, Vol. 19, pp. 69–81, 1984.
150. BEATTY, M. F., and CHOW, A. C., *On the Transverse Vibration of a Rubber String*, Journal of Elasticity, Vol. 13, pp. 317–344, 1983.
151. BAKER, T. J., *The Frequency of Transverse Vibrations of a Stretched India Rubber Cord*, Philosophical Magazine, Vol. 49, pp. 347–351, 1900.
152. VON LANG, V., *Ueber Transversale Töne von Kaulschuk-fäden*, Annalen der Physik und Chemie, Vol. 68, pp. 335–342, 1899.
153. CHEN, P. J., *Selected Topics in Wave Propagation*, Noordhoff International Publishers, Leyden, The Netherlands, 1976.
154. ERICKSEN, J. L., *Introduction to the Thermodynamics of Solids*, Chapman and Hall, London, England, 1991.
155. LEE, S. J., and SHIELD, R. T., *Variational Principles in Finite Elastostatics*, Zeitschrift für angewandte Mathematik und Physik, Vol. 31, pp. 437–472, 1980.
156. SHIELD, R. T., *Equilibrium Solutions in Finite Elasticity*, ASME Journal of Applied Mechanics, Vol. 50, pp. 1171–1180, 1983.
157. RIVLIN, R. S., Editor, *Finite Elasticity*, ASME, AMD Vol. 27, New York, New York, 1977.
158. CARLSON, D. E., and SHIELD, R. T., Editors, *Finite Elasticity*, Proceedings of the IUTAM Symposium on Finite Elasticity, Lehigh University, 1980, Martinus Nijhoff Publishers, The Hague, The Netherlands, 1982.

3

Propagating and Static Exponential Solutions in a Deformed Mooney–Rivlin Material

Ph. Boulanger and M. Hayes

Abstract. A homogeneous isotropic incompressible elastic material whose mechanical behavior is modeled through use of the Mooney–Rivlin form of the stored energy function is maintained in a state of finite static homogeneous biaxial deformation. Superimposed on the finite deformation is an infinitesimal, possibly time dependent, deformation. For purely homogeneous waves the slowness surface consists of two coaxial spheroids. Propagating exponential solutions (PES) are inhomogeneous plane wave solutions for which the planes of constant phase are not the same as the planes of constant amplitude. All PES are obtained systematically through the use of a formulation involving bivectors in which a directional bivector **C** and its attendant directional ellipse is prescribed. Of particular interest is the case when **C** is isotropic because it yields solutions with a displacement field independent of the material constants and of the parameters describing the homogeneous static biaxial deformation. Static exponential solutions (SES) are similar to PES but are time independent. All SES are obtained systematically. It is seen that for such solutions the directional ellipse of **C** must be similar and similarly situated to an elliptical section of either sheet of the slowness surface.

Key Words. Homogeneous isotropic incompressible elastic materials, inhomogeneous plane waves, evanescent waves, propagating exponential solutions (PES), static exponential solutions (SES), bivector, homogeneous

---

Ph. Boulanger • Professor, Département de Mathématique, Université Libre de Bruxelles, Brussels, 1050 Belgium. **M. Hayes** • Professor, Department of Mathematical Physics, University College, Dublin 4, Ireland.

Nonlinear Effects in Fluids and Solids, edited by M. M. Carroll and M. Hayes, Plenum Press, New York, 1996.

plane waves, directional ellipse, principal invariants, slowness bivector, propagation condition, slowness surface, neo-Hookean materials, circularly polarized waves.

1. Introduction

We consider a homogeneous isotropic incompressible elastic body composed of rubber for which it is assumed that the energy function has the Mooney–Rivlin form. The body is first subjected to a static finite homogeneous biaxial deformation and maintained in that state. Superimposed on this deformation is an infinitesimal, possibly time dependent, deformation. The resulting linearized equations are well known and have been the subject of many studies. We cite the work of Hayes and Rivlin (Ref. 1), related work on thermoelastic waves by Flavin and Green (Ref. 2), the stability analysis of Sawyers and Rivlin (Ref. 3), the examination of existence and uniqueness of solution by Hayes and Horgan (Ref. 4), the basic study on impulsive lines of force by Belward (Ref. 5), and papers on waves in prestressed cylinders by Belward (Ref. 6) and Belward and Wright (Ref. 7).

The purpose of this paper is to consider both inhomogeneous plane waves, sometimes called evanescent waves or propagating exponential solutions (PES), and also static exponential solutions.

PES are those waves for which the planes of constant phase are not the same as the planes of constant amplitude. For example, Rayleigh, Love, and Stoneley waves are special combinations of PES. Typically, for PES, the displacement field $\mathbf{u}(\mathbf{x}, t)$ is the real part of the complex expression $\{\mathbf{A} \exp \iota\omega (N\mathbf{C} \cdot \mathbf{x} - t)\}$. Here ω is a prescribed real angular frequency, and $\mathbf{C}$ is a prescribed bivector (i.e., complex vector). The various PES may be obtained systematically (Ref. 8) by finding the complex number N and the bivector amplitude $\mathbf{A}$ for each prescribed ω and $\mathbf{C}$. In the special case of homogeneous plane waves, $\mathbf{C}$ is taken to be a real unit vector along the propagation direction, and N^{-1} is then the propagation speed.

Because $\mathbf{C}$ is allowed to be a bivector, a greater variety of solutions is obtained than when $\mathbf{C}$ is a real vector. In particular, for certain choices of $\mathbf{C}$, we may find $N^{-1} = 0$, and hence a solution of the equilibrium equations is obtained: it is the real part of the expression $\{\mathbf{A} \exp \iota\omega N\mathbf{C} \cdot \mathbf{x}\}$, where now ωN may be an arbitrary complex number. Such solutions are called static exponential solutions (SES).

Here we obtain all possible PES and SES of the linearized equations.

The form of the paper is as follows. In Sections 2 and 3, the basic equations and deformations are recalled. Then, in Section 4, PES and SES are described in detail. The secular equation for N^{-1} is obtained in Section 5. Results for homogeneous plane waves are recalled in Section 6: the slowness surface consists of two sheets which are coaxial spheroids. Next, in Section 7, all PES are systematically obtained. Of particular interest is the case when **C** is isotropic, because it yields solutions with a displacement field independent of the material constants and the parameters describing the homogeneous static biaxial deformation. Finally, in Section 8, all possible SES are obtained. They arise when the directional ellipse of the bivector **C** (Ref. 8) is similar and similarly situated to an elliptical section of either sheet of the slowness surface.

2. Mooney–Rivlin Materials

A Mooney–Rivlin material is a nonlinear incompressible elastic material for which the strain energy per unit volume is given by

$$W = c_1(I - 3) + c_2(II - 3). \tag{1}$$

Here c_1 and c_2 are constants and I, II, III denote the principal invariants of the left Cauchy–Green strain tensor $\mathbf{B} = \mathbf{F}\mathbf{F}^T$ (where $\mathbf{F}$ is the deformation-gradient tensor):

$$I = \mathrm{tr}\mathbf{B}, \qquad II = \tfrac{1}{2}\{(\mathrm{tr}\mathbf{B})^2 - \mathrm{tr}(\mathbf{B}^2)\}, \qquad III = \det \mathbf{B}. \tag{2}$$

The condition of incompressibility is

$$III = 1, \tag{3}$$

and the constitutive equation for the Cauchy stress tensor **T** is

$$\mathbf{T} = -p\mathbf{1} + 2c_1\mathbf{B} - 2c_2\mathbf{B}^{-1}, \tag{4}$$

where p is an arbitrary hydrostatic pressure. In the special case when $c_2 = 0$, the material is said to be neo-Hookean.

3. Small Deformations Superimposed on Large

We assume that a Mooney–Rivlin material is first subjected to a large static homogeneous biaxial deformation given in Cartesian coordinates by

$$x_1 = \lambda X_1, \qquad x_2 = \lambda X_2, \qquad x_3 = \mu X_3, \tag{5}$$

where **X** and **x** denote the particles' positions in the undeformed and deformed state, respectively.

Also, owing to the constraint of incompressibility, the constant extension ratios λ and μ satisfy

$$\lambda^2\mu = 1. \tag{6}$$

The equations governing a superimposed infinitesimal displacement $\mathbf{u}(\mathbf{x}, t)$ from this state of finite deformation are given by (Ref. 4)

$$Ru_{1,11} + (R - S)u_{1,22} + \kappa^2 Ru_{1,33} + Su_{2,12} - q_{,1} = \rho\partial_t^2 u_1, \tag{7a}$$

$$Su_{1,12} + (R - S)u_{2,11} + Ru_{2,22} + \kappa^2 Ru_{2,33} - q_{,2} = \rho\partial_t^2 u_2, \tag{7b}$$

$$R(u_{3,11} + u_{3,22}) + \kappa^2 Ru_{3,33} - q_{,3} = \rho\partial_t^2 u_3, \tag{7c}$$

$$u_{1,1} + u_{2,2} + u_{3,3} = 0, \tag{7d}$$

where

$$R = 2\lambda^2(c_1 + \lambda^2 c_2), \qquad S = 2\lambda^2(\lambda^2 - \mu^2)c_2, \qquad \kappa^2 = \mu^2/\lambda^2, \tag{8}$$

and where $q(\mathbf{x}, t)$ denotes the incremental pressure.

4. Propagating and Static Exponential Solutions

We now seek solutions of the system (7) of the form

$$\mathbf{u} = \{\mathbf{A}\,(\exp \iota\omega)(\mathbf{S}\cdot\mathbf{x} - t)\}^+, \qquad q = \{Q\,(\exp \iota\omega)(\mathbf{S}\cdot\mathbf{x} - t)\}^+. \tag{9}$$

Here ω denotes the prescribed real angular frequency, $\mathbf{S} = \mathbf{S}^+ + \iota\mathbf{S}^-$ the slowness bivector (complex vector), and $\mathbf{A} = \mathbf{A}^+ + \iota\mathbf{A}^-$, $Q = Q^+ + \iota Q^-$ denote the complex amplitudes of the displacement field and incremental pressure. When $\mathbf{S}^-$ is not parallel to $\mathbf{S}^+$, (9) represents an inhomogeneous harmonic plane wave. When $\mathbf{S}^-$ is parallel to $\mathbf{S}^+$, the wave (9) is said to be homogeneous.

For inhomogeneous plane waves, the slowness bivector $\mathbf{S}$ may be written (Ref. 8)

$$\mathbf{S} = N\mathbf{C}, \qquad \mathbf{C} = m\hat{\mathbf{m}} + \iota\hat{\mathbf{n}}, \tag{10}$$

where N is a complex number, $\hat{\mathbf{m}}$ and $\hat{\mathbf{n}}$ are orthogonal unit vectors, and m is a real number. To determine all of the possible slowness bivectors $\mathbf{S}$, the complex number N has to be found for every choice of the bivector $\mathbf{C}$, that is, for every choice of the orthogonal unit vectors $\hat{\mathbf{m}}$, $\hat{\mathbf{n}}$ and of the real number m. Associated with $\mathbf{C}$ is its directional ellipse whose parametric equation is $\mathbf{x} = m\hat{\mathbf{m}} \cos\theta + \hat{\mathbf{n}} \sin\theta$ $(0 \leq \theta \leq 2\pi)$. When N is found, $\mathbf{S}^+$ and $\mathbf{S}^-$ are determined as a pair of conjugate radii of an ellipse similar and similarly situated to the directional ellipse of $\mathbf{C}$ (Ref. 8).

For homogeneous waves, it may be assumed that $\mathbf{C}$ is a real unit vector and we write

$$\mathbf{S} = N\mathbf{C}, \qquad \mathbf{C} = \hat{\mathbf{n}}. \tag{11}$$

In both cases waves of the form (9) may be written as

$$\mathbf{u} = \{\mathbf{A}\,(\exp \iota\omega)\, N(\mathbf{C} \cdot \mathbf{x} - N^{-1}t)\}^+, \tag{12a}$$

$$q = \{Q\,(\exp \iota\omega)\, N(\mathbf{C} \cdot \mathbf{x} - N^{-1}t)\}^+. \tag{12b}$$

They are also called propagating exponential solutions (PES).

SES are solutions of the form

$$\mathbf{u} = \{\mathbf{A}\,(\exp \iota\omega)\, N\mathbf{C} \cdot \mathbf{x}\}^+, \qquad q = \{Q\,(\exp \iota\omega)\; N\mathbf{C} \cdot \mathbf{x}\}^+. \tag{13}$$

The expressions (13) may be seen as the limiting case of the expressions (12) when N^{-1} and ω tend to zero, the product ωN being held fixed. If, for a certain prescribed $\mathbf{C}$, we find $N^{-1} = 0$, then an SES of the form (13) is obtained, ωN being an arbitrary complex number.

5. Propagation Condition–Secular Equation

Introducing the form (12) of the PES into the system (7), we obtain the propagation condition

$$\begin{bmatrix} \mathbf{C}^T\beta\mathbf{C} + SC_1^2 - \rho N^{-2} & SC_1C_2 & 0 & C_1/\omega N \\ SC_1C_2 & \mathbf{C}^T\beta\mathbf{C} + SC_2^2 - \rho N^{-2} & 0 & C_2/\omega N \\ 0 & 0 & \mathbf{C}^T\alpha\mathbf{C} - \rho N^{-2} & C_3/\omega N \\ C_1/\omega N & C_2/\omega N & C_3/\omega N & 0 \end{bmatrix} \begin{bmatrix} A_1 \\ A_2 \\ A_3 \\ \iota Q \end{bmatrix} = 0, \tag{14}$$

where the quadratic forms $\mathbf{C}^T\,\alpha\mathbf{C}$ and $\mathbf{C}^T\,\beta\mathbf{C}$ are defined by

$$\mathbf{C}^T\alpha\mathbf{C} = R(C_1^2 + C_2^2) + \kappa^2 R C_3^2, \tag{15a}$$

$$\mathbf{C}^T\beta\mathbf{C} = (R - S)\,(C_1^2 + C_2^2) + \kappa^2 R C_3^2. \tag{15b}$$

So that waves may propagate, the propagation condition (14) must have nontrivial solutions. This gives the secular equation (by setting to zero the determinant of the system)

$$(\mathbf{C}\cdot\mathbf{C})(\mathbf{C}^T\alpha\mathbf{C} - \rho N^{-2})(\mathbf{C}^T\beta\mathbf{C} - \rho N^{-2}) = 0. \tag{16}$$

When $\mathbf{C}$ is prescribed this is a quadratic equation for N^{-2} (if $\mathbf{C}\cdot\mathbf{C} \neq 0$). On multiplying the first three equations (14) by C_1, C_2, C_3, and on using the fourth equation, $\mathbf{C}\cdot\mathbf{A} = 0$, we note that

$$(\mathbf{C}\cdot\mathbf{C})Q = 0, \tag{17}$$

and hence the incremental pressure is zero if $\mathbf{C}\cdot\mathbf{C} \neq 0$.

Also, on multiplying the first equation (14) by C_2, the second by C_1, and subtracting, we have

$$(\mathbf{C}^T\beta\mathbf{C} - \rho N^{-2})(C_2A_1 - C_1A_2) = 0. \tag{18}$$

6. Homogeneous Waves

For homogeneous waves, $\mathbf{C} = \hat{\mathbf{n}}$ (real unit vector). Thus, $\mathbf{C}\cdot\mathbf{C} = 1$, and hence $Q = 0$. The roots of the secular equation (16) are

$$\rho N_\alpha^{-2}(\hat{\mathbf{n}}) = \hat{\mathbf{n}}^T\alpha\hat{\mathbf{n}}, \qquad \rho N_\beta^{-2}(\hat{\mathbf{n}}) = \hat{\mathbf{n}}^T\beta\hat{\mathbf{n}}, \tag{19}$$

and the corresponding amplitudes $\mathbf{A}_\alpha$, $\mathbf{A}_\beta$ of the displacement field are given, up to a scalar factor, by

$$\mathbf{A}_\alpha = (\hat{n}_1\hat{n}_3, \hat{n}_2\hat{n}_3, -\hat{n}_1^2 - \hat{n}_2^2), \qquad \mathbf{A}_\beta = (-\hat{n}_2, \hat{n}_1, 0). \tag{20}$$

Hence, both waves are linearly polarized.

We here assume that both roots (19) are positive for every propagation direction $\hat{\mathbf{n}}$ ("inherent stability" of the deformed state of the material). Thus, α and β are both positive definite, which means that

$$R > 0, \qquad R - S > 0. \tag{21}$$

The slowness surface corresponding to the roots (19) consists of two spheroids

$$\mathbf{S}^T\alpha\mathbf{S} = \rho, \qquad \mathbf{S}^T\beta\mathbf{S} = \rho, \tag{22}$$

with common axis along the x_3-direction. The spheroids $\mathbf{x}^T\alpha\mathbf{x} = 1$ and $\mathbf{x}^T\beta\mathbf{x} = 1$ (which are similar and similarly situated to the two sheets of the slowness surface) will be called the α- and the β-ellipsoid, respectively.

Finally, we note that there is just one direction $\hat{\mathbf{n}} = \mathbf{k} = (0, 0, 1)$ such that both roots (19) are equal. For this direction, we have

$$\rho N_{\alpha}^{-2}(\mathbf{k}) = \rho N_{\beta}^{-2}(\mathbf{k}) = \kappa^2 R, \tag{23}$$

and the corresponding amplitude $\mathbf{A}$ may be any bivector in the plane orthogonal to $\mathbf{k}$. In particular, when $\mathbf{A} \cdot \mathbf{A} = 0$, the wave is circularly polarized.

In the special case of neo-Hookean materials, $S = 0$, and the two sheets of the slowness surface coalesce: $\alpha = \beta$. In this case the secular equation has always a double root, and circularly polarized homogeneous waves are possible for every propagation direction.

7. Propagating Evanescent Solutions

For inhomogeneous waves, $\mathbf{C} = m\hat{\mathbf{m}} + \iota\hat{\mathbf{n}}$, and thus $\mathbf{C} \cdot \mathbf{C} = m^2 - 1$ may be zero. We consider in turn the case when $\mathbf{C}$ is not isotropic ($m^2 \neq 1$) and the case when $\mathbf{C}$ is isotropic ($m^2 = 1$).

(I) $\mathbf{C} \cdot \mathbf{C} = m^2 - 1 \neq 0$: $\mathbf{C}$ is not isotropic.

In this case, $Q = 0$ and the roots of the secular equation (16) are

$$\rho N_{\alpha}^{-2}(\mathbf{C}) = \mathbf{C}^T\alpha\mathbf{C}, \qquad \rho N_{\beta}^{-2}(\mathbf{C}) = \mathbf{C}^T\beta\mathbf{C}. \tag{24}$$

Here we assume that they are different from zero. The corresponding amplitudes $\mathbf{A}_{\alpha}$, $\mathbf{A}_{\beta}$ of the displacement field are given, up to a scalar factor, by

$$\mathbf{A}_{\alpha} = (C_1C_3, C_2C_3, -C_1^2 - C_2^2), \qquad \mathbf{A}_{\beta} = (-C_2, C_1, 0). \tag{25}$$

We note that

$$\mathbf{A}_{\alpha} \cdot \mathbf{A}_{\alpha} = (\mathbf{C} \cdot \mathbf{C})(C_1^2 + C_2^2), \qquad \mathbf{A}_{\beta} \cdot \mathbf{A}_{\beta} = C_1^2 + C_2^2, \qquad \mathbf{A}_{\alpha} \cdot \mathbf{A}_{\beta} = 0, \tag{26a}$$

$$\mathbf{C} \cdot \mathbf{A}_{\alpha} = 0, \qquad \mathbf{C} \cdot \mathbf{A}_{\beta} = 0. \tag{26b}$$

Hence, in general (for $C_1^2 + C_2^2 \neq 0$), both waves are elliptically polarized. Because $\mathbf{C}$, $\mathbf{A}_{\alpha}$, and $\mathbf{A}_{\beta}$ are orthogonal, the projections of the ellipses of $\mathbf{C}$ and of $\mathbf{A}_{\beta}$ onto the plane of the ellipse of $\mathbf{A}_{\alpha}$ are similar and similarly situated to the ellipse of $\mathbf{A}_{\alpha}$ rotated through a quadrant.

However, for $C_1^2 + C_2^2 = 0$ (with $C_3 \neq 0$), that is, when the projection of the directional ellipse of $\mathbf{C}$ onto the x_1, x_2-plane is a circle, $\mathbf{A}_{\alpha}$ and $\mathbf{A}_{\beta}$ are both isotropic and parallel. In this special case, we have only one circularly polarized propagation mode corresponding to the double root

$$\rho N_\alpha^{-2} = \rho N_\beta^{-2} = \kappa^2 R C_3^2 \tag{27}$$

of the secular equation (16).

(II) $\mathbf{C} \cdot \mathbf{C} = m^2 - 1 = 0$: $\mathbf{C}$ is isotropic.

Clearly, the secular equation (16) is now satisfied whatever the value of N may be. Assuming first that $\rho N^{-2} \neq \mathbf{C}^T \beta \mathbf{C}$, we note from (18) that $A_1/A_2 = C_1/C_2$. It then follows from the propagation condition (14) that the amplitudes are given, up to a scalar factor, by

$$\mathbf{A} = \mathbf{C}, \qquad Q = \iota\omega N(\mathbf{C}^T \alpha \mathbf{C} - \rho N^{-2}). \tag{28}$$

Because $\mathbf{C}$ is isotropic, the displacement field is circularly polarized, but there is now an associated incremental pressure field. Note that in (28), ω and N are arbitrary.

As an example of this type of solution, take $\mathbf{C} = (1, 0, \iota)$ and multiply the amplitudes (28) by an arbitrary real factor a. Then, for N real, (12) yields

$$u_1 = a\cos(kx_1 - \omega t)\exp(-kx_3), \qquad u_2 = 0, \tag{29a}$$

$$u_3 = -a\sin(kx_1 - \omega t)\exp(-kx_3), \tag{29b}$$

$$q = ak\{\rho\omega^2 k^{-2} + (\kappa^2 - 1)R\}\sin(kx_1 - \omega t)\exp(-kx_3), \tag{29c}$$

where ω and $k = \omega N$ are arbitrary (real). It is easily checked that (29) is indeed a solution of the system (7). We note that the displacement field is independent of the material coefficients R, S, κ^2.

Now, take $\rho N^{-2} = \mathbf{C}^T \beta \mathbf{C}$. Corresponding to this choice of N, it is seen from the propagation (14) that two propagation modes are possible: $\mathbf{A}$ and Q may be any linear combinations $\mathbf{A} = \alpha_1 \mathbf{A}_1 + \alpha_2 \mathbf{A}_2$, $Q = \alpha_1 Q_1 + \alpha_2 Q_2$, where α_1 and α_2 are arbitrary constants and

$$\mathbf{A}_1 = \mathbf{C}, \qquad Q_1 = -\iota\omega N S C_3^2, \tag{30}$$

$$\mathbf{A}_2 = (-C_2, C_1, 0), \qquad Q_2 = 0. \tag{31}$$

In particular, we note that $\mathbf{A}$ may thus be any bivector orthogonal to $\mathbf{C}$.

8. Static Exponential Solutions

As for propagating solutions, we consider in turn the case when **C** is not isotropic and the case when **C** is isotropic.

(I) $\mathbf{C} \cdot \mathbf{C} = m^2 - 1 \neq 0$: **C** is not isotropic.

SES occur when the secular equation has a zero root for N^{-2}. As the roots are given by (24) this will be the case when $\mathbf{C}^T\alpha\mathbf{C} = 0$ or $\mathbf{C}^T\beta\mathbf{C} = 0$.

Let $\mathbf{P} = \mathbf{P}^+ + \iota\mathbf{P}^-$ denote a bivector, and **g** a positive definite 3×3 matrix. It has been shown (Ref. 9) that $\mathbf{P}^T\mathbf{g}\mathbf{P} = 0$ if and only if the ellipse of the bivector **P** is similar and similarly situated to a section of the ellipsoid $\mathbf{x}^T\mathbf{g}\mathbf{x} = 1$ by a central plane.

Hence, SES are possible when the ellipse of **C** is similar and similarly situated to a central section of the α-ellipsoid or the β-ellipsoid. Here the circular sections of these spheroids by the x_1, x_2-plane have to be excluded because **C** is assumed not to be isotropic.

(Ia) $\mathbf{C}^T\alpha\mathbf{C} = 0\ (C_3 \neq 0)$.

Corresponding to the root $N_\alpha^{-2} = 0$ of the secular equation, we have a SES with **A** and Q given, up to a scalar factor, by

$$\mathbf{A} = (C_1, C_2, \kappa^2 C_3), \qquad Q = 0. \tag{32}$$

As an example, take $\mathbf{C} = (\kappa, 0, \iota)$ and take $k = \omega N$ to be arbitrary and real. Then (13) yields the static solution

$$u_1 = a\kappa \cos(k\kappa x_1)\exp(-kx_3), \qquad u_2 = 0, \tag{33a}$$

$$u_3 = -a\kappa^2 \sin(k\kappa x_1)\exp(-kx_3), \qquad q = 0. \tag{33b}$$

(Ib) $\mathbf{C}^T\beta\mathbf{C} = 0\ (C_3 \neq 0)$.

Corresponding to the root $N_\beta^{-2} = 0$ of the secular equation, we have a SES with **A** and Q given, upt to a scalar factor, by

$$\mathbf{A} = (-C_2, C_1, 0), \qquad Q = 0. \tag{34}$$

As an example, take $\mathbf{C} = (\kappa, \iota\kappa\sqrt{(2R - S)/(R - S)}, 1)$ and take $k = \omega N$

to be arbitrary real. Then (13) yields the static solution

$$u_1 = a\kappa\sqrt{\frac{2R-S}{R-S}}\exp\left(-k\kappa\sqrt{\frac{2R-S}{R-S}}x_2\right)\sin k(\kappa x_1 + x_3), \tag{35a}$$

$$u_2 = a\kappa\exp\left(-k\kappa\sqrt{\frac{2R-S}{R-S}}x_2\right)\cos k(\kappa x_1 + x_3), \tag{35b}$$

$$u_3 = 0, \qquad q = 0. \tag{35c}$$

(II) $\mathbf{C}\cdot\mathbf{C} = m^2 - 1 = 0$: $\mathbf{C}$ is isotropic.

In this case N is arbitrary, and in order to obtain SES we let N^{-1} tend to zero, ωN being fixed.

First we assume that $\mathbf{C}^T\beta\mathbf{C} \neq 0$, which means that the isotropic $\mathbf{C}$ is not in the plane of the circular section of the β-spheroid, that is, not in the x_1, x_2-plane ($C_3 \neq 0$). Then, as a limiting case of (28), we obtain an SES with $\mathbf{A}$ and Q given, up to a scalar factor, by

$$\mathbf{A} = \mathbf{C}, \qquad Q = \iota\omega N\mathbf{C}^T\alpha\mathbf{C}, \tag{36}$$

where ωN is now an arbitrary complex number. As an example, we note that (29) in which we take $\omega = 0$ is an SES of this type.

Finally, we assume that $\mathbf{C}^T\beta\mathbf{C} = \mathbf{C}\cdot\mathbf{C} = 0$, thus that $\mathbf{C}$ is isotropic in the x_1, x_2-plane. Then, also $\mathbf{C}^T\alpha\mathbf{C} = 0$. In this case, using the propagation condition (14) with $N^{-2} = 0$ and ωN fixed (nonzero), we note that $\mathbf{A}$ may be any bivector orthogonal to $\mathbf{C}$ and thus we have

$$\mathbf{A} = \alpha\mathbf{C} + \beta\mathbf{k}, \qquad Q = 0, \tag{37}$$

where α and β are arbitrary complex constants. As an example, take $\mathbf{C} = (1, \iota, 0)$ and $\mathbf{A} = (a, \iota a, b)$ where a and b are real. Then (13) yields, with $k = \omega N$ real,

$$u_1 = a\cos(kx_1)\exp(-kx_2), \tag{38a}$$

$$u_2 = -a\sin(kx_1)\exp(-kx_2), \tag{38b}$$

$$u_3 = b\cos(kx_1)\exp(-kx_2), \qquad q = 0. \tag{38c}$$

Such a solution is independent of the material coefficients R, S, κ^2. The same solution is thus valid for any Mooney–Rivlin material in any biaxial state of deformation given by (5).

References

1. HAYES, M., and RIVLIN, R. S., *Propagation of a Plane Wave in an Isotropic Elastic Material Subjected to Pure Homogeneous Deformation*, Archive for Rational Mechanics and Analysis, Vol. 8, pp. 15–22, 1961.
2. FLAVIN, J. N., and GREEN, A. E. *Plane Thermoelastic Waves in an Initially Stressed Medium*, Journal of the Mechanics and Physics of Solids, Vol. 9, pp. 179–190, 1961.
3. SAWYERS, K. N., and RIVLIN, R. S., *Instability of an Elastic Material*, International Journal of Solids and Structures, Vol. 9, pp. 607–613, 1973.
4. HAYES, M., and HORGAN, C. O., *On the Dirichlet Problem for Incompressible Elastic Materials*, Journal of Elasticity, Vol. 4, pp. 17–25, 1974.
5. BELWARD, J. A., *Some Dynamic Properties of a Prestressed Incompressible Hyperelastic Material*, Bulletin of the Australian Mathematical Society, Vol. 8, pp. 61–73, 1973.
6. BELWARD, J. A., *The Propagation of Small Amplitude Waves in Prestressed Incompressible Elastic Cylinders*, International Journal of Engineering Science, Vol. 14, pp. 647–659, 1976.
7. BELWARD, J. A., and WRIGHT, S. J., *Small-Amplitude Waves with Complex Wave Numbers in a Prestressed Cylinder of Mooney Material*, Quarterly Journal of Mechanics and Applied Mathematics, Vol. 40, pp. 383–394, 1987.
8. HAYES, M., *Inhomogeneous Plane Waves*, Archive for Rational Mechanics and Analysis, Vol. 85, pp. 41–79, 1984.
9. BOULANGER, PH., and HAYES, M., *Electromagnetic Plane Waves in Anisotropic Media: An Approach Using Bivectors*, Philosophical Transactions of the Royal Society of London, Vol. A330, pp. 335–393, 1990.

4

Circularly Polarized Waves of Finite Amplitude in Elastic Dielectrics

M. M. CARROLL AND M. F. MCCARTHY

Abstract. It is shown that time sinusoidal progressive waves and standing waves can propagate in the direction of a biasing static electric field in isotropic elastic dielectrics. It is assumed that the rates are such that the induced electric displacement field may be regarded as quasistatic. The nature of the electromechanical coupling is examined and a specific solution in terms of Jacobian elliptic functions is obtained by neglecting terms higher than degree three in the electric and mechanical fields.

Key Words. Homogeneous isotropic elastic dieletrics, quasistatic electric fields, simple shear, biasing electric field, standing wave solution, generalized shear modulus, generalized shear compliance, circularly polarized wave, approximate constitutive theory.

1. Introduction

Exact solutions of the nonlinear elastodynamic equations for homogeneous isotropic materials were presented in earlier papers (Refs. 1–4). These motions are time sinusoidal, finite amplitude, circularly polarized progressive waves and standing waves. In the present paper we examine the influence of electrical polarization on these waves in homogeneous isotropic elastic dielectrics. We sim-

---

M. M. Carroll • Dean, Brown School of Engineering, Rice University, Houston, Texas 77005.
M. F. McCarthy • Registrar, University College Galway, Galway, Ireland.

Nonlinear Effects in Fluids and Solids, edited by M. M. Carroll and M. Hayes, Plenum Press, New York, 1996.

plify the problem by assuming that the rates are such that the electric fields may be regarded as quasistatic. With this assumption, progressive wave solutions are obtained for both compressible and incompressible materials, with a uniform static electric field in the direction of propagation. Standing wave solutions are obtained for incompressible materials, and the nature of the standing wave solution for compressible materials, which was treated for elastic materials in Ref. 1, is indicated but not treated in detail.

We use the constitutive equations for isotropic elastic dielectrics in the form given by Singh and Pipkin (Ref. 5). These equations are specialized to describe the response in simple shear, with a biasing electric field applied in the plane of shear and normal to the direction of shear, in Section 3. Propagating wave solutions are presented in Section 4 and the electromechanical coupling is examined, the most notable effect being the appearance of a propagating transverse electric displacement field component in the direction of shear. The standing wave solution is presented in Section 5. In this case the electric field remains longitudinal but becomes nonuniform, and the electric displacement field again acquires a transverse component which rotates with the direction of shear. In Section 6 we examine the standing wave solution at the level of approximation in which terms up to third degree in the amount of shear and the applied field strength are retained. In this case the solution is given in terms of Jacobian elliptic functions.

2. Basic Equations

In describing the response of an isotropic elastic dielectric, we can choose the deformation tensor **B** and electric field vector **e** as independent variables[1] and the Cauchy stress tensor **T** and electric displacement vector **d** as dependent variables. The deformation tensor is defined by $\mathbf{B} = \mathbf{F}\mathbf{F}^T$, where **F** is the deformation gradient tensor. The constitutive equations then have the form

$$\mathbf{T} = \mathbf{T}(\mathbf{B}, \mathbf{e}), \qquad \mathbf{d} = \mathbf{d}(\mathbf{B}, \mathbf{e}), \tag{1}$$

where **T** and **d** are isotropic functions, i.e.,

$$\mathbf{T}(\mathbf{Q}\mathbf{B}\mathbf{Q}^T, \mathbf{Q}\mathbf{e}) = \mathbf{Q}\mathbf{T}(\mathbf{B}, \mathbf{e})\mathbf{Q}^T, \qquad \mathbf{d}(\mathbf{Q}\mathbf{B}\mathbf{Q}^T, \mathbf{Q}\mathbf{e}) = \mathbf{Q}\mathbf{d}(\mathbf{B}, \mathbf{e}), \tag{2}$$

for all orthogonal tensors **Q**. The constitutive equations must be modified for incompressible materials. In this case all admissible deformations are isochoric

[1]We caution that the tensor **B** is not an appropriate independent variable for anisotropic materials. The choice of **e**, rather than **d**, as independent variable is a matter of convenience for the solutions discussed in this paper.

(volume-preserving) so that det $\mathbf{B} = 1$. The Cauchy stress now has an undetermined hydrostatic pressure, so that

$$\mathbf{T} = -p\mathbf{1} + \tilde{\mathbf{T}}(\mathbf{B}, \mathbf{e}), \qquad \mathbf{d} = \tilde{\mathbf{d}}(\mathbf{B}, \mathbf{e}), \tag{3}$$

where $\mathbf{1}$ denotes the unit tensor, p is an arbitrary scalar field, and the domains of $\tilde{\mathbf{T}}$ and $\tilde{\mathbf{d}}$ are restricted to isochoric deformations.

Representation theorems for isotropic tensor and vector functions lead to the following representations of $\mathbf{T}$ and $\mathbf{d}$ (Ref. 5):

$$\mathbf{T} = \alpha\mathbf{1} + \beta\mathbf{B} + \gamma\mathbf{B}^{-1} + \boldsymbol{\epsilon} \times \mathbf{e} + \mathbf{e} \times \boldsymbol{\epsilon}, \tag{4}$$

$$\boldsymbol{\epsilon} = (\alpha_1\mathbf{1} + \beta_1\mathbf{B} + \gamma_1\mathbf{B}^{-1})\mathbf{e}, \tag{5}$$

$$\mathbf{d} = (\alpha_2\mathbf{1} + \beta_2\mathbf{B} + \gamma_2\mathbf{B}^{-1})\mathbf{e}, \tag{6}$$

where the α's, β's, and γ's are scalar functions of the isotropic invariants

$$\mathrm{I}_1 = \mathrm{tr}\mathbf{B}, \qquad \mathrm{I}_2 = \mathrm{tr}\mathbf{B}^*, \qquad \mathrm{I}_3 = \det \mathbf{B}, \tag{7a}$$

$$\mathrm{I}_4 = \mathbf{e} \cdot \mathbf{e}, \qquad \mathrm{I}_5 = \mathbf{e} \cdot \mathbf{B}\mathbf{e}, \qquad \mathrm{I}_6 = \mathbf{e} \cdot \mathbf{B}^{-1}\mathbf{e}, \tag{7b}$$

and $\mathbf{B}^*$ is the adjugate of $\mathbf{B}$. The scalar response functions must meet certain conditions to ensure the existence of an energy density (Ref. 6), but we do not make these conditions explicit here. The constitutive equations for incompressible isotropic elastic dielectrics also have the form (4)–(6), except that α in (4) is now an arbitrary scalar field, $\alpha = -p$, and the remaining α's, β's, and γ's are functions of the five invariants I_1, I_2, I_4, I_5 and I_6.

We assume that there is no applied body force field and no free currents or charges, and also that the rates of the processes considered are such that the electric fields may be regarded as quasistatic. Then the relevant field equations are

$$\mathrm{div}\, \mathbf{T} = \rho\ddot{\mathbf{x}}, \tag{8a}$$

$$\mathrm{curl}\, \mathbf{e} = \mathbf{0}, \tag{8b}$$

$$\mathrm{div}\, \mathbf{d} = 0, \tag{8c}$$

where ρ is the mass density and $\ddot{\mathbf{x}}$ is the acceleration.

3. Simple Shearing and Polarization

Consider an isotropic elastic dielectric which is subjected to a simple shear, so that a particle initially at (X, Y, Z) in a rectangular Cartesian system is now at

(x, y, z), given by

$$x = X + \kappa Z, \qquad y = Y, \qquad z = Z. \tag{9}$$

Suppose also that a uniform static electric field is applied in the z-direction, so that

$$e_x = 0, \qquad e_y = 0, \qquad e_z = E. \tag{10}$$

Here κ (>0) is the amount of shear, E (>0) is the electric field strength, and the electric field is applied in the plane of shear and normal to the direction of shear.

A simple computation from (9) gives the components of the deformation tensor and its inverse as

$$\mathbf{B} = \begin{bmatrix} 1+\kappa^2 & 0 & \kappa \\ 0 & 1 & 0 \\ \kappa & 0 & 1 \end{bmatrix}, \tag{11a}$$

$$\mathbf{B}^{-1} = \begin{bmatrix} 1 & 0 & -\kappa \\ 0 & 1 & 0 \\ -\kappa & 0 & 1+\kappa^2 \end{bmatrix}. \tag{11b}$$

The invariants I_i are given by equations (7), (10), and (11) as

$$I_1 = 3 + \kappa^2, \qquad I_2 = 3 + \kappa^2, \qquad I_3 = 1, \tag{12a}$$

$$I_4 = E^2, \qquad I_5 = E^2, \qquad I_6 = (1+\kappa^2)E^2. \tag{12b}$$

The stress and electric displacement are obtained by substituting from equations (10)–(12) in equations (4)–(6). They may be written as

$$T_{xx} = \sigma_1(\kappa^2, E^2), \quad T_{yy} = \sigma_2(\kappa^2, E^2), \quad T_{zz} = \sigma_3(\kappa^2, E^2), \tag{13a}$$

$$T_{yz} = 0, \quad T_{zx} = \mu(\kappa^2, E^2)\kappa, \quad T_{xy} = 0, \tag{13b}$$

and

$$d_x = \sigma(\kappa^2, E^2)\kappa E, \tag{14a}$$

$$d_y = 0, \tag{14b}$$

$$d_z = \Delta(\kappa^2, E^2)E, \tag{14c}$$

with

$$\sigma_1 = \alpha + \beta(1+\kappa^2) + \gamma, \tag{15a}$$

$$\sigma_2 = \alpha + \beta + \gamma, \tag{15b}$$

$$\sigma_3 = \alpha + \beta + \gamma(1+\kappa^2) + 2\{\alpha_1 + \beta_1 + \gamma_1(1+\kappa^2)\}E^2, \tag{15c}$$

$$\mu = \beta - \gamma + (\beta_1 - \gamma_1)E^2, \tag{15d}$$

$$\delta = \beta_2 - \gamma_2, \tag{15e}$$

$$\Delta = \alpha_2 + \beta_2 + \gamma_2(1+\kappa^2). \tag{15f}$$

The function μ is called the generalized shear modulus. The function δ describes the induced electric displacement component in the direction of shear and, of course, the function $\Delta(0, E^2)$ describes the dielectric response in the absence of deformation.

The response of an incompressible material is similar to that for a compressible one, except that the hydrostatic pressure may be chosen so as to make any one of the normal stresses T_{xx}, T_{yy}, or T_{zz} vanish.

We rewrite equations (13b, center) and (14c) as

$$\tau = \mu(\kappa^2, E^2)\kappa, \tag{16a}$$

$$D = \Delta(\kappa^2, E^2)E. \tag{16b}$$

We assume that equation (16b) can be inverted, for given κ, in the form

$$E = (\kappa^2, D)D. \tag{17}$$

Substitution for E from this equation in (16a) gives the shear response in the form

$$\tau = \tilde{\mu}(\kappa^2, D^2)\kappa. \tag{18}$$

We assume that this can also be inverted, for given D, in the form

$$\kappa = \nu(\tau^2, D^2)\tau, \tag{19}$$

where ν is the generalized shear compliance.

4. Circularly Polarized Progressive Waves

We now consider a circularly polarized harmonic progressive wave of the form

$$x = X + A\sin(kZ - \omega t), \qquad y = Y + A\cos(kZ - \omega t), \qquad z = Z, \tag{20}$$

and a uniform static electric field in the z-direction,

$$e_x = 0, \qquad e_y = 0, \qquad e_z = E. \tag{21}$$

It was shown previously (Ref. 2) that a finite amplitude wave of the form (20) can propagate in every isotropic elastic solid, either compressible or incompressible. We now investigate the possibility of such waves in the presence of a biasing longitudinal electric field.

A routine calculation gives the components of the deformation tensor as

$$\mathbf{B} = \begin{bmatrix} 1+\kappa^2\cos^2\xi & \kappa^2\cos\xi\sin\xi & \kappa\cos\xi \\ \kappa^2\cos\xi\sin\xi & 1+\kappa^2\sin^2\xi & \kappa\sin\xi \\ \kappa\cos\xi & \kappa\sin\xi & 1 \end{bmatrix}, \tag{22}$$

with

$$\kappa = kA, \qquad \xi = \omega t - kz. \tag{23}$$

The deformation tensor $\mathbf{B}$ in (22) is related to that in (11a) by an orthogonal transformation

$$\mathbf{B}(\kappa^2, \xi) = \mathbf{Q}(\xi)\mathbf{B}(\kappa^2, 0)\mathbf{Q}(\xi)^T, \tag{24}$$

with

$$\mathbf{Q}(\xi) = \begin{bmatrix} \cos\xi & -\sin\xi & 0 \\ \sin\xi & \cos\xi & 0 \\ 0 & 0 & 1 \end{bmatrix}, \tag{25}$$

Thus, $\mathbf{Q}(\xi)$ describes a rotation through an angle ξ about the z-axis, and the local deformation in the circularly polarized wave is a shear of amount kA, such that the direction and plane of shear at any location z rotate about the z-axis with angular frequency ω. Since the electric field (10) is invariant under the rotation $\mathbf{Q}(\xi)$, it follows from the isotropy conditions (2) that the stress and the electric displacement associated with the fields (20) and (21) are obtained from those associated with (9) and (10) by rotation $Q(\xi)$. Thus,

$$\mathbf{T} = \begin{bmatrix} \sigma_1\cos^2\xi + \sigma_2\sin^2\xi & (\sigma_1-\sigma_2)\cos\xi\sin\xi & \mu\kappa\cos\xi \\ (\sigma_1-\sigma_2)\cos\xi\sin\xi & \sigma_1\sin^2\xi + \sigma_2\cos^2\xi & \mu\kappa\sin\xi \\ \mu\kappa\cos\xi & \mu\kappa\sin\xi & \sigma_3 \end{bmatrix}, \tag{26}$$

$$d_x = \delta\kappa E \cos \xi, \tag{27a}$$

$$d_y = \delta\kappa E \sin \xi, \tag{27b}$$

$$d_z = \Delta\, E, \tag{27c}$$

where the constitutive response functions are defined by equations (12) and (15), with $\kappa = kA$. Of course, these results could also be obtained directly by substitution from equations (21) and (22) in equations (4)–(7) and making use of the definitions (15).

The equations of motion (8) reduce to

$$(\mu\kappa \cos \xi)' = \rho\omega^2 A \sin \xi, \tag{28a}$$

$$(\mu\kappa \sin \xi)' = -\rho\omega^2 A \cos \xi, \tag{28b}$$

$$\sigma_3' = 0, \tag{28c}$$

where the prime denotes differentiation with respect to z. The first two equations are compatible and reduce to the algebraic equation

$$\omega^2/k^2 = \mu(k^2A^2, E^2)/\rho, \tag{29}$$

which determines the phase velocity of the wave. Since κ and E are constants, equation (28c) is satisfied identically for both compressible and incompressible solids (choosing p = const in the latter case). The electric field equations are also satisfied identically since both E and Δ are constant. Thus, the circularly polarized progressive wave (20) can propagate in the presence of a biasing electric field (21). The interaction effects are: (i) the influence of the electric field on the phase velocity in (29), (ii) the influence of the electric field on the normal stresses in (26), (iii) the influence of the mechanical wave on the static longitudinal electric displacement in (27c), and (iv) the appearance of a propagating transverse electric displacement field in (27a, b), which is in phase with the direction of shearing (or exactly out of phase of $\delta < 0$).

5. Circularly Polarized Standing Waves

Sinusoidal finite amplitude standing waves of the form

$$x = X + \phi(Z) \cos (\omega t) + \psi(Z) \sin (\omega t), \tag{30a}$$

$$y = Y + \phi(Z) \sin (\omega t) - \psi(Z) \cos (\omega t), \tag{30b}$$

$$z = Z \tag{30c}$$

can exist in all incompressible isotropic elastic solids (Refs. 3, 4). The modification of this motion for compressible solids (Ref. 1) is discussed below. We now investigate the effect of a biasing electric field on the standing wave (30) by exploring the possibility of this motion and a nonuniform longitudinal electric field

$$e_x = 0, \qquad e_y = 0, \qquad e_z = E(z). \tag{31}$$

in an incompressible isotropic nonlinear dielectric.

We introduce functions $\kappa(z)$ and $\theta(z)$, defined by

$$\kappa = \sqrt{\phi'^2 + \psi'^2}, \qquad \theta = \arctan(\psi'/\phi'), \tag{32}$$

so that

$$\phi' = \kappa \cos\theta, \qquad \psi' = \kappa \sin\theta. \tag{33}$$

A routine calculation shows that the deformation tensor **B** associated with the motion (30) has the form (22), except that κ is now a function of z and ξ is defined as

$$\xi = \omega t - \theta. \tag{34}$$

Consequently, the stress and electric displacement components are again given by equations (26) and (27), with the understanding that σ_1, σ_2, and σ_3 involve the arbitrary pressure $\alpha = -p$. With the assumption that this does not depend on x or y, the equation of motion (8a) and the electrical field equation (8c) reduce to

$$(\mu\kappa \cos\xi)' = -\rho\omega^2(\phi \cos(\omega t) + \psi \sin(\omega t)), \tag{35a}$$

$$(\mu\kappa \sin\xi)' = -\rho\omega^2(\phi \sin(\omega t) - \psi \cos(\omega t)), \tag{35b}$$

$$\sigma_3' = 0, \tag{35c}$$

$$(\Delta E)' = 0. \tag{35d}$$

Substitution for κ and ξ from equations (33) and (34) in (35a, b) shows that these latter equations are compatible and reduce to

$$(\mu\phi')' = -\rho\omega^2\phi, \qquad (\mu\psi')' = -\rho\omega^2\psi. \tag{36}$$

The arbitrary pressure can be chosen so that (35c) is satisfied. Finally, (35d) may be written as

$$\Delta E = D, \tag{37}$$

with D constant, and this can be inverted in the form (17), i.e.,

$$E = (\kappa^2, D)D. \tag{38}$$

This equation expresses E in terms of κ^2 for given constant D and substitution for E in (36) gives

$$(\tilde{\mu}\phi')' = -\rho\omega^2\phi, \qquad (\tilde{\mu}\psi')' = -\rho\omega^2\psi, \tag{39}$$

with $\tilde{\mu}(\kappa^2, D^2)$ defined in equation (18).

As discussed in Refs. 3 and 4, the change of variables

$$\Phi = \mu\phi', \qquad \Psi = \mu\psi' \tag{40}$$

reduces equations (36) to the form

$$\Phi'' = -\rho\omega^2\nu(\tau^2, D^2)\Phi, \qquad \Psi'' = -\rho\omega^2\nu(\tau^2, D^2)\Psi. \tag{41}$$

Since D is constant, these are the equations of motion for a particle in an attractive central force field, with (Φ, Ψ) as Cartesian coordinates of the particle and z as time. The equations can also be written in polar form,

$$\tau'' - \tau\theta'^2 = -\rho\omega^2\nu\tau, \qquad \tau\theta'' + 2\tau'\theta' = 0, \tag{42}$$

with

$$\Phi = \tau\cos\theta, \qquad \Psi = \tau\sin\theta. \tag{43}$$

First integrals of equations (42) are

$$\frac{1}{2}(\tau'^2 + h^2/\tau^2) + \rho\omega^2\int_0^\tau \nu(s^2, D^2)s\,ds = T, \tag{44}$$

$$\tau^2\theta' = h, \tag{45}$$

where the constants T and h are the energy and angular momentum in the central force orbit. Equations (44) are first order ordinary differential equations in τ and θ, so that the problem is reduced to quadratures (Refs. 3, 4).

We remark that if the electric displacement vector d is chosen as independent variable, then it should be assumed to have the form (27), i.e., a uniform longitudinal component D and a time sinusoidal transverse component in the direction of shear. The longitudinal electric field component is then given directly by equation (38) and the amplitude of the transverse component is found from the condition that the electric field not have a transverse component. Thus, the effect of the mechanical wave is to make the electric field nonuniform and the electric displacement field nonlongitudinal.

The situation is somewhat more complicated for compressible materials. In this case, (35) is a system of four equations in the three unknown functions ϕ, ψ, and E, and it will not admit solutions except in very special cases. In fact (Ref. 4), one must modify the motion (30) by including a nonuniform static longitudinal stretch, by replacing (30c) by

$$z = \eta(Z). \tag{46}$$

This leads to a system of equations, analogous to (35), in the four unknown functions ϕ, ψ, η, and E. Integration of (35c, d) gives two algebraic equations involving constant stress and electric displacement components $T_{zz} = \sigma$ and $d_z = D$. If these equations can be solved for the functions η and E in terms of κ^2, σ, and D, then substitution in the equations analogous to (35a, b) again leads to a problem in central force motion.

6. Approximations

We illustrate the foregoing solutions by considering an approximate constitutive theory in which the amount of shear κ and the longitudinal electric displacement component D are both small of order ϵ, say. We retain terms of order ϵ^3 and neglect terms of order ϵ^4.

Equations (16) now take the form

$$\tau = (\mu_0 + \mu_1\kappa^2 + \mu_2E^2)\kappa, \tag{47}$$

$$D = (\Delta_0 + \Delta_1\kappa^2 + \Delta_2E^2)E. \tag{48}$$

This latter equation can be inverted to give

$$E = (\epsilon_0 + \epsilon_1\kappa^2 + \epsilon_2D^2)D, \tag{49}$$

with

$$\epsilon_0 = 1/\Delta_0, \qquad \epsilon_1 = -\Delta_1/\Delta_2^2 \qquad \epsilon_2 = -\Delta_2/\Delta_0^4, \tag{50}$$

and substitution for E from (49) in (47) gives

$$\tau = (\overline{\mu}_0 + \mu_1\kappa^2)\kappa, \qquad \overline{\mu}_0 = \mu_0 + \mu_2 D^2/\Delta_0^2. \tag{51}$$

This can be inverted to give

$$\kappa = (\nu_0 + \nu_1\tau^2)\tau, \tag{52}$$

with

$$\nu_0 = 1/\overline{\mu}_0 = \frac{1}{\mu_0}(1 - \mu_2 D^2/\mu_0\Delta_0^2), \qquad \nu_1 = -\frac{\mu_1}{\mu_0^4}(1 - 4\mu_2 D^2/\mu_0\Delta_0^2). \tag{53}$$

Substituting from equation (52) in equation (44) gives

$$\tau' = \sqrt{2T - h^2/\tau^2 - \rho\omega^2(\nu_0\tau^2 + \tfrac{1}{2}\nu_1\tau^4)}\,. \tag{54}$$

Writing

$$\begin{aligned} &2T\tau^2 - h^2 - \rho\omega^2(\nu_0\tau^4 + \tfrac{1}{2}\nu_1\tau^6) \\ &= \tfrac{1}{2}\rho\omega^2\nu_1(a - \tau^2)(\tau^2 - b)(\tau^2 - c), \end{aligned} \tag{55}$$

and assuming that the constants T and h are such that this cubic expression in τ^2 has three real zeros a, b, and c, with $a > b > 0 > c$, leads to a periodic solution (Ref. 4),

$$\tau^2 = a - (a-b)\text{sn}^2 m(z - z_0), \tag{56}$$

where sn denotes the Jacobian elliptic function with modulus k and

$$m^2 = \frac{1}{2}\rho\omega^2\nu_1(\text{a} - \text{c}), \qquad k^2 = \frac{a - b}{a - c} \tag{57}$$

Substituting from equation (56) in equation (45) gives the phase angle θ as an elliptic integral of the third kind (Gradshteyn and Ryzhik, Ref. 7)

$$\theta = \theta_0 + \frac{h}{ma}\,\Pi(\zeta, b/a - 1, \text{k}), \qquad \zeta = \sin^{-1}\text{sn}m(z - z_0). \tag{58}$$

The motion is then given by equations (30), (40), and (43) as

$$x = X - \frac{1}{\rho\omega^2}\{\tau'\cos(\omega t-\theta) + \frac{h}{\tau}\sin(\omega t-\theta)\}, \tag{59a}$$

$$y = Y - \frac{1}{\rho\omega^2}\{\tau'\sin(\omega t-\theta) + \frac{h}{\tau}\cos(\omega t-\theta)\}, \tag{59b}$$

$$z = Z. \tag{59c}$$

The electric displacement field has the form

$$d_x = \frac{\delta_0\nu_0}{\Delta_0} D\tau\cos(\omega t-\theta), \qquad d_y = \frac{\delta_0\nu_0}{\Delta_0}\tau\sin(\omega t-\theta), \quad d_z = D, \tag{60}$$

where δ_0 is the value of the function δ in (15e) at $\kappa = 0$ and $E = 0$.

References

1. CARROLL, M. M., *Finite Amplitude Standing Waves in Compressible Elastic Solids,* Journal of Elasticity, Vol. 8, pp. 323–328, 1978.
2. CARROLL, M. M., *Some Results on Finite Amplitude Elastic Waves,* Acta Mechanica, Vol. 3, pp. 167–181, 1967.
3. CARROLL, M. M., *Plane Elastic Standing Waves of Finite Amplitude,* Journal of Elasticity, Vol. 7, pp. 411–424, 1977.
4. CARROLL, M. M., *Reflection and Transmission of Circularly Polarized Elastic Waves of Finite Amplitude,* Journal of Applied Mechanics, Vol. 46, pp. 867–872, 1979.
5. SINGH, M., AND PIPKIN, A. C., *Controllable States of Elastic Dielectrics,* Archive for Rational Mechanics and Analysis, Vol. 21, pp. 169–210, 1966.
6. TOUPIN, R. A., *The Elastic Dielectric,* Journal of Rational Mechanics and Analysis, Vol. 5, pp. 849–915, 1956.
7. GRADSHTEYN, L. S., AND RYZHIK, I. W., *Table of Integrals, Series, and Products,* Academic Press, New York, New York, 1965.

5

Static Theory of Point Defects in Nematic Liquid Crystals

J. L. Ericksen

Abstract. In recent years, important advances have been made in developing the first mathematically rigorous, nonlinear theory of defects in materials. It deals with the static theory of point defects in nematic liquid crystals. This is a brief exposition of important results, along with some discussion of possible formations of droplet problems. Also included are very interesting and hitherto unpublished data from Pat Cladis's laboratory, on the motion of two point defects in a capillary, a fairly simple physical example of a kind of dynamical problem not yet treated by theorists.

Key Words. Point defects, theory of liquid crystals.

1. Introduction

For studies of point and line defects in materials, the small-molecule nematic liquid crystals, commonly used in display devices, are attractive from the viewpoint of both experiment and theory.

Experimentally, one can resolve individual defects and details of configurations near them, with the relatively simple optical (polarizing) microscopes. Often, such configurations are not static but, in many cases, they change very slowly with time, so it is rather easy to observe many details, before they change

---

J. L. Ericksen • Professor Emeritus, Department of Aerospace Engineering and Mechanics and School of Mathematics, University of Minnesota, Minneapolis, Minnesota 55455; *currently* Consultant, Florence, Oregon 97439.

Nonlinear Effects in Fluids and Solids, edited by M. M. Carroll and M. Hayes, Plenum Press, New York, 1996.

appreciably. From numerous observations, some very old, much has been learned about their structure and behavior.

From the viewpoint of theory, rather well-tested nonlinear equations are available, covering both static and dynamic phenomena, providing a basis for analyzing stationary or moving defects. Such theory is rather formidable, but, now and then, a worker has managed to do some relevant analysis. Recently, there has been a spurt of activity, stimulated by the 1984–85 activities of the Institute for Mathematics and its Applications (IMA). Among other things, some expert analysts were introduced to and became fascinated by some of the problems. Bringing to bear kinds of expertise which had been lacking, some have been making important contributions. Briefly, my purpose is to describe some of these, relating to the static theory of point defects.

The reader might find useful Ref. 1, the proceeding of an IMA workshop, designed to acquaint mathematicians with aspects of liquid crystal research. It contains expositions by experts on the basic equations and problems of current interest involving defects, among other things. To make the discussion reasonably self-contained, we do overlap this, a bit.

2. Basic Theory

We will be concerned with nematic liquid crystals at rest, these being optically uniaxial liquids. Commonly, the optic axis varies with position, being influenced by such things as wall effects and electromagnetic fields. The primary function of static theory is to describe and predict such orientation patterns. To describe this axis, we introduce a unit vector n, called the director, so one relevant equation is

$$\mathbf{n} \cdot \mathbf{n} = 1. \tag{1}$$

From its interpretation, $\mathbf{n}$ and $-\mathbf{n}$ are physically equivalent. One should bear this in mind but, for simplicity, I will gloss some subtleties associated with this.

Commonly, static problems are formulated in terms of minimizing energy. At least in part, this involves the bulk energy of the liquid crystal. For this, it is common to use the Oseen–Frank theory, which has done well in describing a great variety of phenomena. According to it, the stored energy per unit volume W is a function of the form

$$\begin{aligned} 2W = {} & K_1(\nabla \cdot \mathbf{n})^2 + K_2(\mathbf{n} \cdot \operatorname{curl} \mathbf{n})^2 \\ & + K_3\|\mathbf{n} \times \operatorname{curl} \mathbf{n}\|^2 + (K_2 + K_4)[\operatorname{tr}(\nabla\mathbf{n})^2 - (\nabla \cdot \mathbf{n})^2], \end{aligned} \tag{2}$$

where the K's are constants or, more realistically, functions of the temperature, considered as constant. This results from assuming that W is quadratic in $\nabla\mathbf{n}$, with coefficients depending on n, and insisting that it be invariant under rather obvious symmetry transformations. Also, the moduli are restricted by the condition that $W \geq 0$, reflecting the assumption that n = const is a configuration of minimum energy. Assuming that this holds in a strict sense, one can bound W as indicated by

$$L_1 \|\nabla n\|^2 \leq W \leq L_2 \|\nabla n\|^2, \tag{3}$$

with positive constants L_1 and L_2. Mathematically, it is of some import to have this, and it is agreeable with physical experience. Often, electromagnetic field energies need to be accounted for and, sometimes, interfacial energies are included but, for simplicity, we will ignore them for the present. Then, the governing equations are of the form (1) and

$$\frac{\partial W}{\partial n} - \nabla \cdot \frac{\partial W}{\partial \nabla n} = \lambda n, \tag{4}$$

with the scalar function λ a Lagrange multiplier, associated with the constraint (1).

For qualitative or semiquantitative analyses, workers sometimes use the special case

$$K_1 = K_2 = K_3 = \mathrm{K}, K_4 = 0 \Rightarrow 2W = \mathrm{K}\|\nabla n\|^2, \tag{5}$$

to simplify analyses. This special case is degenerate, this form of W being invariant under a much larger group than applies, for other values of the K's. Predictions obtained with the special case are, in some cases, roughly correct, but occasionally, quite unrepresentative.

For any choice of material constants,[1]

$$n = x/\|x\| \tag{6}$$

satisfies (3), providing a simple example of a solution singular at a point, a point defect, with topological degree 1. For the special case (5), this generates a collection of solutions of the form

$$n = Qx/\|x\|, \tag{7}$$

with Q any constant orthogonal transformation. For other values of the K's,

[1]Unit vector fields satisfying (4) for all choices of material constants have been characterized by Marris (Refs. 2, 3).

these are not solutions, for $Q \neq \pm 1$. Observed point defects locally resemble (7) with particular choices of Q. Observations known to me could be fit fairly well with $Q = 1$ or a 180° rotation, perhaps combined with $Q = -1$, for whatever this is worth. Singularities of the kind indicated are mild enough to permit energy integrals to converge, and there are other reasons to think that this is characteristic of point defects.

Similarly, there are simple universal solutions representing line defects, e.g.,

$$n = (x_1/r, x_2/r, 0), \qquad r = \sqrt{x_1^2 + x_2^2}. \tag{8}$$

Here, the singularity is strong enough to cause the energy integral to diverge, which is typical of line defects. It would take a long discussion to properly describe current thoughts about the various kinds of line defects, some of which are discussed in Ref. 1. Later, we will say a bit more, about their relation to point defects.

3. Dirichlet Problems

At least roughly, many of the problems considered in practice fit a familiar mathematical pattern. This is to find the minimizers of

$$E = \int_B W(n, \nabla n), \tag{9}$$

where B is a given domain, subject to boundary conditions of the Dirichlet type,

$$n = n_0, \text{ on } \partial B, \tag{10}$$

with n_0 a given unit vector field. Of course, (1) is to be satisfied, almost everywhere. In practice, B is often considered to be infinite, reduced to finite size by assumptions of one-dimensionality or periodicity, but we will ignore such complications.

One needs to appreciate that B and n_0 can be very smooth but, if the topological degree of n_0 is nonzero and B is a ball, topologically, there is no unit vector field matching the boundary conditions and continuous throughout B. So, defects can be forced to occur, for such reasons. Thus, function spaces considered need to contain quite rough vector fields, pretty much all for which (1), (9), and (10) make sense.

Without belaboring matters technical, there is, in this kind of setting, existence theory, assuring us that minimizers exist. There is also some regularity theory, indicating that, where minimizers are continuous, they are analytic. For the special cases (5), only isolated point singularities can occur in minimizers of (9). Theorems available for more general forms of W allow the set of singularities to be a bit more complicated, but it is not certain that this possibility is real. It is also known that minimizers can exhibit singularities when this is not necessary, topologically. For a ball, say, one can promote this by assigning n_0 so that it varies rapidly, on part of the boundary. One can also do so, by taking B to be of suitable shape, with boundary conditions simple enough to provide a design for a realistic experiment. Such results, along with pertinent references, are covered in the article by Hardt and Kinderlehrer, in Ref. 1. This provides a sound basis for a rigorous static theory of point defects occurring in the interior of a liquid crystal. As is also noted there, the term multiplied by $K_2 + K_4$ in (2) can be transformed to a surface integral which is determined by n_0. So, for these Dirichlet problems, one can ignore the term. Related to this is the fact that no way is known to determine K_4 by experiment.

4. Stable Point Defects

Associated with the minimization problem is an idea which is proving to be very useful, that of a stable point defect. Suppose we are given a unit vector field $n(x)$, with an isolated point singularity at $x = x_0$. It is unimportant whether it has other singularities. Consider the restriction of it to $B_R(x_0)$, the ball of radius R with center at x_0. Then, if R is small enough, no other singularities will be contained in the ball. Let n_0 be the vector field obtained, as values of $n(x)$ on $\partial B_R(x_0)$. Now consider the minimization problem for $B = B_R(x_0)$, and the indicated boundary values. Provided that, for sufficiently small R, the given field is a minimizer, we regard the defect as stable. We know that, for the ball, there is at least one minimizer, some way for an unstable defect to adjust, locally, to become stable. So, intuitively, defects are either stable or very unstable.

For a defect at x_0 to be stable, it is clearly necessary that (4) be satisfied near x_0. The universal solution (6) is then a simple candidate, but it is not so easy to determine whether it is stable, in this sense. F.-H. Lin (Ref. 4) found a way of proving that it is, provided

$$K_1 \leq \min(K_2, K_3). \tag{11}$$

Then, S.-Y. Lin (Ref. 5) proved that it is not when

$$K_1 > K_2 = K_3, \qquad \text{with } K_1/K_2 \text{ sufficiently large.} \tag{12}$$

Independently, Hélein (Ref. 6) obtained a better result, showing that it is not, if

$$8(K_2 - K_1) + K_3 < 0. \tag{13}$$

When it is not, there is a minimizer for the ball, matching the radial boundary conditions. Clearly, any rotation R about an axis through the origin takes the ball to itself, leaving the boundary values radial, if one interprets this as indicated by

$$n(x) \rightarrow R^{\mathrm{T}} n(Rx),$$

which maps minimizers to minimizers. It is not hard to show that, if $n(x)$ is mapped to itself by all such R, it must be of the form (6). So, the minimizers not of this form are far from unique. S.-Y. Lin (Ref. 5) did supercomputer calculations, to determine minimizers, with $K_1/K_2 = 2$ and 3, $K_2 = K_3$ replacing the ball by a less symmetric cube. This experiment indicated that (6) is unstable in both cases, as is implied by (13). The minimizers found are topologically similar, with the straight vector lines replaced by curves. One point to be noted is that the capability of doing reliable computations of this kind now exists. Hélein (Ref. 6) presents a simple, explicit minimizer for the degenerate case $K_2 = K_3 = 0$.

At least for the special form of W given by (5), it is known that there are solutions of (4), singular at a point, with any integer as their topological degree. Rather simple examples can be obtained by a construction involving stereographic projection and rational functions. Assuming that a similar variety of solutions exists for more realistic forms of W, there is need to understand why it is that those observed have only ± 1 as their degrees. Indeed, this supplied motivation for developing the notion of a stable point defect. An important step was made by Hardt, Kinderlehrer, and Lin (Ref. 7), who showed that the degree of a stable defect is bounded, the bound depending at most on the values of the material constants. A numerical experiment was performed and reported in the article by Cohen *et al.* (in Ref. 1). Using (5), they took a known solution for a point defect with degree 2, using it to determine boundary values on a cube, and calculated a minimizer. They found that the minimizer involved two point defects of degree 1, the original defect being unstable. This information was passed on to Brezis, Coron, and Lieb, then working together on related problems in France. For the special form (5), they (Ref. 8) managed to prove that the degree of a stable defect can only be ± 1. The situation for the general forms of W is still not resolved, but impressive progress has been made in setting this issue.

Some other results obtained are somewhat negative, but I find them quite interesting. First, I was familiar with an interesting remark by Brinkman and Cladis (Ref. 9), made in a review article which contains some nice pictures of defects in liquid crystals. In their words,

> The total energy stored in the elastic field around a point defect grows linearly with the radius of the volume enclosed. Consequently, two defects have an interaction energy which grows linearly with separation. In this respect, they act like quarks interacting through a gluon field.

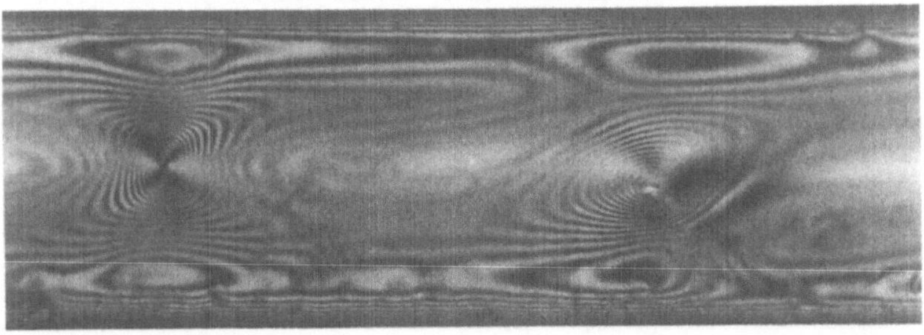

Fig. 1. Photograph taken at a particular time, employing polarized light, indicating the optical variations near the two points.

During the 1984–85 IMA year, Haim Brezis visited, and David Kinderlehrer and I discussed with him prospects for clearly defining and solving a problem of this kind. What emerged from the discussion was the idea of considering vector fields defined throughout space, smooth everywhere except at two points P_1 and P_2, where they have point defects with degrees $+1$ and -1, respectively. Then, try to minimize the energy, in this function space. Brezis, Coron, and Lieb (Ref. 8), using (5), did manage to analyze this, in fact a more general problem. First, they managed to calculate the infimum of the energy, determining that it is proportional to the distance between P_1 and P_2, in accord with the quotation. However, they also showed that the infimum is not attained! It then seems rather clear that problems of this kind are inherently dynamical. Still more illuminating is a result proved by Hardt, Kinderlehrer, and Lin (Ref. 10). Briefly, it shows that one cannot have a sequence of static solutions involving stable point defects which approach each other and, in the limit, cancel each other. Essentially, this means that the phenomena of creation or annihilation of point defects, which are observed, are outside the range of static theory.

The latter conclusion is, I think, quite compatible with observations. Point defects are likely to be generated in the process of filling a capillary, for example. Pat Cladis has kindly supplied me with the pictures reproduced below, indicating what happens to a pair, with degrees $+1$ and -1, in a capillary, treated to make n become normal, at the wall. Figure 1 is a photograph taken at a particular time, employing polarized light, indicating the optical variations near the two points. Figure 2 is a sketch of the corresponding (axisymmetric) orientation pattern. Figure 3 is a record of data[2] showing how the distance between the points

[2]Pat Cladis informs me that these data were taken by a summer employee, Mayola Walters, who worked with her at Bell Laboratories.

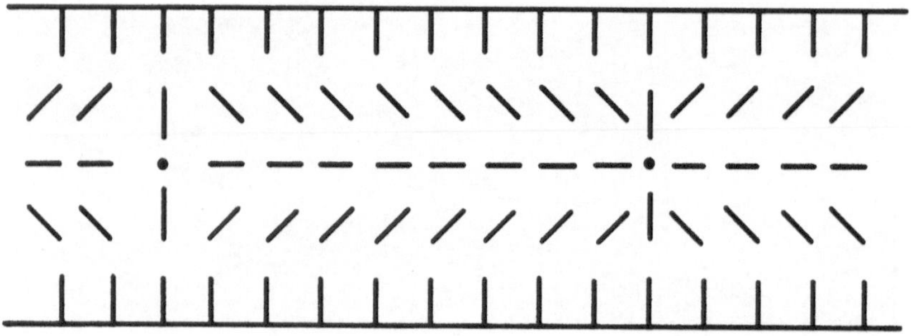

Fig. 2. Sketch of the corresponding (axisymmetric) orientation pattern.

changes with time. Starting a bit less than 4 diameters apart, the two start moving toward each other, at a rate of about a diameter per thousand minutes, slow enough to make it somewhat reasonable to consider using static theory to analyze the behavior at early times. However, when the distance between them becomes a bit less than a diameter, the relative velocity starts to increase dramatically, until they come together and annihilate each other. Other evidence of this kind supports the view that dynamical theory must be used for analyzing creation and annihilation. Naive scaling analyses based on dynamical theory suggest that if t, the time to annihilation, is very small, the distance d between the

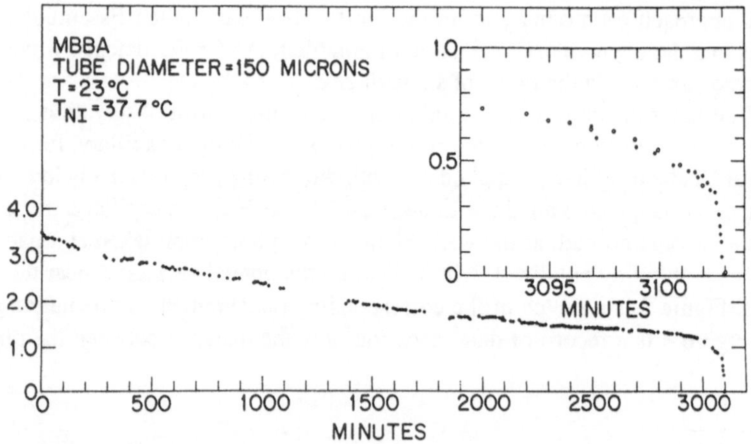

Fig. 3. Record of data showing how the distance between the points changes with time.

two should be related to this by

$$d \cong \text{const}\,\sqrt{t}\,. \tag{14}$$

Better data would be needed to check this, but it does not look so unreasonable, from the data given.

With the normal orientation at the capillary wall, one might think that, in equilibrium, one would wind up with a configuration like that given by (8), involving a line defect. Instead, there is an "escape into the third dimension," as Brinkman and Cladis (Ref. 9) put it. That is, the director swings out of the plane, to become tangent to the center line, and eliminate the line defect. Singularities do occur, in the form of the two point defects, but only as a transient effect. The director does vary quite rapidly near the center line, being close to the planar form a short distance away. In older times, workers mistook such configurations for some involving true line defects. As was mentioned earlier, true line defects also occur and are of interest, so there is need to improve theory for these.

What we have been discussing are point defects occurring in the interior of liquid crystals. Rather often, they are observed on the boundary, where the liquid crystal contacts another material. One could try to analyze these by considering the boundary values n_0 in (10) to admit singularities. However, rather different formulations are more reasonable, physically. In some cases, observations suggest that n makes some definite angle with the normal to the interface, but is free to take any direction on the cone of directions defined by this condition. Usually, but not always, the angle is 90°. It can be 0°, in which case (10) applies. Such boundary conditions are likely to occur where a liquid crystal contacts untreated glass or another fluid. Saupe (Ref. 11) discusses likely configurations, for cases where the angle is 90°. One can apply various surface treatments to glass to induce boundary conditions better described by (10), and this is exploited in making devices and in designing experiments. I am told that, in unpublished work, existence and regularity theory have been extended to cover these conical boundary conditions, but I have not seen the details. Except in the degenerate case where the angle is 0°, the projection of n on the plane tangent to the interface has a fixed nonzero length. Topologically, it is impossible for such a vector field to be continuous everywhere on the boundary of a region topologically equivalent to a ball. Various observations of droplets do indicate the surface defects suggested by such reasoning. Volovik and Lavrentovich (Ref. 12) report observations of an interesting case. It involves a liquid crystal droplet suspended in another liquid. The angle is sometimes 0°, sometimes 90°, depending on temperature. It is clear from their discussion that they believe that such orientation boundary conditions apply quite well to their observations, even though the director field varies considerably. I know of no reason to doubt that this is true, for this and other droplet problems. Their droplets are nearly spherical. Other droplets observed are more clearly tactoids, shaped like footballs or cigars, with point defects likely to be present at the ends. If we accept the hint, we might formulate a droplet problem as follows. Consider a fixed mass of liquid crystal. Or, what is the same for an incompressible fliud, fix its volume V. It will occupy some (unknown) region B, subject to the constraint

$$\text{volume of } B = V. \tag{15}$$

If ν is the unit normal to ∂B, we are to satisfy a boundary condition of the conical form,

$$(n \cdot \nu)^2 = \text{a}, \qquad \text{almost everywhere on } \partial B, \tag{16}$$

with a some given constant. Then, subject to these constraints and (1), we should seek minimizers of the total energy

$$E = \int_B W + E_s, \tag{17}$$

where E_s represents surface energy. For this, we could adopt the conventional estimate

$$E_s = \text{const} \times (\text{area of } \partial B). \tag{18}$$

As far as I know, no one has considered problems of this kind.

In a study now rather old, Chandrasekhar (Ref. 13) considered a rather different formulation, now being reconsidered by some analysts. Again, (15) and (17) apply, but not (16) or (18). Replace (18) by

$$E_s = \int_{\partial B} W_s, \tag{19}$$

with W_s some function of $(n \cdot \nu)^2$, and again seek the energy minimizers. When liquid crystal workers consider such surface energies, they usually assume the form

$$W_s = b + c(n \cdot \nu)^2, \qquad b > 0, \qquad b + c > 0, \tag{20}$$

with b and c constants. This can have minima only for $n \cdot \nu = 0$ or ± 1, which covers more commonly observed cases, but not all of the observations. Even with this simple choice, these problems are formidable. To make some headway,

Chandrasekhar borrowed ideas from the theory of crystal growth. Briefly, put n = const, which certainly minimized the bulk energy. Then pick B to minimize the surface energy. This can be done rather easily, using the Wulff construction, for any given function W_s. Among other things, one can get tactoids, at least qualitatively like the shapes observed. Some but not all formal reasoning[3] suggests that such estimates might be good, asymptotically, as $V \rightarrow 0$. On the other hand, the description is clearly not good for the droplets of finite size observed by Volovik and Lavrentovich (Ref. 12), for example. In this is a hint that sound asymptotic analysis might produce some surprises. The previous formulation could be viewed as a very different kind of hopeful approximation. In (16), pick a so as to minimize W_s. With $(n \cdot v)^2$ so fixed on ∂B, W_s will be constant, in accord with (18), whether it is given by (20) or some more general function. In a rough way, one could understand how this might tend to give good approximations, if Ws has a very sharp minimum, increasing rapidly as the argument changes from the minimizing value. If there is some clear experimental evidence concerning this function, I do not know of it. Since (20) is a rather popular guess, it would be interesting to know what it predicts, for various values of the adjustable constants, and how well this agrees with observations. What is predicted, using (20) and the simplifying assumption n = const, is discussed in some detail by Virga (Ref. 14).

In a rather sketchy way, this describes some of the problems relating to point defects, and recent progress, improving related theory. There is a large literature on defects in liquid crystals, not easily summarized. Some of this is covered rather nicely, by Kléman (Ref. 15).

Acknowledgments

I wish to thank Pat Cladis for permission to use her illustrations and David Kinderlehrer for helpful discussions and criticisms.

References

1. ERICKSEN, J. L., and KINDERLEHRER, D., *Theory and Applications of Liquid Crystals*, IMA Volumes in Mathematics and Its Applications, Springer-Verlag, Berlin, Germany, Vol. 5, 1987.

[3]It is a common prejudice that volume energies will become small with respect to surface energies as the volume shrinks to zero. Assume that point defects occur, and you are likely to come to the opposite conclusion.

2. MARRIS, A. W., *Universal Solutions in the Hydrostatics of Nematic Liquid Crystals*, Archive for Rational Mechanics and Analysis, Vol. 67, pp. 251–303, 1978.
3. MARRIS, A. W., *Addition to Universal Solutions in the Hydrostatics of Nematic Liquid Crystals*, Archive for Rational Mechanics and Analysis, Vol. 69, pp. 323–380, 1979.
4. LIN, F.-H., *A Remark on the Map* $x/|x|$, Comptes Rendus de l'Académie des Sciences, Paris, France, Vol. 305, pp. 529–531, 1987.
5. LIN, S.-Y., *Numerical Analysis for Liquid Crystal Problems*, PhD Thesis, University of Minnesota, 1987.
6. HÉLEIN, F., *Minima de l'Énergie Libre Fonctionelle des Cristaux Liquides*, Comptes Rendus de l'Académie des Sciences, Paris, France, Vol. 305, pp. 565–568, 1987.
7. HARDT, R., KINDERLEHRER, D., and LIN, F.-H., *Existence and Partial Regularity of Static Liquid Crystal Configurations*, Communications in Mathematical Physics, Vol. 105, pp. 547–550, 1986.
8. BREZIS, H., CORON, J. M., and LIEB, E., *Estimations d'Énergie pour des Applications de* R^3 *a Valeurs dans* S^2, Comptes Rendus de l'Académie des Sciences, Paris, France, Vol. 303, pp. 207–210, 1986.
9. BRINKMAN, W. F., and CLADIS, P. E., *Defects in Liquid Crystals*, Physics Today, Vol. 35, pp. 48–54, 1982.
10. HARDT, R., KINDERLEHRER, D., and LIN, F.-H., *A Remark About the Stability of Smooth Equilibrium Configurations of Static Liquid Crystals*, Molecular Crystals and Liquid Crystals, Vol. 139, pp. 189–194, 1986.
11. SAUPE, A., *Disclinations and Properties of the Director Field in Nematic and Cholesteric Liquid Crystals*, Molecular Crystals and Liquid Crystals, Vol. 21, pp. 211–238, 1973.
12. VOLOVIK, G. E., and LAVRENTOVICH, O. D., *Topological Dynamics of Defects: Boojums in Nematic Drops*, Journal of Experimental and Theoretical Physcis, Vol. 85, pp. 1159–1166, 1983.
13. CHANDRASEKHAR, S., *Surface Tension of Liquid Crystals*, Liquid Crystals, Edited by G. H. Brown, G. J. Dienes, and M. M. Labes, Gordon and Breach, New York, New York, pp. 531–540, 1966.
14. VIRGA, E. G., *Forme di Equilibrio di Piccole Gocce di Cristallo Liquido*, Istituto di Analisi Numerica, Pavia, Italy, Vol. 562, pp. 1–8, 1987.
15. KLÉMAN, M., *Points, Lines and Walls*, John Wiley and Sons, New York, New York, 1983.

6

Function Spaces and Fading Memory

G. FICHERA

Abstract. Functional analysis is a powerful tool which has been extensively used in problems arising in continuum mechanics. However, the use of functional analytic methods requires that an *a priori* choice be made of the function space in which to operate. Many basic concepts of the theories under consideration are strictly connected to the choice of space.

It is first shown, by two examples, how fundamental concepts of analysis have completely different contents when the underlying function space is changed. Then it is shown how the concept of fading memory for elastic materials depends heavily on the choice of a suitable function space so that this concept becomes completely vacuous because of the impossibility of choosing, on the basis of purely physical consideration, an appropriate function space.

Key Words. Functional analysis, function space, fading memory, complex Banach space, Cauchy problem, Laplace equation, topology, elastic solid, influence function, stress relaxation tensor, fading memory principle.

1. Introduction

In the last 40 years, functional analysis has been extensively used in problems arising from continuum mechanics. There is no doubt that powerful methods of the former have led to the solution of several difficult problems of the latter. However, there is one point which deserves to be considered carefully.

The use of functional analytic methods requires an a *priori* choice of the

G. Fichera • Professor, Department of Mathematics, University of Rome-1, 00199 Rome, Italy.

Nonlinear Effects in Fluids and Solids, edited by M. M. Carroll and M. Hayes, Plenum Press, New York, 1996.

function space in which to operate; on the other hand, many basic concepts of the theories under consideration are strictly connected to the choice of this space. Even in the domain of pure mathematics, some fundamental definitions have quite a different content when passing from one function space to another. Of course, the same is true in mathematical physics. However, there is a basic difference: while in pure mathematics the researcher has only to acknowledge this change of meaning of the relevant concepts when changing the function spaces, in theories connected with the physical world the choice should be unambiguous and the function space to be chosen should be the outcome of precise physical experiments. But is that always possible?

In this paper, we first show by two examples how fundamental concepts of analysis have completely different contents when the underlying function space is changed. Afterward, we shall show how the concept of fading memory for elastic materials depends heavily on the choice of a suitable function space, so that this concept, although very successful in the recent literature, becomes completely vacuous because of the impossibility of choosing, on the basis of purely physical considerations, an appropriate function space.

2. Stability of the Solution of a Volterra Integral Equation

Let S be a complex Banach space, and let $\mathcal{T}(S)$ be the space of bounded linear mappings K of S into itself. The space $\mathcal{T}(S)$ of these mappings is a Banach space endowed with the norm

$$||| K ||| = \sup_{\| u \| = 1} \| Ku \|.$$

Let us consider the interval $I \equiv [t_0, +\infty)$ of the real axis, and let $C^0(I, S)$ be the space of functions which are valued in S and continuous, in the strong topology of S, over I.

Denote by $K(t, \tau)$ a mapping of $\mathcal{T}(S)$ defined and continuous for every $(t, \tau) \in I \times I$. If $\varphi(t) \in C^0(I, S)$, the Volterra integral equation

$$u(t) = \varphi(t) + \int_{t_0}^{t} K(t,\tau)\, u(\tau)\, d\tau \tag{1}$$

has one and only one solution. The solution $u(t)$ of (1) is said to be asymptotically stable (for $t \rightarrow +\infty$) whenever

$$\lim_{t \rightarrow +\infty} \| u(t) \| = 0. \tag{2}$$

As an example, let us consider the Volterra integral equation

$$U(x, t) = e^{-x^2} + \int_{-1}^{t} \left(\frac{2}{\tau} - 6\tau^5 x^2\right) U(x, \tau)\, d\tau. \tag{3}$$

We assume as space S the space of the L^1 functions $v(x)$ in $(-1, 1)$ with the norm

$$\| v \| = \int_{-1}^{1} |v(x)|\, dx.$$

Consider $U(x, t)$ as a function $u(t)$ defined for $t \in [1, +\infty)$ and valued in S. Set

$$Ku = \left(\frac{2}{\tau} - 6\tau^5 x^2\right) U(x, \tau), \qquad \varphi = e^{-x^2}.$$

Equation (1) has the following solution:

$$u(t) \equiv U(x, t) = t^2\, e^{-t^6 x^2}. \tag{4}$$

Hence,

$$\| u(t) \| = t^2 \int_{-1}^{1} e^{-t^6 x^2}\, \mathrm{dx} = \frac{1}{t} \int_{-t^3}^{t^3} e^{-\xi^2}\, d\xi \leq \frac{1}{t} \int_{-\infty}^{+\infty} e^{-\xi^2}\, d\xi = \frac{\sqrt{\pi}}{t}.$$

This proves that condition (2) is satisfied and $u(t)$ is asymptotically stable.

Let us now assume like S the space of continuous functions $v(\mathrm{x})$ in $[-1, 1]$ with the norm

$$\| v \| = \max_{[-1, 1]} |v(x)|.$$

From (4), we deduce that

$$\| u(t) \| = \max_{-1 \leq x \leq 1} [t^2 e^{-t^6 x^2}] = t^2,$$

$$\lim_{t \to +\infty} \| u(t) \| = +\infty.$$

In conclusion, assuming as space S the space $L^1(-1, 1)$, it is true that the solution of (3) is asymptotically stable. If S is the space $C^0[-1, 1]$, this is false.

3. Cauchy Problem for the Laplace Equation

According to a classical definition given by Hadamard a problem for a partial differential equation is well set if: (i) the solution exists, (ii) it is unique, (iii) it depends continuously on the data.

Consider the Cauchy problem for harmonic functions,

$$\frac{\partial^2 u}{\partial x^2} + \frac{\partial^2 u}{\partial y^2} = 0, \tag{5}$$

$$u(x, 0) = \varphi_0(x), \qquad u_y(x, 0) = \varphi_1(x), \qquad 0 \leq x \leq \pi. \tag{6}$$

Suppose that $\varphi_0(x)$ and $\varphi_1(x)$ are analytic functions of x in the closed interval $[0, \pi]$. This means that, for every $x_0 \in [0, \pi]$, $\varphi_k(x)$, $k = 1, 2$, can be developed in a power series in $x - x_0$ converging in a neighborhood of x_0. Then, for the classical Cauchy–Kowalewska theorem, it is possible to find $\sigma > 0$ such that, for $0 \leq x \leq \pi$, $-\sigma \leq y \leq \sigma$, a unique solution of (5) exists satisfying the conditions (6).

Following Hadamard (Ref. 1, pp. 33–34), assume that

$$\varphi_0^{(n)}(x) = 0, \qquad \varphi_1^{(n)}(x) = e^{-\sqrt{n}} \sin(nx), \tag{7}$$

where n is a positive integer.

We have

$$u^{(n)} = \frac{1}{n} e^{-\sqrt{n}} \sin(nx) \sinh(ny).$$

Let us introduce in the space S of the data $\Phi \equiv (\varphi_0, \varphi_1)$ the following norm:

$$\| \Phi \| = \sum_{h=0}^{m} \left(\max_{[0, \pi]} \left| \frac{d^h \varphi_0}{dx^h} \right| + \max_{[0, \pi]} \left| \frac{d^h \varphi_1}{dx^h} \right| \right), \tag{8}$$

where m is an arbitrary nonnegative integer.

With the choice (7), we have

$$\| \Phi^{(n)} \| = e^{-\sqrt{n}} \sum_{h=0}^{m} n^h.$$

Hence,

$$\lim_{n \to \infty} \| \Phi^{(n)} \| = 0.$$

Introduce in the space U of the solutions u the norm

$$\| u \| = \max_{[0, \pi] \times [-\sigma, \sigma]} |u(x, y)|. \tag{9}$$

It is easily seen that, no matter how small σ has been chosen, one has

$$\lim_{n \to \infty} \| u^{(n)} \| = +\infty.$$

This proves that condition (iii) for the problem (5), (6) to be well set is not satisfied. One can read in Hörmander's outstanding book (see Ref. 2, p. 115): "It was emphasized by Hadamard that the Cauchy–Kowalewska theorem is of limited value because the solution needs not to depend continuously on φ_0 and φ_1, even if these functions are given the C^∞ topology and a very weak topology is used for u."

In our opinion, this criticism of the Cauchy–Kowalewska theorem seems to be unfair, since by assuming a more suitable norm in the space of the data one gets the continuous dependence of the solution of the problem (5), (6) on the data.

In fact, since the theorem of Cauchy–Kowalewska is valid if φ_0 and φ_1 are analytic functions of x in $[0, \pi]$, by introducing the complex variable $z = x + iy$ one can consider φ_0 and φ_1 as holomorphic functions of z in a rectangle R_σ: $-\sigma \le x \le \pi + \sigma$, $-\sigma \le y \le \sigma$, $\sigma > 0$ of the z-plane. Let us assume that the space S of the data is the space of pairs $\Phi = (\varphi_0, \varphi_1)$ of functions which are holomorphic in $R_\sigma - \partial R_\sigma$ and continuous in R_σ.

If u is a harmonic function in $R_\sigma - \partial R_\sigma$ and continuous in R_σ, we denote by v the harmonic conjugate of u such that $v(0, 0) = 0$. Let $f(z) = u(x, y) + iv(x, y)$. If a solution of (5), (6) exists, then for $0 \le x \le \pi$,

$$f'(x + i0) = \varphi_0'(x + i0) - i\,\varphi_1(x + i0).$$

Hence, for $z \in R_\sigma$,

$$f(z) = \varphi_0'(z) - i\,\varphi_1(z),$$

which implies that

$$f(z) = \varphi_0(z) - i\int_0^z \varphi_1(\zeta)\, d\zeta. \tag{10}$$

The function $u = R\,f(z)$ is the solution of (5), (6).

If we introduce in S the norm

$$\|\Phi\| = \max_{R_\sigma} |\varphi_0(z)| + (\pi = \sigma) \max_{R_\sigma} |\varphi_1(z)|, \tag{11}$$

and assumes for u the norm (9), we have from (10)

$$\|u\| \leq \|\Phi\|,$$

which proves the continuous dependence of the solution u on the data. Hence, the Cauchy problem (5), (6) which is not well set in the topologies defined by (8), (9), becomes well set if we introduce the topologies (9), (11).

After the example studied in Section 2, this is another example proving how the choice of a function space and the relevant topology can influence a basic concept of mathematical analysis.

4. Concept of Fading Memory for an Elastic Solid

Let us consider the constitutive equations of a linearly elastic solid Ω with memory. Let $x \equiv (x_1, x_2, x_3)$ be a point of $\mathbb{R}^3$ and assume for simplicity that the body is homogeneous and that the translation invariance axiom holds. We have

$$\sigma_{ih}(x, t) = G_{ihjk}(0)\, \epsilon_{jk}(x, t) + \int_{-\infty}^{t} \dot{G}_{ihjk}(t - \tau)\epsilon_{jk}(x, \tau)d\tau. \tag{12}$$

As usual, σ_{ih} are the stress components and ϵ_{jk} the strain components in the rectangular Cartesian frame of $\mathbb{R}^3$. It is assumed that

$$G_{ihjk}(s) = G_{jkih}(s) = G_{hijk}(s), \qquad s \geq 0,$$

and moreover

$$G_{ihjk}(0)\, \epsilon_{ih}\, \epsilon_{jk} > 0.$$

Coleman and Noll in an important paper remark (see Ref. 3, p. 243) that "the memory of a simple material fades in time." With this statement, they mean that the elastic body has the tendency to forget the history of its far past, i.e., that strains suffered by the body in the past become less and less influential in (12) when t increases. This idea already appears in the work of Volterra, although expressed by different analytic means (see Ref. 4, p. 102).

To express mathematically their concept, Coleman and Noll introduce an influence function $h(s)$ with the following properties:

$$\text{(i)} \quad h(s) > 0, \qquad \text{for all } s \geq 0;$$

$$\text{(ii)} \quad \lim_{s \to +\infty} s^p h(s) = 0, \qquad \text{for some } p > 0.$$

Moreover, they assume that the stress relaxation tensor $G_{ihjk}(s)$ satisfies the condition

$$\text{(iii)} \quad \int_0^{+\infty} |\dot{G}_{ihjk}(s)|^2 [h(s)]^{-2}\, ds < +\infty.$$

It is evident how $h(s)$ influences the memory of the body: the larger p, the faster the fading of the memory.

It is obvious that this concept of an influence function is a purely mathematical one. In fact, the wise principle, according to which in any problem of mathematical physics the constants and the functions which are given as data must be provided or at least confirmed by precise physical experiments, is violated: no physical experiment carried out in a necessarily finite interval of time will be able to determine the asymptotic behavior of an influence function $h(s)$ when $s \to \infty$. As a consequence, one is forced to test the admissibility of an influence function $h(s)$ by other means.

Let us denote by $u(x, t)$ the displacement vector and let us restrict ourselves to consider only the boundary condition

$$u(x, t) = 0, \qquad x \in \partial\Omega, \qquad -\infty < t \leq T. \tag{13}$$

For the quasistatic problem, we have the following integro-differential system:

$$G_{ihjk}(0)\, u_{j/hk}(x, t) + \lambda \int_{-\infty}^{t} \dot{G}_{ihjk}(t - \tau) u_{j/hk}(x, \tau) d\tau = f_i(x, t), \tag{14}$$

where $f_i(x, t)$ are determined by the components of the body forces acting on the solid. The system (14) must be solved assuming $\lambda = 1$ and considering the boundary condition (13).

A necessary condition for $h(s)$ to be admissible is that, for any stress relaxation tensor $G_{ihjk}(t)$ satisfying (iii), the problem (13), (14) should have one and only one solution.

Simple examples showing that in general this is not the case were provided in Ref. 5.

Several authors have proposed additional conditions to be satisfied by the stress relaxation tensor in order to avoid the counterexamples given in Ref. 5. However, even with these further restrictions proposed for $G_{ihjk}(t)$, examples were found proving that the existence and/or uniqueness of the solution of (13), (14), with $\lambda = 1$, fail to hold (see Refs. 6, 7).

Assuming that Ω is bounded and has a smooth boundary, let us consider the BVP of classical elasticity

$$G_{ihjk}(0)\, w_{j/hk} = v_i, \qquad \text{in } \Omega, \tag{15}$$

$$w_{i|\partial\Omega} = 0. \tag{16}$$

Suppose that $v(x) = (v_1(x), v_2(x), v_3(x))$ is a vector-valued function with components in the space $L^2(\Omega)$. Then, the problem (15), (16) has only one solution belonging to the Sobolev space $H^2(\Omega)$ (see Ref. 8, pp. 370 and 381). Let us denote by $w = \Gamma v$ this solution, where Γ is the Green operator of the problem (15), (16).

Problem (13), (14), when $u(x, t)$ is understood as a vector-valued function of t with components in $H^2(\Omega)$ and λ is a complex parameter, is perfectly equivalent to the functional equation

$$v + \lambda\mathcal{T}v = f, \tag{17}$$

where the linear operator $\mathcal{T}$ has the following components:

$$\mathcal{T}_i v = \int_{-\infty}^{t} \dot{G}_{ihjk}(t - \tau)[\Gamma_j v]_{/hk}\, d\tau.$$

Of course, $v(x, t)$ must be sought in a function space of vector-valued functions of t with components in $L^2(\Omega)$ such that the operator $\mathcal{T}$ is defined in this space and maps the space into itself.

Suppose that we have introduced a norm in the space S of a 3-vector-valued function $v(t)$, $-\infty \leq t \leq T$, with components in $L^2(\Omega)$ and, through functional completion, consider S to be a Banach space.

It seems reasonable to define a fading memory principle (FMP) through an admissibility condition for the stress relaxation tensors $G_{ihjk}(t)$ somewhat more sophisticated than condition (iii).

In Ref. 6 the following axiomatic FMP is proposed: the stress relaxation tensor $G_{ihjk}(t)$ is admissible with respect to FMP if:

(a) the operator $\mathcal{T}$ is defined for every $v \in S$, maps S into S, and is bounded;
(b) if $G_{ihjk}(t)$ is admissible, existence and uniqueness hold for (17) in the space S;
(c) if $G_{ihjk}(t)$ is admissible, $\lambda\, G_{ihjk}(t)$ is admissible for every λ with $|\lambda| \leq 1$.

While the meaning of conditions (a) and (b) is obvious, condition (c) means that, if an FMP is satisfied for $G_{ihjk}(t)$, then this FMP must be satisfied for the not stronger memory defined by $\lambda\, G_{ihjk}(t)$

Denote by $\| v \|$ the norm of a vector v of S and set

$$||| \mathcal{T} ||| = \sup_{\| v \| = 1} \| \mathcal{T} v \|, \qquad \rho_{\mathcal{T}} = \lim_{k \to \infty} ||| \mathcal{T}^k |||^{1/k}.$$

The existence of the last limit is well known: $\rho_{\mathcal{T}}$ is the spectral radius of the operator $\mathcal{T}$.

In Ref. 6, the following theorem is proved.

Theorem 1. The stress relaxation tensor $G_{ihjk}(t)$ is admissible with respect to FMP if and only if

$$\rho_{\mathcal{T}} < 1. \tag{18}$$

It is evident that condition (18) is more sophisticated than condition (iii) of Coleman and Noll. The influence function $h(s)$ can be used to define a norm in S.

Let $h(s)$ be a continuous influence function with $p > \frac{1}{2}$. One can define the following norm in S:

$$\| v \|^2 = \sup_{-\infty < t \leq T} \int_{-\infty}^{t} [h(t \quad \tau)]^2 d\tau \int_{\Omega} |v(x, \tau)|^2 dx. \tag{19}$$

Since

$$\sum_{j,h,k}^{1,3} \int_\Omega |(\Gamma_j v)_{/hk}|^2 dx \leq c \int_\Omega |v|^2 dx,$$

with c only depending on Ω and on G_{ihjk} (0) (see Ref. 8, p. 370), one has

$$\int_\Omega |\mathcal{T} v|^2 dx \leq c \left(\int_0^{+\infty} |\dot{G}(s)|^2 [h(s)]^{-2} ds \right) \| v \|^2,$$

where

$$|\dot{G}(s)|^2 = \sum_{ihjk}^{1,3} |\dot{G}_{ihjk}(s)|^2.$$

Hence,

$$\| \mathcal{T} v \|^2 = \sup_{-\infty < t \leq T} \int_{-\infty}^{t} [h(t - \tau)]^2 d\tau \int_\Omega |\mathcal{T} v|^2 dx$$

$$\leq c \left(\int_0^{+\infty} [h(s)]^2 ds \int_0^{+\infty} |\dot{G}(s)|^2 [h(s)]^{-2} ds \right) \| v \|^2.$$

Of course, infinitely many choices different from (19) are possible for a norm in S which is compatible with the given influence function $h(s)$.

One sees that the condition (iii) permits only to satisfy in infinitely many ways hypothesis (a) of our axiomatic FMP, while (b) needs not be satisfied.

Unfortunately, the axiomatic FMP which is proposed in Ref. 6 depends strongly on the topology of the space S. In fact, in Ref. 6, it is shown that $\rho_{\mathcal{T}}$ can have a completely different value if one changes the topology of S. Hence, even if one supplements the condition (iii) of Coleman and Noll with more stringent conditions like (b) and (c) of our axiomatic FMP, the FMP depends heavily on the topology adopted in S with the following implication: a statement concerning fading memory which is true for a certain topology in S can become false when changing the topology.

On the other hand, no physical experiments carried out in a necessarily finite interval of time can suggest the topology to be adopted in S.

References

1. HADAMARD, J., *Lectures on Cauchy's Problem in Linear Partial Differential Equations*, Yale University Press, New Haven, Connecticut, 1923; Reprinted by Dover Publications, New York, New York, 1952.
2. HÖRMANDER, L., *Linear Partial Differential Equations*, 2nd Revised Edition, Die Grundleheren der Mathematischen Wissenschaften in Einzerdarstellungen, Springer, Berlin, Germany, Vol. 116, 1964.
3. COLEMAN, B., AND NOLL, W., *Foundations of Linear Viscoelasticity*, Reviews of Modern Physics, Vol. 33, pp. 239–249, 1961.
4. VOLTERRA, V., *Leçons sur les Fonctions de Lignes*, Gauthier–Villars, Paris, France, 1943.
5. FICHERA, G., *Avere una Memoria Tenace Crea Gravi Problemi*, Archive for Rational Mechanics and Analysis, Vol. 70, pp. 101–112, 1979.
6. FICHERA, G., *Sul Principio della Memoria Evanescente*, Rendiconti del Seminario Matematico di Padova, Vol. 68, pp. 245–259, 1982.
7. FICHERA, G., *On Linear Viscoelasticity*, Mechanics Research Communications, Vol. 12, pp. 241–242, 1985.
8. FICHERA, G., *Existence Theorems in Elasticity*, Handbuch der Physik, Springer, Berlin, Germany, Vol. VI a/2, pp. 347–389, 1972.

7

On Thermomechanical Formulation of Theories of Continuous Media

A. E. GREEN AND P. M. NAGHDI†

Abstract. The first part of this paper is concerned with some basic developments within the scope of the classical (three-dimensional) thermomechanics. Specifically, setting aside the invariance conditions under superposed rigid body motions but starting with an energy balance which explicitly accounts for the rate of change of thermal energy (or heat) in any part of the body (in addition to the usual heat flux and heat supply), both the basic balance equations of the purely mechanical theory and entropy balance are derived. In utilizing these derived results, the second part of the paper deals with developments of certain features of thermomechanical theories of shell-like and rodlike bodies in which the use of entropy balance plays a significant role when temperature variations are accounted for also along the shell thickness and in the cross section of the rod.

Key Words. Thermomechanical theory of deformable continua, thermoelastic theories of shells and rods, Cosserat surfaces, Cosserat curves, specific Gibbs free energy, internal stress, invariance under superposed rigid body motions, second law of thermodynamics, closure property, entropy inequality, empirical temperature, empirical heat energy, thermomechanical process, convected coordinates.

†Deceased.

A. E. Green • Professor, Mathematical Institute, University of Oxford, Oxford OX1 3LB, England. **P. M. Naghdi** • Professor, Department of Mechanical Engineering, University of California, Berkeley, Berkeley, California 94720.

Nonlinear Effects in Fluids and Solids, edited by M. M. Carroll and M. Hayes, Plenum Press, New York, 1996.

1. Introduction

This paper is concerned firstly with some basic developments in Section 2 within the scope of the (three-dimensional) thermomechanical theory of deformable continua as further discussed in Section 1.1, and secondly with developments of certain features of thermoelastic theories of shells and rods obtained by an approximation procedure from the three-dimensional equations. The latter approximate procedure is motivated by corresponding theories of elastic shells and rods constructed by a direct approach on the basis of (three-dimensional) continuum models known as Cosserat surfaces and Cosserat curves, respectively.

Frequently, in the interest of simplicity, the development of theories of shells and rods (either by the direct approach or from the three-dimensional equations) are based on the lowest level of kinematical approximations and may lead to undesirable features (as compared to known exact solutions). Such undesirable features can be overcome if the development of the constitutive equations are recast in terms of the specific Gibbs free energy (rather than the specific Helmholtz free energy). It is this particular aspect of the theories of shells and rods that is discussed in Sections 3 and 4 and is further elaborated on in Section 1.2.

1.1. Background on the Three-Dimensional Thermomechanical Theory. Classical continuum thermomechanics has a long history. In recent years the methods of development of the basic equations for the theory have gradually crystallized into a pattern followed by many writers, but with differences in presentation and notations. Kinematics is introduced first, followed by ideas of body and surface forces acting on bodies together with equations of mass conservation, balance of momentum and balance of moment of momentum; the latter balance laws, in turn, give rise to the concept of internal stress in the body. These purely mechanical aspects of the theory are then supplemented by the concepts of temperature, heat and heat flux followed by energy and an equation for energy balance. These basic kinematical and dynamical equations apply to all types of single phase theories of continua. To complete the theory, constitutive equations are required which define the particular type of material under discussion, together with concepts of invariance under superposed rigid body motions which place some restrictions on these equations. Finally, ideas arising from the Second Law of Thermodynamics, together with the concept of entropy are used to place further restrictions on constitutive equations.

The restrictions on constitutive equations referred to at the end of the preceding paragraph are usually effected with the use of the Clausius–Duhem or similar inequalities, where often the concept of entropy first appears. An alternative procedure has been developed by Green and Naghdi (Refs. 1, 2) in which an entropy balance is postulated.[1] The local equation resulting from the balance of entropy for the special case of an inviscid fluid has the same form as the energy equation (but not for other materials), and hence does not provide an independent equation in the case of inviscid fluids. Moreover, for the purpose of eliminating the heat supply and the body force terms, substitution of the local equation for the balance of entropy and the equations of motion into the local balance of energy results in a reduced energy equation. The latter is then regarded as an identity for all thermomechanical processes and is used to place restrictions on constitutive equations. The Second Law of Thermodynamics is subsequently interpreted directly by a number of inequalities (Ref. 3) which can be used to place further restrictions on constitutive equations.

It is known that by starting with the energy equation and using invariance conditions under superposed rigid body motions, one can derive the remaining conservation laws of the mechanical theory for mass, momentum, and moment of momentum from the balance of energy. In Section 2 we follow the development presented in Green and Naghdi (Ref. 4) and by setting aside invariance conditions but starting with an energy balance, both the balance equations of the mechanical theory and entropy balance are deduced. The mechanical equations have a slightly more general form than those usually postulated, and this generality is then removed with the help of invariance conditions. The latter part of the derivation in Section 2 has some similarity with that used by Green and Rivlin (Refs. 5–7), by Green, Naghdi, and Rivlin (Ref. 8), and by Naghdi (Ref. 9, pp. 487–490), but now a clearer interpretation is given when new quantities are introduced.

1.2. Background on the Developments of Thermoelastic Theories of Shells and Rods. Few exact solutions have been obtained for specific materials and problems, and usually it is necessary to adopt some form of approximation for either the field equations or the constitutive equations, or both. Advantage may sometimes be taken of some geometrical property of the body, particularly if it is shell-like or rodlike in shape. The remainder of the paper (Sections 3 and 4) is addressed to these two categories, namely, shell-like and rodlike bodies. In this connection, we begin by first recalling that a satisfactory development of the foundation of the theory of elastic shells (and to a lesser extent also the theory of elastic rods) has progressed rather slowly and has taken nearly a century to arrive at the current understanding and state of the subject.[2]

[1]The introduction of balance of entropy is motivated by the structure of the balance of energy in the special case of inviscid fluids.

[2]Indeed this statement is true even in the context of the classical linear theory based on the so-called Kirchhoff–Love assumptions, which began in 1888 with the work of A. E. H. Love.

Most of the difficulties in the classical developments of shells (and to some extent rods) stem from difficulties in the foundation of the subject and also some undesirable features in the derivations of the basic shell and rod equations (including the constitutive equations) from the corresponding equations in the three-dimensional theory by approximations. Partly because of these past conceptal difficulties, the present writers and others have preferred to construct theories of shells and rods by a direct approach based on (three-dimensional) continuum models called Cosserat (or directed) surfaces for shells and Cosserat (or directed) curves for rods. Within the scope of general hierarchical theories of this kind, the simplest developments employ a single director for shells and two directors for rods.[3] The chief advantages of such a direct approach for shells and rods are: (1) flexibility in the development of constitutive equations and (2) the closure property, i.e., the fact that the direct procedure automatically provides for closure at every hierarchical level of shell and rod theories. In fact, the absence of closure is one of the flaws in the classical formulations of shells and rods by approximation from three-dimensional equations even in the presence of infinitesimal deformation. Despite such drawbacks, systematic developments from three-dimensional equations have not been ignored by the present writers and generally have been used as one way to identify the constitutive coefficients and the inertia coefficients in the direct approach. It may also be noted here that after a dynamical theory by the direct approach is constructed for shells (and similarly for rods), then a corresponding system of equations can also be derived from the three-dimensional equations with one-to-one correspondence between the two systems. It is in this light that we discuss thermomechanical theories of shells and rods in Sections 3 and 4 of the present paper. The thermodynamical equations in Section 3 and 4 are strongly motivated by the corresponding derivations from the equations of the three-dimensional theory, but the resulting theory has more freedom in the choice of constitutive equations. Sometimes, but not always, it is convenient to be guided by the approximate results from three-dimensional theory but in special cases the choice is helped by obtaining results corresponding to exact three-dimensional solutions of particular problems.

The equations which are derived from the equations of motion and the constitutive equations in the three-dimensional theory are not necessarily unique and often do not have closure.[4] However, once the kinematics and the basic

[3]An account of hierarchical theories of shells and rods can be found in Naghdi (Ref. 10), where the original references are also cited.

[4]In this connection, it should be recalled that the structure of the equations for shells (and rods) depends on the choice of weighting functions in the definitions of various force and couple resultants, as well as on the nature of the assumed approximate expressions for the components of the velocity (or the displacement) field in the three-dimensional theory.

equations of the direct formulation of shells (and rods) are known, then with the use of an approximate expression for the three-dimensional displacement (which is arranged in terms of displacements in the direct formulation), a fully equivalent system can also be derived from the equations of the three-dimensional theory. For example, with reference to the simplest bending theory of shells, a part of the approximate expression for the displacement in the three-dimensional equations can be readily identified with that of a single director in the direct approach (see, e.g., Ref. 9, Section 7). Similarly, in the case of the simplest bending theory of rods, a part of the approximate expression for the displacement in the three-dimensional equations can be identified with two directors in the direct approach.

Keeping the above background in mind, one procedure which starts with equations of the three-dimensional theory introduces approximations and expresses the constitutive equations in terms of the specific Helmholtz free energy function ψ^* in the three-dimensional theory (see Refs. 11–14). Some undesirable constitutive features, however, arise for elastic materials when the approximate representation for the displacements are such that it is equivalent to the use of one director for shells and two for rods. For example, when one director is used for shells, the three-dimensional specific Helmholtz free energy function ψ^* is calculated in terms of a (three-dimensional) strain measure, and the specific Helmholtz free energy function ψ for the shell would then be found from ψ^* by integration across the shell thickness. The expressions for the stress-resultants (membrane forces) and the stress-couples (bending moments)—corresponding respectively to forces and director forces in the direct approach—then follows from ψ. If the resulting equations are subsequently applied to the simple problem of flexure of a flat plate, the results will not agree very well with the known exact three-dimensional solution in the linear theory for pure bending of a plate. A similar situation presents itself in the case of simple flexure of a beam of rectangular cross section. One way of avoiding this difficulty is to add one more director for shells and three more for rods, but this leads to a more complicated theory in both cases. For this reason an alternative method of approximation with the help of the three-dimensional Gibbs function G^* has been used by Green and Naghdi (Refs. 13, 15, 16), Naghdi (Ref. 9), and Green, Naghdi, and Wenner (Ref. 17). This ensures that with one director for shells and two for rods, satisfactory results can be obtained for flexure problems. The earlier thermodynamical aspects of the approximations (Refs. 9, 13) in the three-dimensional theory involved the energy equation and an entropy inequality. But, now that the entropy balance is available along with the three-dimensional derivations from the energy equation given in Section 2, derivations of various equations for shells and rods are reconsidered in Sections 3 and 4, respectively. This is carried out on the basis of the developments of Section 2, and a detailed discussion is given concerning the approximate derivation of constitutive equations from the equations of the three-dimensional theory.

2. General Thermomechanical Theory

Consider a body and its motion which is continuously mapped into configurations in a three-dimensional Euclidean space. In particular, suppose that particles (material points) of the body in some fixed reference configuration $\boldsymbol{\kappa}_0$ are identified by a position vector $\mathbf{X}$ and that corresponding points in the current configuration $\boldsymbol{\kappa}$ at time t are denoted by a position vector $\mathbf{x}$. A motion of the body is then specified by a continuous invertible mapping,

$$\mathbf{x} = \boldsymbol{\chi}(\mathbf{X}, t). \tag{1}$$

The particle velocity $\mathbf{v}^*$ at time t is defined by[5]

$$\mathbf{v}^* = \dot{\mathbf{x}}, \tag{2}$$

where a superposed dot denotes material time differentiation holding $\mathbf{X}$ fixed. The usual kinematical variables which occur in the theory may now be defined in terms of $\mathbf{x}$ and $\mathbf{v}^*$. Sufficient conditions (such as differentiability) are always assumed for the mathematical validity of the theory. Also, the mass density at each material point in the current configuration is denoted by ρ^* and is a function of $\mathbf{X}$ and t.

Let the body in its current configuration $\boldsymbol{\kappa}$ occupy a material volume $\mathcal{R}^*$ bounded by a smooth closed surface $\partial\mathcal{R}^*$. In line with concepts of force, work, and energy, we suppose that the body is acted on by surface (or contact) forces $\bar{\mathbf{t}}$ per unit area over $\partial\mathcal{R}^*$ and by body forces $\mathbf{b}^*$ per unit mass throughout $\mathcal{R}^*$ which are defined by rates of work of $\bar{\mathbf{t}}$ and $\mathbf{b}^*$ equal to $\bar{\mathbf{t}} \cdot \mathbf{v}^*$ and $\mathbf{b}^* \cdot \mathbf{v}^*$ for all choice of $\mathbf{v}^*$, respectively, per unit area of $\partial\mathcal{R}^*$ and per unit mass in $\mathcal{R}^*$. Consider any subset S_t of the body in its configuration $\boldsymbol{\kappa}$ at time t occupying a volume $\mathcal{P}^* \subseteq \mathcal{R}^*$ bounded by a closed surface $\partial\mathcal{P}^*$. The externally applied forces $\bar{\mathbf{t}}$, $\mathbf{b}^*$ are assumed to give rise to a surface (or contact) force $\mathbf{t}$ per unit area of $\partial\mathcal{P}^*$, defined by a corresponding rate of work $\mathbf{t} \cdot \mathbf{v}^*$ for all $\mathbf{v}^*$. The surface force $\mathbf{t}$ is external to $\mathcal{P}^*$. In addition, the external forces may give rise to internal body forces in $\mathcal{P}^*$ but we do not need to consider these at present. Also, a kinetic energy $(1/2)\mathbf{v}^* \cdot \mathbf{v}^*$ per unit mass is defined throughout $\mathcal{R}^*$.

Next, the ideas of empirical temperature T and heat energy are assumed. A scalar $\theta^* > 0$ is defined throughout $\mathcal{R}^*$ which depends on T and on the kinematical and thermal properties of the body restricted by the condition $\partial\theta^*/\partial T > 0$.

[5]We use the symbol $\mathbf{v}^*$ (rather than $\mathbf{v}$) for the particle velocity in the three-dimensional theory in order to reserve the corresponding symbol without a superposed star "*" for later use. Similar notations with a superposed star "*" are used in this section for other quantities such as mass density ρ^* and a measure of temperature θ^*.

Then, θ^* is invertible with respect to T and is adopted as a measure of temperature throughout $\mathcal{R}^*$. There are external rates of supplies of heat to the body consisting of a surface flux of heat $-\bar{h}^*$ per unit area over $\partial\mathcal{R}^*$ and volume rate of supply of heat r* per unit mass throughout $\mathcal{R}^*$. These are accompanied by an external surface flux of entropy defined by $-\bar{k}^* = \bar{h}^*/\theta^*$ per unit area over $\partial\mathcal{R}^*$ and an external volume rate of supply of entropy $s^* = r^*/\theta^*$ per unit mass throughout $\mathcal{R}^*$. It is assumed that these give rise to a surface flux of heat $-h^*$ into $\mathcal{P}^*$ and measured per unit area of $\partial\mathcal{P}^*$, together with a surface flux of entropy $-k^*$ defined by $k^* = h^*/\theta^*$ per unit area. Internal volume rate of supply of heat and entropy may also be generated in $\mathcal{P}^*$, but these are not considered here. Finally, corresponding to the kinetic energy density in $\mathcal{P}^*$ arising from mechanical concepts there is a density of heat ζ^* together with an entropy density η^* defined by $\eta^* = \zeta^*/\theta^*$, per unit mass in $\mathcal{R}^*$.

It is now convenient to define the various work and heat supply contributions to any part (or subset) of the body S_t occupying a volume $\mathcal{P}^*$. The kinetic energy K and the thermal energy H in S_t are (see also Ref. 4)

$$K(S_t) = \int_{\mathcal{P}^*} \frac{1}{2}\rho^*\mathbf{v}^* \cdot \mathbf{v}^* dv, \tag{3a}$$

$$H(S_t) = \int_{\mathcal{P}^*} \rho^*\zeta^* dv = \int_{\mathcal{P}^*} \rho^*\theta^*\eta^* dv. \tag{3b}$$

The total rate of work on S_t due to the rate of work R_b by the body force $\mathbf{b}^*$ and the rate of work R_c by the contact force $\mathbf{t}$ is

$$R(S_t) = R_b(S_t) + R_c(S_t), \tag{4a}$$

$$R_b(S_t) = \int_{\mathcal{P}^*} \rho^*\mathbf{b}^* \cdot \mathbf{v}^* dv, \qquad R_c(S_t) = \int_{\partial\mathcal{P}^*} \mathbf{t} \cdot \mathbf{v}^* da. \tag{4b}$$

Similarly, the total rate of heat Q supplied to S_t is

$$Q(S_t) = \int_{\mathcal{P}^*} \rho^* r^* dv - \int_{\partial\mathcal{P}^*} h^* da$$

$$= \int_{\mathcal{P}^*} \rho^*\theta^* s^* dv - \int_{\partial\mathcal{P}^*} \theta^* k^* da. \tag{5}$$

Finally, the total rate at which internal energy (both mechanical work and thermal) is generated in S_t is

$$W(S_t) = \int_{\mathcal{P}^*} \rho^* w^* dv, \tag{6}$$

where w^* is the density of such energy.

We now state the first law of thermodynamics (or balance of energy) for any part S_t as follows:

(i) The total rate of work of external body forces in S_t plus the rate of work of tractions on ∂S_t, plus energy due to external volume supply of heat in S_t and external surface flux of heat across ∂S_t, minus the total rate of change of kinetic energy and heat in S_t plus the total rate at which internal energy (mechanical and thermal) is generated in S_t is zero.

(ii) The total heat energy Q and mechanical work R supplied to or extracted from S_t in a cycle is zero.

A cycle referred to in (ii) is a thermomechanical process, during a closed interval of time $\mathcal{I} = [t_1, t_2]$ with $t_1 < t_2$, involving various kinematical, kinetical, and thermal variables, which define the state of any part of the body and which assume the same values at the beginning and end of the cycle at times t_1 and t_2. For example, the body could be completely at rest at the beginning and end of the cycle, with the same values of the kinematical and kinetical variables.

In adopting a balance of energy for S_t in the form stated in (i) and (ii), it is tacitly assumed that the theory under discussion is a local theory, i.e., any internal forces or rate of supplies of heat generated in S_t, as well as quantities such as $\mathbf{t}$ and k^*, do not depend on kinematical quantities outside S_t. Some modifications are necessary in order to discuss a nonlocal theory.

In the remainder of this section we follow closely the derivation given in Green and Naghdi (Ref. 4). Thus, in view of the statement (i), the balance of energy may be stated in the form

$$-\frac{d}{dt}\{K(S_t) + H(S_t)\} + R(S_t) + Q(S_t) + W(S_t) = 0, \tag{7}$$

which may also be written as

$$-\frac{d}{dt}\int_{\mathcal{P}^*} \left(\frac{1}{2}\mathbf{v}^* \cdot \mathbf{v}^* + \eta^*\theta^*\right)\rho^* dv + \int_{\mathcal{P}^*} (s^*\theta^* + \mathbf{b}^* \cdot \mathbf{v}^* + w^*)\rho^* dv$$

$$+ \int_{\partial\mathcal{P}^*} (\mathbf{t} \cdot v^* - k^*\theta^*) da = 0. \tag{8}$$

Observing that the time rates of change of kinetic energy $\dot{K}$ and of thermal energy $\dot{H}$ take the same values at the beginning ($t = t_1$) and end ($t = t_2$) of any cycle during the time interval $\mathcal{J} = [t_1, t_2]$, from the statement (ii) and an expression resulting from integration of (7) with respect to t between the limits t_1 and t_2, we obtain

$$\oint_{t_1}^{t_2} W(S_t)dt = 0, \tag{9}$$

which holds for any cycle during the time interval $\mathcal{J}$. If the time integral in (9) is taken over any thermomechanical process which is not a cycle during the closed time interval $[t_1, t]$, the integral is nonzero and will not depend on the path but only on the state of the body at time t (and t_1), so that

$$\int_{t_1}^{t} W(S_t)dt = -\psi(S_t), \tag{10}$$

and by (6) we may write

$$\psi(S_t) = \int_{\mathcal{P}^*} \rho^*\psi^* dv. \tag{11}$$

The scalar function ψ^* in (11), as becomes evident later, is known as the specific Helmholtz free energy. From (10) it is seen that

$$W(S_t) = -\frac{d}{dt}\psi(S_t) \tag{12}$$

and then by virtue of (6) and (11) we also have

$$-\int_{\mathcal{P}^*} \rho^* w^* dv = \frac{d}{dt}\int_{\mathcal{P}^*} \rho^*\psi^* dv \tag{13}$$

for all parts $\mathcal{P}^*$ in the current configuration κ. Hence, we may deduce the local result

$$\rho^* J w^* = -\overline{\rho^* J \psi^*}^{\cdot}, \tag{14}$$

where

$$J = \det \mathbf{F}, \qquad \mathbf{F} = \frac{\partial \boldsymbol{\chi}}{\partial \mathbf{X}}. \tag{15}$$

With the help of (12) or (13), the balance of energy in the forms (7) and (8) can be reduced to

$$\bar{E} = -\frac{d}{dt}\{K(S_t) + H(S_t) + \psi(S_t)\} + R(S_t) + Q(S_t) = 0, \tag{16}$$

$$\bar{E} = -\frac{d}{dt}\int_{\mathcal{P}^*}\left[\frac{1}{2}\mathbf{v}^*\cdot\mathbf{v}^* + \psi^* + \eta^*\theta^*\right]\rho^* dv + \int_{\mathcal{P}^*}(s^*\theta^* + \mathbf{b}^*\cdot\mathbf{v}^*)\rho^* dv + \int_{\partial\mathcal{P}^*}(\mathbf{t}\cdot\mathbf{v}^* - k^*\theta^*)da = 0. \tag{17}$$

Setting

$$E(S_t) = \psi(S_t) + H(S_t), \qquad E(S_t) = \int \rho^*\epsilon^* dv, \qquad \epsilon^* = \psi^* + \eta^*\theta^*, \tag{18}$$

the balance equations (16) or (17) can be rewritten in its usual form with ϵ^* being the specific internal energy and $s^*\theta^*$, $k^*\theta^*$ replaced by r^*, h^*, respectively. As noted previously (Ref. 4), in the present approach it is the Helmholtz free energy and not the internal energy which first occurs in the energy balance because there is a counterpart to the kinetic energy in the expression for H due to heat density $\omega^* = \eta^*\theta^*$ [see (3b)].

By applying (17) to an elementary tetrahedron and using the usual smoothness assumptions, we obtain

$$(\mathbf{t} - \mathbf{T}\mathbf{n})\cdot\mathbf{v}^* - (k^* - \mathbf{p}^*\cdot\mathbf{n})\theta^* = 0, \tag{19}$$

where $\mathbf{n}$ is an outward unit normal to $\partial\mathcal{P}^*$, $\mathbf{T}$ is a second-order tensor called the stress tensor, and $\mathbf{p}^*$ is an entropy flux vector which is related to the usual heat flux vector $\mathbf{q}^*$ by $\mathbf{q}^* = \mathbf{p}^*\theta^*$. With the help of (19) the energy balance (17) may be further reduced to the form

$$\bar{E} = \int_{\mathcal{P}^*}\Big\{-(\dot{\rho}^* + \rho^*\operatorname{div}\mathbf{v}^*)\left[\frac{1}{2}\mathbf{v}^*\cdot\mathbf{v}^* + \psi^* + \eta^*\theta^*\right] + (-\rho^*\dot{\mathbf{v}}^* + \rho^* b^* + \operatorname{div}\mathbf{T})\cdot\mathbf{v}^* - \frac{1}{2}\rho^*\boldsymbol{\lambda}^*\cdot\mathbf{w} + \mathbf{T}\cdot\mathbf{D} + (-\rho^*\dot{\eta}^* + \rho^* s^* - \operatorname{div}\mathbf{p}^*)\theta^* - \mathbf{p}^*\cdot\operatorname{grad}\theta^* - \rho^*(\dot{\psi}^* + \eta^*\dot{\theta}^*)\Big\}dv = 0, \tag{20}$$

where the "div" operator is with respect to the place $\mathbf{x}$, $\mathbf{D}$ is the rate of deformation tensor, $\mathbf{w}$ = curl $\mathbf{v}^*$ is the vorticity vector, and

$$-\rho^*\boldsymbol{\Gamma}^* = \mathbf{T} - \mathbf{T}^T, \qquad \boldsymbol{\lambda}^* \times \mathbf{c} = \boldsymbol{\Gamma}^*\mathbf{c}, \qquad \mathbf{w} \times \mathbf{c} = 2\mathbf{W}\mathbf{c} \tag{21}$$

for all arbitrary vectors $\mathbf{c}$. Also, in the reduced energy equation (20), the notation $\mathbf{T}^T$ stands for the transpose of the second-order tensor, $\mathbf{T}$, $\boldsymbol{\Gamma}^*$ is a skew-symmetric second-order tensor with $\boldsymbol{\lambda}^*$ being the corresponding axial vector, and $\mathbf{W}$ is the spin or the vorticity tensor.

We now define a vector $\mathbf{f}^*$ and scalars ξ^* and m by the equations

$$m = \dot{\rho}^* + \rho^* \text{div } \mathbf{v}^*, \tag{22}$$

$$\rho^*\mathbf{f}^* = \rho^*\dot{\mathbf{v}}^* + m\,\mathbf{v}^* - \rho^*\mathbf{b}^* - \text{div } \mathbf{T}, \tag{23}$$

$$\rho^*\xi^* = \rho^*\dot{\eta}^* + m\,\eta^* - \rho^*s^* + \text{div } \mathbf{p}^*. \tag{24}$$

Using (22), (23), and (14), equation (20) becomes

$$\bar{E} = \int_{\mathcal{P}^*} \left\{ \frac{1}{2} m\,\mathbf{v}^* \cdot \mathbf{v}^* - m\,\psi^* - \rho^*\mathbf{f}^* \cdot \mathbf{v}^* - \frac{1}{2}\rho^*\boldsymbol{\lambda}^* \cdot \mathbf{w} + \mathbf{T} \cdot \mathbf{D} - \rho^*\xi^*\theta^* - \mathbf{p}^* \cdot \text{grad } \theta^* - \rho^*(\dot{\psi}^* + \eta^*\dot{\theta}^*) \right\} dv = 0. \tag{25}$$

Recalling that $\bar{E}$ represents the sum of all energies in the equation (17) for the balance of energy in statement (i) due to work done by external body and contact forces R on $\mathcal{P}^*$ and external rate of supplies of heat Q to $\mathcal{P}^*$, minus the rate of change of kinetic energy, then the expression (25) enables us to give an interpretation and meaning to m, $\mathbf{f}^*$, and ξ^* which at present are defined by (22), (23), and (24). Thus, the terms $= \mathbf{T} \cdot \mathbf{D}$ and $-\mathbf{p}^* \cdot \text{grad } \theta^*$ in (25) represent rate of work of the internal stress $\mathbf{T}$ and the rate of internal generation of heat due to entropy flux vector $\mathbf{p}^*$, while $-\rho^*\mathbf{f}^* \cdot \mathbf{v}^*$ and $-1/2$ $\rho^*\boldsymbol{\lambda}^* \cdot \mathbf{w}$ readily suggest the interpretations that $\mathbf{f}^*$ is an internal force and $\boldsymbol{\lambda}^*$ an internal couple, and $-\rho^*\xi^*\theta^*$ means that ξ^* is an internal generation of entropy. Further the terms containing m imply that m is an internal generation of mass.

In a complete theory it is necessary to specify constitutive equations for $\mathbf{T}$, $\mathbf{p}^*$, $\mathbf{f}^*$, ξ, m, as well as for ψ^* and η^* (or θ^*). Then, (22)–(24) represent field equations for balances of force and momentum, and entropy, respectively, together with a balance of mass, and these equations cease to be merely definitions of m, $\mathbf{f}^*$, ξ^*. It should be noted that in this approach there is, at this stage, a nonzero internal force $\mathbf{f}^*$ and couple $\boldsymbol{\lambda}^*$ although only stress vectors were applied to the surface. Later it is seen that they are zero. However, if the theory had been a nonlocal theory these quantities would not be zero and would have been present in the original energy balance (8) since they are affected by material outside $\mathcal{P}^*$, unless $\mathcal{P}^*$ is the whole body (see, e.g., Ref. 18).

Since (25) must hold for every part $\mathcal{P}^*$, assuming that the integrand is continuous, we deduce the local equation

$$\begin{aligned}&\tfrac{1}{2}m\,\mathbf{v}^*\cdot\mathbf{v}^* - m\,\psi^* - \rho^*\mathbf{f}^*\cdot\mathbf{v}^* - \tfrac{1}{2}\,\rho^*\boldsymbol{\lambda}^*\cdot\mathbf{w} + \mathbf{T}\cdot\mathbf{D}\\ &- \rho^*\xi\theta^* - \mathbf{p}^*\cdot\operatorname{grad}\theta^* - \rho^*(\dot{\psi}^* + \eta^*\dot{\theta}^*) = 0.\end{aligned} \tag{26}$$

Given constitutive equations for $\mathbf{T}$, $\mathbf{p}^*$, η^*, m, $\mathbf{f}^*$, ξ^*, the equations (22)–(24) with specified values for the external force and external supply of entropy and suitable boundary and initial conditions are sufficient to determine the velocity (or the displacement) and temperature fields. Moreover, as in Ref. 1 the balance of energy (26) must be regarded as an identity which provides some restrictions on constitutive equations.

It will be observed that the balance of force and momentum in (23) contains the extra internal force $\mathbf{f}^*$ and is, therefore, somewhat different from that in the usual approach which postulates a momentum balance at the outset. Moreover, the entropy balance (24) appears here naturally instead of being a separate postulate motivated by the energy equation for an inviscid fluid (1). Most of the existing literature on thermomechanics does not admit an entropy balance but postulates an entropy inequality as a representation of the Second Law of Thermodynamics. Here, restrictions arising from this law have been set aside.

So far ideas of invariance under superposed rigid body motions have not been introduced. If these are now invoked, then it is known that from the energy equation (17) (Refs. 5–9, 18) the field equations for mass conservation and momentum may be recovered in the forms (22) and (23) with

$$m = 0, \qquad \mathbf{f}^* = \mathbf{0}. \tag{27}$$

Moreover, the field equations for moment of momentum can also be obtained and yield

$$\boldsymbol{\Gamma}^* = 0 \quad \text{or} \quad \mathbf{T} = \mathbf{T}^T. \tag{28}$$

Alternatively, these results may now be found from (26). In addition, following Naghdi (Ref. 9), integral balances of mass conservation, momentum, and moment of momentum follow from (17), again with $m = 0$, $\mathbf{f}^* \equiv \mathbf{0}$.

From the application of the invariance conditions to (19), we obtain the results

$$\mathbf{t} = \mathbf{T}\mathbf{n}, \qquad \text{hence } k^* = \mathbf{p}^*\cdot\mathbf{n}. \tag{29}$$

At this point restrictions which may arise from the Second Law of Thermodynamics should be discussed but these are not considered here. Instead reference is made to the discussion of this aspect of the subject by Green and Naghdi (Ref. 3) after suppressing the electrodynamic effects.

It should be noted that if comparisons are made in particular applications with, for example, values of the entropy flux from other approaches, particularly with those found from kinetic theory, it is important to be sure that similar quantities are being compared. In any case, it is really only possible to compare results obtained from the complete theory and, in particular, with those found from experiment.

3. Thermomechanical Theory of Shells and Plates

It is convenient to introduce a system of curvilinear coordinates θ^i, $i = 1, 2, 3$, in the reference configuration of the body and regard these as a convected system throughout the motion. Then

$$\mathbf{X} = \mathbf{X}(\theta^i), \qquad \mathbf{x} = \mathbf{x}(\theta^i, t), \tag{30a}$$

$$\mathbf{G}_i = \partial \mathbf{X}/\partial \theta^i, \qquad \mathbf{g}_i = \partial \mathbf{x}/\partial \theta^i, \tag{30b}$$

$$\mathbf{G}^i \cdot \mathbf{G}_j = \delta^i_j, \qquad \mathbf{g}^i \cdot \mathbf{g}_j = \delta^i_j, \qquad G_{ij} = \mathbf{G}_i \cdot \mathbf{G}_j, \qquad G^{ij} = \mathbf{G}^i \cdot \mathbf{G}^j, \tag{30c}$$

$$g_{ij} = \mathbf{g}_i \cdot \mathbf{g}_j, \qquad g^{ij} = \mathbf{g}^i \cdot \mathbf{g}^j, \qquad g^{1/2} = [\mathbf{g}_1, \mathbf{g}_2, \mathbf{g}_3], \tag{30d}$$

$$G^{1/2} = [\mathbf{G}_1 \mathbf{G}_2 \mathbf{G}_3], \tag{30e}$$

where $\mathbf{g}_i$, $\mathbf{g}^i$ are covariant and contravariant base vectors, respectively, g_{ij} and g^{ij} are covariant and contravariant metric tensors, respectively, in the configuration at time t, and δ^i_j is the Kronecker delta. Corresponding quantities in the reference configuration are $\mathbf{G}_i$, $\mathbf{G}^i$, G_{ij}, G^{ij}. Also

$$\mathbf{t}^i = \mathbf{T}\mathbf{g}^i, \qquad \mathbf{t}^i_{\mathrm{R}} = \mathbf{T}_{\mathrm{R}}\mathbf{G}^i, \qquad \mathbf{T} = \mathbf{t}^i \otimes \mathbf{g}_i, \qquad \mathbf{T}_{\mathrm{R}} = \mathbf{t}^i_{\mathrm{R}} \otimes \mathbf{G}_i, \tag{31a}$$

$$\mathbf{T}^i = g^{1/2}\mathbf{t}^i = G^{1/2}\mathbf{t}^i_{\mathrm{R}}, \qquad g^{1/2}\mathrm{div}\,\mathbf{T} = G^{1/2}\,\mathrm{Div}\,\mathbf{T}_{\mathrm{R}} = \partial \mathbf{T}^i/\partial \theta^i, \tag{31b}$$

$$p^i = g^{1/2}\mathbf{p}^* \cdot \mathbf{g}^i = G^{1/2}\mathbf{p}^*_{\mathrm{R}} \cdot \mathbf{G}^i, \qquad g^{1/2}\,\mathrm{div}\,\mathbf{p}^* = G^{1/2}\,\mathrm{Div}\,\mathbf{p}^*_{\mathrm{R}} = \partial p^i/\partial \theta^i. \tag{31c}$$

Introducing the notation $\theta^3 = z$, the body is assumed to be bounded by the surfaces

$$z = z_2(\theta^1, \theta^2), \qquad z = z_1(\theta^1, \theta^2), \qquad z_1 < z_2, \tag{32}$$

which are smooth and nonintersecting. In the region $z_1 < z < z_2$ the position vectors **X** and **x**, and the temperature θ^*, are represented by the series

$$\mathbf{X} = \sum_{N=0}^{\infty} \lambda_N(z)\mathbf{D}_N, \qquad \mathbf{D}_0 = \mathbf{R}, \qquad \mathbf{D}_N = \mathbf{D}_N(\theta^\alpha), \tag{33a}$$

$$\mathbf{x} = \sum_{N=0}^{\infty} \lambda_N(z)\mathbf{d}_N, \qquad \mathbf{d}_0 = \mathbf{r}, \qquad \mathbf{d}_N = \mathbf{d}_N(\theta^\alpha, t), \tag{33b}$$

$$\theta^* = \sum_{N=0}^{\infty} \mu_N(z)\theta_N, \qquad \theta_N = \theta_N(\theta^\alpha, t), \qquad \theta_0 = \theta, \tag{33c}$$

$$\lambda_0 = 1, \qquad \mu_0 = 1, \tag{33d}$$

where Greek indices take the values 1, 2. The functions $\lambda_N(z)$, $\mu_N(z)$ are assumed to be chosen so that the series in (33) represent the deformation of the body completely in the region and are smoothly differentiable as many times as required in the analysis. Also, r in (33b, center) may be regarded as the position vector at some reference value of z (say $z = z_0$), defining a two-dimensional surface whose base vectors, metric tensors, and unit normal are

$$\mathbf{a}_\alpha = \mathbf{a}_\alpha(\theta^\beta, t) = \partial\mathbf{r}/\partial\theta^\alpha, \tag{34a}$$

$$\mathbf{a}^\alpha \cdot \mathbf{a}_\beta = \delta^\alpha_\beta, \tag{34b}$$

$$a^{1/2} = [\mathbf{a}_1\mathbf{a}_2\mathbf{a}_3], \tag{34c}$$

$$a^{1/2}\mathbf{a}_3 = \mathbf{a}_1 \times \mathbf{a}_2, \tag{34d}$$

$$a_{\alpha\beta} = \mathbf{a}_\alpha \cdot \mathbf{a}_\beta, \tag{34e}$$

$$a^{\alpha\beta} = \mathbf{a}^\alpha \cdot \mathbf{a}^\beta. \tag{34f}$$

Corresponding quantities in the reference configuration are denoted by capital letters, e.g., $\mathbf{A}_\alpha$, $A_{\alpha\beta}$, and so on. The particle velocity in the three-dimensional theory is then given by

$$\mathbf{v}^* = \sum_{N=0}^{\infty} \lambda_N(z)\mathbf{w}_N, \qquad \mathbf{v} = \mathbf{w}_0, \tag{35a}$$

$$\mathbf{w}_N = \mathbf{w}_N(\theta^\alpha, t) = \dot{\mathbf{d}}_N, \tag{35b}$$

where a superposed dot again denotes material time differentiation.

Now let ζ^i, $i = 1, 2, 3$, be a fixed system of curvilinear coordinates and let points in the region occupied by the body in the current configuration be specified by a position vector $\bar{\mathbf{x}} = \bar{\mathbf{x}}(\zeta^i)$ with corresponding base vectors and metric tensors

$$\bar{\mathbf{g}}_i = \partial\bar{\mathbf{x}}/\partial\zeta^i, \qquad \bar{\mathbf{g}}^i \cdot \bar{\mathbf{g}}_k = \delta^i_k, \qquad \bar{g}_{ik} = \bar{\mathbf{g}}_i \cdot \bar{\mathbf{g}}_k, \tag{36a}$$

$$\bar{g}^{ik} = \bar{\mathbf{g}}^i \cdot \bar{\mathbf{g}}^k, \qquad \bar{g}^{1/2} = [\bar{\mathbf{g}}_1\bar{\mathbf{g}}_2\bar{\mathbf{g}}_3]. \tag{36b}$$

Select a fixed surface which, with its corresponding base vectors, metric tensors, and unit normal, is specified by

$$\bar{\mathbf{r}} = \bar{\mathbf{r}}(\zeta^i), \qquad \bar{\mathbf{a}}_\alpha = \partial\bar{\mathbf{r}}/\partial\zeta^i, \qquad \bar{\mathbf{a}}^\alpha \cdot \bar{\mathbf{a}}_\beta = \delta^\alpha_\beta, \qquad \bar{a}^{1/2} = [\bar{\mathbf{a}}_1\bar{\mathbf{a}}_2\bar{\mathbf{a}}_3], \tag{37a}$$

$$\bar{a}^{1/2}\bar{\mathbf{a}}_3 = \bar{\mathbf{a}}_1 \times \bar{\mathbf{a}}_2, \qquad \bar{a}_{\alpha\beta} = \bar{\mathbf{a}}_\alpha \cdot \bar{\mathbf{a}}_\beta, \qquad \bar{a}^{\alpha\beta} = \bar{\mathbf{a}}^\alpha \cdot \bar{\mathbf{a}}^\beta. \tag{37b}$$

The surfaces (32) which bound the body are now specified by

$$\zeta = \zeta_1(\zeta^1, \zeta^2, t), \qquad \zeta = \zeta_2(\zeta^1, \zeta^2, t), \qquad \zeta^3 = \zeta. \tag{38}$$

In terms of the fixed coordinates ζ^i, the velocity and temperature of a particle at time t may be represented by

$$\mathbf{v}^* = \bar{\mathbf{v}}^*(\zeta^i, t) = \bar{v}^{*i}\bar{\mathbf{g}}_i = \sum_{N=0}^{\infty} \bar{\lambda}_N(\zeta)\bar{\mathbf{w}}_N, \tag{39a}$$

$$\bar{\mathbf{v}} = \bar{\mathbf{w}}_0, \qquad \bar{\mathbf{w}}_N = \bar{\mathbf{w}}_N(\zeta^\alpha, t), \tag{39b}$$

$$\theta^* = \sum_{N=0}^{\infty} \bar{\mu}_N(\zeta)\bar{\theta}_N, \qquad \bar{\theta}_N = \bar{\theta}_0, \qquad \bar{\theta}_N = \bar{\theta}_N(\zeta^\alpha, t). \tag{39c}$$

The surfaces (38) are material surfaces which move with the body so that

$$\frac{\partial\zeta_1}{\partial t} = \bar{v}^{*3} - \bar{v}^{*\alpha}\frac{\partial\zeta_1}{\partial\zeta^\alpha}\ (\zeta = \zeta_1), \tag{40a}$$

$$\frac{\partial\zeta_2}{\partial t} = \bar{v}^{*3} - \bar{v}^{*\alpha}\frac{\partial\zeta_2}{\partial\zeta^\alpha}\ (\zeta = \zeta_2). \tag{40b}$$

Later, the convected coordinates θ^i are chosen so that at time t, the θ^i-curves coincide with the fixed ζ^i-curves, i.e., $\theta^i = \zeta^i + b^i$ where b^i are constants. Also, the moving surface represented by the vector $\mathbf{r}$ in (33b, center) and associated base vectors and unit normal given by (34a, d), coincides with the fixed surface (37) at time t. With this choice of θ^i, we have

$$\mathbf{g}_i = \bar{\mathbf{g}}_i, \qquad g_{ij} = \bar{g}_{ij}, \qquad g^* = \bar{g}^{1/2}, \qquad \mathbf{a}_\alpha = \bar{\mathbf{a}}_\alpha, \qquad a_{\alpha\beta} = \bar{a}_{\alpha\beta}, \qquad \text{etc.}, \tag{41a}$$

$$\lambda_N(z) = \bar{\lambda}_N(\zeta), \qquad \mu_N(z) = \bar{\mu}_N(\zeta). \tag{41b}$$

Now, in order to construct the desired theory, we may begin with the energy balance (17) and the reduced form (26) in which use is also made of (27) and (28) from invariance considerations. Next make use of the Lagrangian form of representation (33) and (35), substitute into (17) and (25), and perform the integration with respect to z. This follows the procedure used by Green and Naghdi (Refs. 12, 13). The Eulerian forms will be derived later. Thus, we obtain

$$\begin{aligned}
\bar{E} = &-\frac{d}{dt}\int_{\mathscr{P}} \left(\frac{1}{2}\sum_{M=0}^{\infty}\sum_{N=0}^{\infty} y_{MN}\mathbf{w}_M\cdot\mathbf{w}_N + \psi + \sum_{N=0}^{\infty}\tilde{\eta}_N\theta_N\right)\rho\, d\sigma \\
&+\int_{\mathscr{P}}\sum_{N=0}^{\infty}(s_N\theta_N + \boldsymbol{\ell}_N\cdot\mathbf{w}_N)\rho\, d\sigma + \int_{\partial\mathscr{P}}\sum_{N=0}^{\infty}(\mathbf{m}_N\cdot\mathbf{w}_N - k_N\theta_N)\, ds \\
= &\int_{\mathscr{P}}\left\{\sum_{N=0}^{\infty}\left(\mathbf{k}_N\cdot\mathbf{w}_N + \mathbf{M}_N^{\alpha}\cdot\frac{\partial\mathbf{w}_N}{\partial\theta^\alpha} - \mathbf{p}_N\cdot\tilde{\mathbf{g}}_N\right)\right. \\
&\left. - \rho\left\{\dot{\psi} + \sum_{N=0}^{\infty}(\tilde{\eta}_N\dot{\theta}_N + \xi_N\theta_N)\right\}\right\} d\sigma = 0,
\end{aligned} \tag{42}$$

where $\mathscr{P}$ is an arbitrary part of the surface specified by (33b, center) and (34) bounded by a smooth closed curve $\partial\mathscr{P}$ whose outward unit normal in the surface is

$$\boldsymbol{\upsilon} = \upsilon_\alpha \mathbf{a}^\alpha = \upsilon^\alpha \mathbf{a}_\alpha, \tag{43}$$

and

$$\tilde{\mathbf{g}}_N = \text{grad}_s\, \theta_N = \mathbf{a}^\alpha\, \partial\theta_N/\partial\theta^\alpha. \tag{44}$$

The various functions in (42) are defined by the following integrals:

$$\rho a^{1/2} = \int_{z_1}^{z_2} \rho^* g^{1/2} dz, \tag{45a}$$

$$\rho a^{1/2} y_{MN} = \int_{z_1}^{z_2} \rho^* g^{1/2} \lambda_M(z) \lambda_N(z) dz, \tag{45b}$$

$$\mathbf{M}_N^\alpha a^{1/2} = \int_{z_1}^{z_2} \mathbf{T}^\alpha \lambda_N(z) dz, \qquad \mathbf{k}_N a^{1/2} = \int_{z_1}^{z_2} \mathbf{T}^3 \lambda'_N(z) dz, \tag{46a}$$

$$\mathbf{M}_0^\alpha = \mathbf{N}^\alpha, \qquad \mathbf{k}_0 = \mathbf{0}, \qquad \mathbf{m}_N = \mathbf{M}_N^\alpha \nu_\alpha, \tag{46b}$$

$$\rho \boldsymbol{\ell}_N a^{1/2} = \int_{z_1}^{z_2} \rho^* g^{1/2} \mathbf{b}^* \lambda_N(z) dz + [\mathbf{t}\lambda_N(z)(g g^{33})^{1/2}]_{z=z_1} + [\mathbf{t}\lambda_N(z)(g g^{33})^{1/2}]_{z=z_2}, \tag{47}$$

$$\rho s_N a^{1/2} = \int_{z_1}^{z_2} \rho^* g^{1/2} s^* \mu_N(z) dz - [k^* \mu_N(z)(g g^{33})^{1/2}]_{z=z_1} - [k^* \mu_N(z)(g g^{33})^{1/2}]_{z=z_2}, \tag{48}$$

$$\mathbf{p}_N \cdot \mathbf{a}^\alpha a^{1/2} = \int_{z_1}^{z_2} P^\alpha \mu_N(z) dz, \tag{49}$$

$$\rho \tilde{\eta}_N a^{1/2} = \int_{z_1}^{z_2} \rho^* g^{1/2} \eta^* \mu_N(z) dz, \tag{50}$$

$$\rho \xi_N a^{1/2} = \int_{z_1}^{z_2} \rho^* g^{1/2} \xi^* \mu_N(z) dz + \int_{z_1}^{z_2} P^3 \mu'_N(z) dz, \tag{51}$$

$$\rho \psi a^{1/2} = \int_{z_1}^{z_2} \rho^* g^{1/2} \psi^* dz, \tag{52}$$

where in (46) and (51), $\lambda'_N(z) = d\lambda_N(z)/dz$, $\mu'_N(z) = d\mu_N(z)/dz$. With the usual continuity assumptions (42) yields the equation

$$-\rho\left\{\dot{\psi} + \sum_{N=0}^{\infty} (\tilde{\eta}_N\dot{\theta}_N + \xi_N\theta_N)\right\} + \sum_{N=0}^{\infty}\left(\mathbf{k}_N \cdot \mathbf{w}_N + \mathbf{M}_N^\alpha \cdot \frac{\partial \mathbf{w}_N}{\partial \theta^\alpha} - \mathbf{p}_N \cdot \tilde{\mathbf{g}}_N\right) = 0. \tag{53}$$

Using the definitions (31), (35), and (45)–(51) the following equations may be obtained from (22), (23), and (24) in which $m = 0$, $\mathbf{f}^* = \mathbf{0}$, namely,

$$\dot{\rho} + \rho \operatorname{div}_s \mathbf{v} = 0 \quad \text{or} \quad \rho a^{1/2} = \rho_R A^{1/2}, \tag{54}$$

$$\rho \sum_{M=0}^{\infty} y_{MN}\dot{\mathbf{w}}_N = \rho \boldsymbol{\ell}_N - \mathbf{k}_N + \operatorname{div}_s \mathbf{M}_N, \tag{55}$$

$$\rho\dot{\tilde{\eta}} = \rho(s_N + \xi_N) - \operatorname{div}_s \mathbf{p}_N, \qquad N = 0, 1, 2, \ldots, \tag{56}$$

where the vector fields $\mathbf{M}_N$ and the surface divergence operators $\operatorname{div}_s \mathbf{M}_N$ and $\operatorname{div}_s \mathbf{v}$ are defined by

$$\mathbf{M}_N = \mathbf{M}_N^\alpha \otimes \mathbf{a}_\alpha, \qquad a^{1/2} \operatorname{div}_s \mathbf{M}_N = \partial(a^{1/2}\mathbf{M}_N^\alpha)/\partial\theta^\alpha, \qquad \operatorname{div}_s \mathbf{v} = \mathbf{a}^\alpha \cdot \partial\mathbf{v}/\partial\theta_\alpha.$$

Also, from (28), (31), and (35), it is seen that

$$\sum_{N=0}^{\infty}\left(\mathbf{d}_N \times \mathbf{k}_N + \frac{\partial \mathbf{d}_N}{\partial \theta^\alpha} \times \mathbf{M}_N^\alpha\right) = \mathbf{0}. \tag{57}$$

The meanings of the functions $\boldsymbol{\ell}_N$, $\mathbf{k}_N$, $\mathbf{M}_N$, η_N, s_N, ξ_N, $\mathbf{p}_N$ in (55)–(57) follow from the formulas (46)–(51) and the corresponding meanings attached to $\mathbf{b}^*$ in Section 2 so that (55) represent the equations of motion for shell-like bodies in terms of stress resultants derived here from the three-dimensional equations. Similarly, (57) is the moment of momentum equation, and (56) are equations of entropy balance. In this form the equations represent those for the theory of shells based on Cosserat surfaces.

Once constitutive equations have been assumed for

$$\mathbf{M}_N^\alpha, \mathbf{k}_N, \mathbf{p}_N, \psi, \tilde{\eta}_N, \xi_N, \tag{58}$$

the remaining reduced energy equation (53) becomes an identity for all velocity and temperature fields. This places restrictions on the constitutive equations. The constitutive equations may be derived from the corresponding equations for

$$\mathbf{T}, \mathbf{p}^*, \psi^*, \eta^*, \xi^*, \tag{59}$$

and the formulas (46)–(52). Apart from the entropy balance equations, the equations derived here are similar to those given by Green and Naghdi (Refs. 12, 13). In the earlier papers instead of entropy balances, a series of energy equations were derived from the three-dimensional theory, which were then supplemented by entropy inequalities. Entropy balances were introduced in later papers of Green and Naghdi (Refs. 19–21) for a direct development of a theory of shells or a theory of water waves for finite depths. To complete the discussion, jump conditions are required across surfaces of discontinuity and these may be obtained from (21) and (23) with the help of (45)–(52), but are omitted here.

The foregoing theory, which is derived here from the exact three-dimensional equations, can be put in one-to-one correspondence with the corresponding development by direct approach. It is now in a form which can be used for special classes of continua, particularly those which have a reference configuration, such as elastic shells. Equations for an elastic shell-like body in terms of the specific Helmholtz free energy function have been obtained previously (Refs. 13, 15). In this general form the equations were still difficult to use for applications to special problems so that methods of approximation were considered. For example, if the theory is limited to a surface and two or more directors, the resulting constitutive equations which are then expressed in terms of the elastic constants of the linear three-dimensional theory seem to yield a satisfactory approximation for most applications. Motivated by simplicity, however, most approximate theories are for a surface with one director corresponding to $\lambda_1 = z$, $\lambda_N = 0$, $N > 2$, an approximate theory being particularly appropriate for thin elastic shell-like bodies. However, in the context of linear elasticity, it was found that the natural form of the theory in terms of the Helmholtz free energy function was unsatisfactory since, for the very simple static problem of flexure of a thin plate, it yielded unsatisfactory results compared with the known exact solutions in the three-dimensional theory. As this was one of the basic problems in the applications, it is evident that such an approximation was unacceptable. Of course, whatever the approximate theory, it cannot make satisfactory predictions for all problems. An alternative three-dimensional approximation for elastic materials using the specific Gibbs free energy function has been used, e.g., by Green, Naghdi, and Wenner (Ref. 22), Naghdi (Ref. 9), and Green and Naghdi (Ref. 15). If the calculations for an elastic shell are completed with this function a set of constitutive equations is obtained for the approximate theory consisting of a surface and single director which, in the linearized case, provides satisfactory solutions for most of the simple problems, including the flexure of a plate. It may not, however, provide good approximate results for some dynamical problems.

Partly for the foregoing reasons, the present authors have preferred to adopt the direct approach to shell theory, using a finite number of directors, and regarding it as a model for the complete theory, so that more flexibility is possible in the choice of constitutive equations. The basic dynamical equations, however, are supported by the corresponding (approximate) equations in the three-dimensional theory, whatever type of continua is under discussion.

The above theory for shells and plates has been presented in Lagrangian form, but for certain applications involving fluid media it is more convenient to have a Eulerian form as developed by Green and Naghdi (Refs. 20, 21). The difficulties arise when some terms in the basic equations involve time derivatives. As in the paragraph after (40), choose the coordinate θ^i so that at time t the θ^i-curves coincide with the first system ζ^i and the moving surface given by (34) coincides with the fixed surface (37) at time t. Then, the time derivative terms in (7) become

$$-L = \frac{\partial}{\partial t}\int_{\mathcal{P}^*}\left(\frac{1}{2}\,\mathbf{v}^*\cdot\mathbf{v}^* + \psi^* + \eta^*\theta^*\right)\rho^* dv$$

$$= \frac{\partial}{\partial t}\int_{\bar{\mathcal{P}}^*}\left(\frac{1}{2}\,\mathbf{v}^*\cdot\mathbf{v}^* + \psi^* + \eta^*\theta^*\right)\rho^* dv$$

$$+ \int_{\partial\bar{\mathcal{P}}^*}\left(\frac{1}{2}\,\mathbf{v}^*\cdot\mathbf{v}^* + \psi^* + \eta^*\theta^*\right)\mathbf{v}^*\cdot\mathbf{n}da, \tag{60}$$

where $\bar{\mathcal{P}}^*$ is a fixed curve bounded by a fixed closed surface $\partial\bar{\mathcal{P}}^*$ such that $\mathcal{P}^*$ coincides with $\bar{\mathcal{P}}^*$ at time t. The rest of the development follows lines similar to that used for the Lagrangian form of the theory earlier in this section, but now in relation to the fixed surface (37). This manner of obtaining the Eulerian forms of the theory is analogous to the procedure used in a three-dimensional context by Oldroyd (Ref. 23).

4. Thermomechanical Theory of Rods

We again employ a system of convected curvilinear coordinates θ^i, $i = 1, 2, 3$, in the reference configuration of the body and for convenience set $\theta^3 = \theta$. The general formulas summarized in (30) and (31) remain valid here, but instead of the description (32) the body is now assumed to be rod-like along the θ-direction and bounded by a lateral surface

$$H(\theta^1, \theta^2, \theta) = 0. \tag{61}$$

In the region bounded by the surface (61), the position vectors **X** and **x** and the temperature θ^* are now represented by

$$\mathbf{X} = \sum_{N=0}^{\infty} \lambda_N(\theta^1, \theta^2)\mathbf{D}_N, \qquad \lambda_0 = 1, \qquad \mathbf{D}_0 = \mathbf{R}, \qquad \mathbf{D}_N = \mathbf{D}_N(\theta), \tag{62a}$$

$$\mathbf{x} = \sum_{N=0}^{\infty} \lambda_N(\theta^1, \theta^2)\mathbf{d}_N, \qquad \mathbf{d}_0 = \mathbf{r}, \qquad \mathbf{d}_N = \mathbf{d}_N(\theta, t), \tag{62b}$$

$$\theta^* = \sum_{N=0}^{\infty} \mu_N(\theta^1, \theta^2)\theta_N, \qquad \theta_N = \theta_N(\theta, t), \qquad \mu_0 = 1. \tag{62c}$$

Although some of the symbols are the same as those used in Section 3, the meanings are different and the notations are distinct apart from (30), (31), and (36). The position vector **r** in (62b, center) may be regarded as representing points on a one-dimensional curve in the current configuration of the body at time t, with tangent vector defined by

$$\mathbf{a}_3 = \partial\mathbf{r}/\partial\theta, \qquad a_{33} = \mathbf{a}_3 \cdot \mathbf{a}_3. \tag{63}$$

At any point on this curve and with the help of (30), the base vectors and metric tensors in the configuration at time t take the values

$$\mathbf{g}_i = \mathbf{a}_i(\theta, t), \qquad \mathbf{g}^i = \mathbf{a}^i(\theta, t), \qquad g_{ij} = a_{ij}, \tag{64a}$$

$$g^{ij} = a^{ij}, \qquad g = a. \tag{64b}$$

Corresponding quantities in the reference configuration of the body are denoted by capital letters, e.g., **R**, $\mathbf{A}_3$, $\mathbf{G}_i = \mathbf{A}_i(\theta)$, etc. The particle velocity $\mathbf{v}^*$ is then given by

$$\mathbf{v}^* = \sum_{N=0}^{\infty} \lambda_N(\theta^1, \theta^2)\mathbf{w}_N, \qquad \mathbf{v} = \mathbf{w}_0, \qquad \mathbf{w}_N = \mathbf{w}_N(\theta, t) = \dot{\mathbf{d}}_N. \tag{65}$$

Select a fixed system of curvilinear coordinates ζ^i and corresponding base vectors, etc., as defined in (36), also a fixed curve which, with its corresponding tangent vector, is given by

$$\bar{\mathbf{r}} = \bar{\mathbf{r}}(\zeta), \qquad \zeta^3 = \zeta, \qquad \bar{\mathbf{a}}_3 = \partial\bar{\mathbf{r}}/\partial\zeta, \qquad \bar{a}_{33} = \bar{\mathbf{a}}_3 \cdot \bar{\mathbf{a}}_3. \tag{66}$$

At points on this curve the base vectors and metric tensors in (36) assume the values

$$\bar{\mathbf{g}}_i = \bar{\mathbf{a}}_i(\zeta), \qquad \bar{\mathbf{g}}^i = \bar{\mathbf{a}}^i(\zeta), \qquad \bar{g}_{ij} = \bar{a}_{ij}, \tag{67a}$$

$$\bar{g}^{ij} = \bar{a}^{ij}, \qquad \bar{g} = \bar{a}. \tag{67b}$$

In terms of the fixed coordinates ζ^i, the velocity $\mathbf{v}^*$ and temperature θ^* at time t may be represented by

$$\mathbf{v}^* = \bar{\mathbf{v}}^*(\zeta, t) = \bar{v}^{*i}\bar{\mathbf{g}}_i = \sum_{N=0}^{\infty} \bar{\lambda}_N(\zeta^1, \zeta^2)\bar{\mathbf{w}}_N, \tag{68a}$$

$$\theta^* = \sum_{N=0}^{\infty} \bar{\mu}_N(\zeta^1, \zeta^2)\theta_N, \qquad \bar{\mathbf{w}}_N = \bar{\mathbf{w}}_N(\zeta, t), \tag{68b}$$

$$\bar{\theta}_N = \dot{\bar{\theta}}_N(\zeta, t), \qquad \bar{\mathbf{w}}_0 = \bar{\mathbf{v}}_0. \tag{68c}$$

The lateral surface (61) which bounds the body is now specified by

$$\bar{H}(\zeta^1, \zeta^2, \zeta, t) = 0. \tag{69}$$

This is a material surface which moves with the body so that, at the surface (69),

$$\frac{\partial \bar{H}}{\partial t} + \bar{v}^{*i}\frac{\partial \bar{H}}{\partial \zeta^i} = 0. \tag{70}$$

Later the convected coordinates θ^i are chosen so that at time t, the θ^i-curves coincide with the fixed ζ^i-curves and the moving curve (63) coincides with the fixed curve (66).

As in Section 3, our starting point is again the energy balance (17) along with (26)–(28). With the use of the Lagrangian form of representations (62) and (65), substitute into (17) and (26) and perform an integration along the curve between $\bar{\theta}_1$ and $\bar{\theta}_2$. This follows a similar procedure to that used by Green and Naghdi (Refs. 12, 13); and in the present context, with a different notation, yields

$$\bar{E} = -\frac{d}{dt}\int_{\bar{\theta}_1}^{\bar{\theta}_2} \left\{\frac{1}{2}\sum_{M=0}^{\infty}\sum_{N=0}^{\infty} y_{MN}\mathbf{w}_M\cdot\mathbf{w}_N + \psi + \tilde{\eta}_N\theta_N\right\}\rho ds$$

$$+ \int_{\zeta_1}^{\zeta_2} \sum_{N=0}^{\infty}(s_N\theta_N + \boldsymbol{\ell}_N\cdot\mathbf{w}_N)\rho ds + \left[\sum_{N=0}^{\infty}(\mathbf{m}_N\cdot\mathbf{w}_N - k_N\theta_N)\right]_{\zeta_1}^{\zeta_2}$$

$$= \int_{\bar{\theta}_1}^{\bar{\theta}_2} \left[\sum_{N=0}^{\infty} \left(\mathbf{k}_N \cdot \mathbf{w}_N + \mathbf{m}_N \cdot \frac{\partial \mathbf{w}_N}{\partial \theta} - k_N \frac{\partial \theta_N}{\partial \theta} \right) \right.$$

$$\left. - \rho \left\{ \dot{\psi} + \sum_{N=0}^{\infty} (\tilde{\eta}_N \dot{\theta}_N + \xi_N \theta_N) \right\} \right] ds = 0, \tag{71}$$

where $ds = a_{33}^{1/2}\, ds$ and $\bar{\theta}_1 < \theta < \bar{\theta}_2$ is an arbitrary part of the curve (63). The functions in (71) are defined by

$$\lambda = \rho a_{33}^{1/2} = \iint \rho^* g^{1/2} dA, \qquad dA = d\theta^1 d\theta^2, \tag{72a}$$

$$\lambda y_{MN} = \iint \rho^* g^{1/2} \lambda_M \lambda_N dA, \tag{72b}$$

$$\mathbf{m}_N = \iint \mathbf{T}^3 \lambda_N dA, \tag{73a}$$

$$\mathbf{k}_N = \iint \mathbf{T}^\beta \frac{\partial \lambda_M}{\partial \zeta^\beta} dA, \tag{73b}$$

$$\lambda \ell_N = \iint \rho^* g^{1/2} \mathbf{b}^* \lambda_N dA + \oint \lambda_N \{ (\mathbf{T}^1 - \bar{\nu}^1 \mathbf{T}^3) d\theta^2 - (\mathbf{T}^2 - \bar{\nu}^2 \mathbf{T}^3) d\dot{\theta} \}. \tag{74}$$

The line integral in (74) is along the curve where ζ = const intersects (69), $\bar{\nu}^\alpha = \bar{\nu} \cdot \mathbf{g}^\alpha$,

$$\bar{\nu} = \bar{\nu}^\alpha \mathbf{g}_\alpha + \mathbf{g}_3 \tag{75}$$

is a vector tangential to (69), and

$$\lambda \tilde{\eta}_N = \iint \rho^* g^{1/2} \eta^* \mu_N dA, \tag{76}$$

$$\lambda s_N = \iint \rho^* g^{1/2} s^* \mu_N dA - \oint \mu_N \{ (P' - \bar{\nu}^1 \mathbf{P}^3) d\theta^2 - (P^2 - \bar{\nu}^2 P^3) d\theta^1 \}, \tag{77}$$

$$\lambda \xi_N = \iint \rho^* g^{1/2} \xi^* \mu_N dA + \iint P^\alpha \frac{\partial \mu_N}{\partial \theta^\alpha} dA, \tag{78}$$

$$k_N = \iint P^3 \mu_N \, dA, \tag{79}$$

$$\lambda\psi = \iint \rho^* g^{1/2} \psi^* dA. \tag{80}$$

From (71) it follows that

$$-\rho\left\{\dot{\psi} + \sum_{N=0}^{\infty} (\tilde{\eta}_N \dot{\theta}_N + \xi_N \theta_N)\right\}$$

$$+ \sum_{N=0}^{\infty} (\mathbf{k}_N \cdot \mathbf{w}_N + \mathbf{m}_N \cdot \frac{\partial \mathbf{w}_N}{\partial \theta} - k_N \frac{\partial \theta_N}{\partial \theta}) = 0. \tag{81}$$

With the help of the definitions (72)–(80) the following equations may be obtained from (22), (23), and (24) in which $m = 0$, $\mathbf{f}^* = 0$, namely,

$$\dot{\rho} + \rho \operatorname{div}_s \mathbf{v} = \dot{\rho} + \rho \mathbf{a}^3 \cdot \partial \mathbf{v}/\partial \theta = 0 \quad \text{or} \quad \rho a_{33}^{1/2} = \rho_R A_{33}^{1/2}, \tag{82}$$

$$\lambda \sum_{M=0}^{\infty} y_{MN} \dot{\mathbf{w}}_M = \lambda \boldsymbol{\ell}_N - \mathbf{k}_N - \partial \mathbf{m}_N/\partial \theta, \tag{83}$$

$$\lambda \dot{\eta}_N = \lambda(s_N + \xi_N) - \partial k_N/\partial \theta, \qquad N = 0, 1, 2, \ldots. \tag{84}$$

Also, from (28), (62), and (73) it is seen that

$$\sum_{N=0}^{\infty} (\mathbf{d}_N \times k_N + \frac{\partial \mathbf{d}_N}{\partial \theta} \times \mathbf{m}_N) = 0. \tag{85}$$

The meanings of the various functions in (82)–(85) follow from the formulas (72)–(80) and the corresponding definitions given to $\mathbf{b}^*$ in Section 2, so that equations (82)–(85) represent the equations of motion in terms of stress resultants, entropy balance, and moment of momentum for rodlike bodies derived from the three-dimensional equations. In this form the equations represent those of the theory of rods based on Cosserat curves. Once constitutive equations have been assumed for

$$\mathbf{m}_N, \mathbf{k}_N, \mathbf{p}_N, \psi, \tilde{\eta}_N, \xi_N, \tag{86}$$

the remaining reduced energy equation (81) becomes an identity for all velocity and temperature fields. This places restrictions on the constitutive equations. The constitutive equations may be derived from the corresponding equations for the variables (59) and the formulas (72)–(80). Apart from the entropy balance, the equations derived here are similar to those given by Green and Naghdi (Refs. 12, 13) but with a different notation. In the earlier papers instead of entropy balances a series of energy equations were derived from the three-dimensional theory, which was then supplemented by entropy inequalities. Entropy balances were used in later papers concerned with a direct development of a theory of rods (Refs. 16, 24).

At the end of Section 3 the application of the theory of shells to specific bodies, the relation with direct theories, and methods of approximation were discussed particularly in relation to a theory with a single director. A parallel discussion may be carried out for a theory of rods. In this theory the special case of a rod modeled by a curve and two directors is of particular interest and, for elasticity, has the problem that approximation from the Helmholtz function yields equations which are unsatisfactory for flexure problems. Then it seems to be more appropriate to use the Gibbs function (see, e.g., Refs. 16, 17). A complete discussion is omitted here since it can be effected in parallel to that contained in Section 3. Finally, for similar reasons to those explained for shells, the present authors have found it more satisfactory to use a direct theory of rods.

The Eulerian form for the inertia terms may be found in a way which is similar to that used for shells. Choose the coordinate system θ^i so that at time t, the θ^i-curves coincide with the fixed system ζ^i and the moving curve given by (63) coincides with the fixed curve (66) at time t. The inertia terms are then obtained from the second form in (60). Details of this are omitted since they will be given in a different context elsewhere.

Acknowledgment

The work of the second author was supported by the U.S. Office of Naval Research under Contract N0014-84-K-0264, Work Unit 4324-436 with the University of California, Berkeley.

References

1. Green, A. E., and Naghdi, P. M., *On Thermodynamics and the Nature of the Second Law*, Proceedings of the Royal Society of London, Vol. A357, pp. 253–270, 1977.
2. Green, A. E., and Naghdi, P. M., *The Second Law of Thermodynamics and Cyclic Processes*, Journal of Applied Mechanics, Vol. 45, pp. 487–492, 1978.
3. Green, A. E., and Naghdi, P. M., *Aspects of the Second Law of Thermodynamics in the Presence of Electromagnetic Effects*, Quarterly Journal of Mechanics and Applied Mathematics, Vol. 37, pp. 179–193, 1984.

4. GREEN, A. E., and NAGHDI, P. M., *A Demonstration of Consistency of an Entropy Balance with Balance of Energy*, Journal of Applied Mathematics and Physics, Vol. 42, pp. 159–168, 1991.
5. GREEN, A. E., and RIVLIN, R. S., *Simple Force and Stress Multipoles*, Archive for Rational Mechanics and Analysis, Vol. 16, pp. 325–353, 1964.
6. GREEN, A. E., and RIVLIN, R. S., *Multipolar Continuum Mechanics*, Archive for Rational Mechanics and Analysis, Vol. 17, pp. 113–147, 1964.
7. GREEN, A. E., and RIVLIN, R. S., *On Cauchy's Equations of Motion*, Zeitschrift fur Angewandte Mathematik und Physik, Vol. 15, pp. 290–292, 1964.
8. GREEN, A. E., NAGHDI, P. M., and RIVLIN, R. S., *Directors and Multipolar Displacements in Continuum Mechanics*, International Journal of Engineering Science, Vol. 2, pp. 611–620, 1965.
9. NAGHDI, P. M., *The Theory of Shells and Plates*, Handbuch der Physik, Springer, Berlin, Germany, Vol. VIa/2, pp. 425–640, 1972.
10. NAGHDI, P. M., *Finite Deformation of Elastic Rods and Shells*, Proceedings of the IUTAM Symposium on Finite Elasticity, Edited by D. E. Carlson and R. T. Shield, Martinus Nijhoff Publishers, The Hague, The Netherlands, pp. 47–103, 1982.
11. GREEN, A. E., LAWS, N., and NAGHDI, P. M., *A Linear Theory of Straight Elastic Rods*, Archive for Rational Mechanics and Analysis, Vol. 25, pp. 285–298, 1967.
12. GREEN, A. E., LAWS, N., and NAGHDI, P. M., *Rods, Plates and Shells*, Proceedings of the Cambridge Philosophical Society, Vol. 64, pp. 895–913, 1968.
13. GREEN, A. E., and NAGHDI, P. M., *Non-Isothermal Theory of Rods, Plates and Shells*, International Journal of Solids and Structures, Vol. 6, pp. 209–244, 1970.
14. ANTMAN, S., and WARNER, W. H., *Dynamical Theory of Hyperelastic Rods*, Archive for Rational Mechanics and Analysis, Vol. 23, pp. 135–162, 1966.
15. GREEN, A. E., and NAGHDI, P. M., *On Electromagnetic Effects in the Theory of Shells and Plates*, Philosophical Transactions of the Royal Society of London, Vol. A309, pp. 559–610, 1983.
16. GREEN, A. E., and NAGHDI, P. M., *Electromagnetic Effects in the Theory of Rods*, Philosophical Transactions of the Royal Society of London, Vol. A314, pp. 311–352, 1985.
17. GREEN, A. E., NAGHDI, P. M., and WENNER, M. L., *On the Theory of Rods: I. Derivations from the Three-Dimensional Equations*, Proceedings of the Royal Society of London, Vol. A337, pp. 451–483, 1974.
18. GREEN, A. E., and NAGHDI, P. M., *On Nonlocal Continuum Mechanics*, Mathematical Proceedings of the Cambridge Philosophical Society, Vol. 83, pp. 307–319, 1978.
19. GREEN, A. E., and NAGHDI, P. M., *On Thermal Effects in the Theory of Shells*, Proceedings of the Royal Society of London, Vol. A365, pp. 161–190, 1979.
20. GREEN, A. E., and NAGHDI, P. M., *A Nonlinear Theory of Water Waves for Finite and Infinite Depths*, Philosophical Transactions of the Royal Society of London, Vol. A320, pp. 37–70, 1986.
21. GREEN, A. E., and NAGHDI, P. M., *Further Developments in a Nonlinear Theory of Water Waves for Finite and Infinite Depths*, Philosophical Transactions of the Royal Society of London, Vol. A324, pp. 47–72, 1987.
22. GREEN, A. E., NAGHDI, P. M., and WENNER, M. L., *Linear Theory of Cosserat Surface and Elastic Plates of Variable Thickness*, Proceedings of the Cambridge Philosophical Society, Vol. 69, pp. 227–254, 1971.

23. Oldroyd, J. E., *On the Formulation of Rheological Equations of State*, Proceedings of the Royal Society of London, Vol. A200, pp. 523–541, 1950.
24. Green, A. E., and Naghdi, P. M., *On Thermal Effects in the Theory of Rods*, International Journal of Solids and Structures, Vol. 14, pp. 829–853, 1979.

Supplementary Reference

25. Green, A. E., and Naghdi, P. M., *On the Derivation of Discontinuity Conditions in Continuum Mechanics*, International Journal of Engineering Science, Vol. 2, pp. 621–624, 1965.

8

Steady Flow of Slightly Viscoelastic Fluids[1]

W. E. Langlois

Abstract. A study is presented of the flow of fluids for which the stress-deformation relation departs only slightly from that of a Newtonian fluid. Incompressibility is assumed throughout and the analysis is limited to flows for which the nonlinear inertia terms in the equations of motion may be completely ignored. For flows of very viscous fluids, a method is presented whereby the effect of a small departure from the Newtonian constitutive equation can be examined. The fluid is assumed to be in a state of steady flow with fixed sources or sinks present within the volume occupied by the fluid or on fixed portions of its boundaries which are assumed to consist solely of rigid walls at rest or in steady motion parallel to themselves. General expressions are obtained for axisymmetric flows. Application is made to steady convergent flow in a right-circular cone.

Key Words. Viscoelastic fluid, kinematic matrices, incompressible, secondary flow, stream function, convergent flow, hypergeometric equation.

Preface. It was my good fortune to write my doctoral dissertation under Professor Rivlin's guidance. Quite apart from the subject matter itself, I learned much about effective ways to carry out and report theoretical research. This often proved useful in areas quite removed from viscoelastic flow. The thesis was one of many documents on nonlinear continuum mechanics to issue from the Division of Applied Mathematics at Brown during the late 1950's—a direct conse-

[1]This paper is a condensed version of Ref. 1.

W. E. Langlois • Staff Member, IBM Almaden Research Laboratory, San Jose, California 95120.

Nonlinear Effects in Fluids and Solids, edited by M. M. Carroll and M. Hayes, Plenum Press, New York, 1996.

quence of Professor Rivlin's presence and influence. It was his original intent to publish at least one paper based on the thesis. However, those were busy times, and the paper was never written. Professor Rivlin often expressed malaise with that situation, to me and to others. In spite of repeated disclaimers to the effect "Ronald, I've published N papers, N + 1 wouldn't make any difference," he somehow felt he had done me a disservice. Publication of this volume gives me the opportunity to put the matter to rest. Although the results are no longer new, the work has archival value, and this seems an appropriate medium.

1. Introduction

Rivlin and Ericksen (Ref. 2) have considered a viscoelastic fluid which is isotropic in its state of rest and for which the stress components t_{ij}, $i, j,$ = 1, 2, 3, in a Cartesian coordinate system, at a point x_i of the fluid, are expressible as polynomials in the gradients of velocity, acceleration, second acceleration, . . . , $(R - 1)$th acceleration. They have shown that the stress matrix $\| t_{ij} \|$ can be expressed as a matrix polynomial in R kinematic matrices $\| A_{ij}^{(\gamma)} \|$, $\gamma = 1, 2, \ldots, R$, with coefficients, which are scalar invariants of these R matrices, expressible as polynomials in the elements of the matrices. The R kinematic matrices $\| A_{ij}^{(\gamma)} \|$ are defined by

$$A_{ij}^{(1)} = \frac{\partial v_i}{\partial x_j} + \frac{\partial v_j}{\partial x_i}, \qquad i, j = 1, 2, 3, \tag{1}$$

$$A_{ij}^{(\gamma+1)} = \frac{\partial A_{ij}^{(\gamma)}}{\partial t} + v_p \frac{\partial A_{ij}^{(\gamma)}}{\partial x_p} + A_{ip}^{(\gamma)} \frac{\partial v_p}{\partial x_j} + A_{pj}^{(\gamma)} \frac{\partial v_p}{\partial x_i},$$
$$i, j = 1, 2, 3 \quad \text{and} \quad \gamma = 1, 2, \ldots, R = 1, \tag{2}$$

where v_i, i = 1, 2, 3, are the velocity components in the x_i coordinate system and t is time.

In the case when the flow is two-dimensional, so that

$$v_1 = v_1(x_1, x_2, t), \qquad v_2 = v_2(x_1, x_2, t), \qquad v_3 = 0, \tag{3}$$

it has been pointed out by Rivlin (Ref. 3) that the two-dimensional stress matrix $\overline{\mathbf{T}} = \| t_{ij} \|$, $i, j,$ = 1, 2, can be expressed in terms of R two-dimensional kinematic matrices $\overline{\mathbf{A}}_\gamma = \| A_{ij}^{(\gamma)} \|$, i, j = 1, 2 and $\gamma = 1, 2, \ldots, R$, by an expression of the form

$$\overline{\mathbf{T}} = \bar{\alpha}_0 \overline{\mathbf{I}} + \sum_{\gamma=1}^{R} \bar{\alpha}_\gamma \overline{\mathbf{A}}_\gamma, \tag{4}$$

while

$$t_{31} = t_{32} = 0. \tag{5}$$

In equation (4), $\bar{\mathbf{I}}$ is the 2×2 unit matrix and $\bar{\alpha}_0, \bar{\alpha}_1, \ldots, \bar{\alpha}_R$ are polynomials in the scalar invariants tr $\bar{\mathbf{A}}_\gamma$, $\gamma = 1, 2, \ldots, R$, and tr $\bar{\mathbf{A}}_\gamma \bar{A}_\delta$, $\gamma, \delta = 1, 2, \ldots, R$. If the fluid is incompressible, it is possible to take $\bar{\alpha}_0 = -\bar{p}$, where $\bar{p}$ is arbitrary, so that equation (4) becomes

$$\bar{\mathbf{T}} = -p\,\bar{\mathbf{I}} + \sum_{\gamma=1}^{R} \bar{\alpha}_\gamma \bar{\mathbf{A}}_\gamma. \tag{6}$$

In the same paper, Rivlin indicated that if the three-dimensional stress matrix $\| t_{ij} \|$ can be expressed as a matrix polynomial in two kinematic matrices, say $\| A_{ij}^{(\mathrm{M})} \| = \mathbf{A}_\mathrm{M}$ and $\| A_{ij}^{(\mathrm{N})} \| = \mathbf{A}_\mathrm{N}$, then $\| t_{ij} \|$ can be expressed by a relation of the form

$$\begin{aligned} \| t_{ij} \| = {} & \alpha_0 \mathbf{I} + \alpha_1 \mathbf{A}_\mathrm{M} + \alpha_2 \mathbf{A}_\mathrm{N} + \alpha_3 \mathbf{A}_\mathrm{M}^2 + \alpha_4 \mathbf{A}_\mathrm{N}^2 \\ & + \alpha_5 (\mathbf{A}_\mathrm{M}\mathbf{A}_\mathrm{N} + \mathbf{A}_\mathrm{N}\mathbf{A}_\mathrm{M}) + \alpha_6 (\mathbf{A}_\mathrm{M}^2 \mathbf{A}_\mathrm{N} + \mathbf{A}_\mathrm{N} \mathbf{A}_\mathrm{M}^2) \\ & + \alpha_7 (\mathbf{A}_\mathrm{N}^2 \mathbf{A}_\mathrm{M} + \mathbf{A}_\mathrm{M} \mathbf{A}_\mathrm{N}^2) + \alpha_8 (\mathbf{A}_\mathrm{M}^2 \mathbf{A}_\mathrm{N}^2 + \mathbf{A}_\mathrm{N}^2 \mathbf{A}_\mathrm{M}^2), \end{aligned} \tag{7}$$

where $\mathbf{I}$ is the 3×3 unit matrix and where $\alpha_0, \alpha_1, \ldots, \alpha_8$ are expressible as polynomials in tr $\mathbf{A}_\mathrm{M}$, tr $\mathbf{A}_\mathrm{N}$, tr $\mathbf{A}_\mathrm{M}^2$, tr $\mathbf{A}_\mathrm{N}^2$, tr $\mathbf{A}_\mathrm{M}^3$, tr $\mathbf{A}_\mathrm{N}^3$, tr $\mathbf{A}_\mathrm{M}\mathbf{A}_\mathrm{N}$, tr $\mathbf{A}_\mathrm{M}^2\mathbf{A}_\mathrm{N}$, tr $\mathbf{A}_\mathrm{M}\mathbf{A}_\mathrm{N}^2$, and tr $\mathbf{A}_\mathrm{M}^2\mathbf{A}_\mathrm{N}^2$. If the fluid is incompressible, then α_0 may be replaced by an arbitrary quantity $-p$.

Green and Rivlin (Ref. 4) have considered incompressible fluids for which the stress components are expressible as polynomials in the velocity gradients only. For these fluids, $\| t_{ij} \|$ can be expressed in the form

$$\| t_{ij} \| = -p\mathbf{I} + \alpha \mathbf{A}_1 + \beta \mathbf{A}_1^2, \tag{8}$$

where α and β are polynomials in tr $\mathbf{A}_1^2$ and tr $\mathbf{A}_1^3$ (tr $\mathbf{A}_1$ is zero since the fluid is incompressible) and where p is arbitrary. They have studied flows of these fluids in straight tubes of various cross sections. Ericksen (Ref. 5) had shown earlier that rectilinear flow in such pipes can be maintained without body forces only under special circumstances. In agreement with Ericksen, Green and Rivlin found that it can be so maintained if the tube is circular or if $\beta = K\alpha$, where K is a constant.

For a tube of elliptical cross section and for a fluid such that

$$\| t_{ij} \| = -p\mathbf{I} + \mu \mathbf{A}_1 + \epsilon\,(c + b \operatorname{tr} \mathbf{A}_1^2)\mathbf{A}_1^2, \tag{9}$$

where μ, c, and b are constants and where ϵ is a small, dimensionless constant, they were able to calculate the flow with neglect only of terms of order ϵ^2. In addition to the flow down the tube, they found a secondary flow in the cross section. This flow consists of four separate vortices and is symmetric about each of the axes of the elliptical cross section.

In the present paper we shall study in further detail flows of fluids for which the stress-deformation relation departs only slightly from that of a Newtonian fluid. Incompressibility of the material will be assumed throughout the paper and the analysis will be limited to flows for which nonlinear inertia terms in the equations of motion may be completely neglected.

In Section 2 we shall prove a theorem which will provide us with a method of studying the effect of a small departure of the stress-deformation relation from that of a Newtonian fluid upon flows of a very viscous fluid. It will be shown that if the fluid described above is in a state of steady flow with fixed sources or sinks present within the volume occupied by the fluid or on fixed portions of its boundaries, which are assumed to consist solely of rigid walls at rest or in steady motion parallel to themselves, then an approximate solution to the governing equations and boundary conditions can be found by the following procedure:

Step 1. A solution (v_i', p') to the equations of motion for an incompressible Newtonian fluid is calculated neglecting inertia subject to the boundary conditions of the situation;

Step 2. The solution (v_i', p') calculated in Step 1 is introduced into the equations of motion for the incompressible viscoelastic fluid described above and the additional body forces required to support this solution are calculated neglecting inertia;

Step 3. A solution (v_i'', p'') to the equations of motion for an incompressible Newtonian fluid under the action of the body forces calculated in Step 2 is calculated neglecting inertia, subject to the boundary conditions of vanishing velocity at the walls, of zero efflux from the sources, and of zero flux across surfaces across which the volume flow is prescribed;

Step 4. Then $(v_i = v_i' - v_i'', p = p' - p'')$ is an approximate solution to the equations of very slow motion for an incompressible slightly viscoelastic fluid subject to the same boundary conditions as used in Step 1. The error involved is of the second order in a dimensionless parameter which measures the departure of the stress-deformation relation for the material from that of a Newtonian fluid.

The remainder of the paper will be concerned with the application of this theorem to specific flow problems.

2. Theorem Concerning the Flow of Slightly Viscoelastic Fluids

Consider a mass of an incompressible, isotropic, slightly viscoelastic fluid, for which the stress components t_{ij}, $i, j = 1, 2, 3$, in a Cartesian coordinate system x_i, at a point x_i of the fluid, are expressible as polynomials in the gradients of velocity, acceleration, second acceleration, . . . , $(R - 1)$th acceleration, to be in a state of steady flow.

Let us consider a domain D, fixed in space, which may be simply or multiply connected. Let us assume that on portions $\Sigma_1, \Sigma_2, \ldots, \Sigma_N$ of the boundaries of the domain D, the velocity v_i of the fluid is specified, while over the remaining portions $S_1, S_2, \ldots, S_M$ the resultant fluxes of fluid $F_1, F_2, \ldots, F_M$ are specified. We then have, on $\Sigma_1, \Sigma_2, \ldots, \Sigma_N$,

$$v_i = V_i, \qquad i = 1, 2, 3, \tag{10}$$

where V_i, $i = 1, 2, 3$, are specified functions of position and, on $S_1, S_2, \ldots, S_M$,

$$\iint\limits_{S_\alpha} v_i n_i dS_\alpha = F_\alpha, \qquad \alpha = 1, 2, \ldots, M \tag{11}$$

where n_i are the components, in the coordinate system x_i, of the outward-drawn unit vector normal to S_α, $\alpha = 1, 2, \ldots, M$.

We assume further that there are present within the domain D fixed point-sources $\sigma_1, \sigma_2, \ldots, \sigma_P$ having constant fluxes $Q_1, Q_2, \ldots, Q_P$, respectively. Then

$$\iint\limits_{A_\gamma} v_i n_i dA_\gamma = Q_\gamma, \qquad \gamma = 1, 2, \ldots, P \tag{12}$$

where A_γ is any surface enclosing σ_γ and no other sources.

We also consider the case when portions $S'_1, S'_2, \ldots, S'_K$ of $S_1, S_2, \ldots, S_M$ form fixed rigid boundaries of the fluid mass, and certain of the point-sources $\sigma_1, \sigma_2, \ldots, \sigma_P$ lie on $S'_1, S'_2, \ldots, S'_K$. We then have $v_i = 0$ on $S_1, S_2, \ldots, S'_M$. Also, if the point-source σ_γ lies on S'_β, then the relation (12) is still valid if A_γ denotes a surface such that $A_\gamma + S''_\beta$ is a closed surface, where S''_β is a portion of S'_β upon which σ_γ and no other sources lie and such that $A_\gamma + S''_\beta$ encloses no sources.

At each point of the fluid, the three equations of motion,

$$\frac{\partial t_{ij}}{\partial x_j} + \rho f_i = \rho v_P \frac{\partial v_i}{\partial x_P}, \qquad i = 1, 2, 3, \tag{13}$$

and the continuity equation,

$$\frac{\partial v_i}{\partial x_i} = 0, \tag{14}$$

must be satisfied. In equations (13), f_i, $i = 1, 2, 3$, are the components in the coordinate system x_i of body force per unit mass of fluid and ρ is the (constant) density of the fluid.

If the flow is such that the nonlinear inertia terms may be neglected, equations (13) become

$$\frac{\partial t_{ij}}{\partial x_j} + \rho f_i = 0\,, \qquad i = 1, 2, 3. \tag{15}$$

Since Rivlin and Ericksen (Ref. 2) have shown that the stress matrix $\| t_{ij} \|$ can be expressed as a matrix polynomial in the kinematic matrices $\| A_{ij}^{(\gamma)} \|$, defined by equations (1) and (2), with coefficients, which are scalar invariants of these matrices, expressible as polynomials in the elements of the matrices, and since we are considering fluids which are nearly Newtonian, the stress-deformation relation may be taken as

$$t_{ij} = -p\delta_{ij} + \mu A_{ij}^{(1)} + \epsilon P_{ij}, \qquad i, j = 1, 2, 3, \tag{16}$$

where ϵ is a small, dimensionless constant, p is arbitrary, μ is constant, and P_{ij} is a symmetric tensor polynomial in the $A_{ij}^{(\gamma)}$, $\gamma = 1, 2, \ldots, R$, with coefficients which are scalar invariants of the $A_{ij}^{(\gamma)}$, $\gamma = 1, 2, \ldots, R$, the coefficient of $A_{ij}^{(1)}$ vanishing when all of the invariants vanish, and with no term of degree zero. For a Newtonian fluid, $\epsilon = 0$.

Substituting from equations (16) into equations (15) and using the incompressibility condition (14), we obtain

$$\mu \frac{\partial^2 v_i}{\partial x_j \partial x_j} + \epsilon \frac{\partial P_{ij}}{\partial x_j} + \rho f_i = \frac{\partial p}{\partial x_i}, \qquad i = 1, 2, 3. \tag{17}$$

Let (v_i', p') be a solution to the equations of motion neglecting inertia for the steady flow of an incompressible, isotropic Newtonian fluid, with viscosity μ, subject to the boundary conditions (10)–(12). The equations of motion and incompressibility condition are then

$$\mu \frac{\partial^2 v_i'}{\partial x_j \partial x_j} + \rho f_i = \frac{\partial p'}{\partial x_i}, \qquad i = 1, 2, 3, \tag{18}$$

$$\frac{\partial v_i'}{\partial x_i} = 0. \tag{19}$$

The boundary conditions are

$$v_i' = V_i, \qquad \text{on } \Sigma_\beta, i = 1, 2, 3 \text{ and } \beta = 1, 2, \ldots, N, \tag{20}$$

$$\iint_{S_\alpha} v_i n_i dS_\alpha = F_\alpha, \qquad \text{on } S_\alpha, \alpha = 1, 2, \ldots, M, \tag{21}$$

$$\iint_{A_\gamma} v_i n_i dA_\gamma = Q_\gamma, \qquad \text{on } A_\gamma, \gamma = 1, 2, \ldots, P, \tag{22}$$

where $A\gamma$ is the surface surrounding the point-source σ_γ.

If the solution ($v_i = v_i'$, $p = p'$) is introduced into equations (17), expressions are found for the body forces f_i' which must be applied to a fluid which obeys the stress-deformation relation (16) in order to maintain the velocity field v_i' and the pressure field p'. The result is

$$f_i' = f_i - \frac{\epsilon}{\rho}\frac{\partial P_{ij}'}{\partial x_j}, \qquad i = 1, 2, 3, \tag{23}$$

where P_{ij}' is the expression calculated for P_{ij} when $v_i = v_i'$, $i = 1, 2, 3$.

Now, let ($\epsilon v_i''$, $\epsilon p''$) be a solution to the equations

$$\mu \frac{\partial^2}{\partial x_j \partial x_j}(\epsilon v_i'') + \rho(f_i' - f_i) = \frac{\partial}{\partial x_i}(\epsilon p''), \qquad i = 1, 2, 3, \tag{24}$$

$$\frac{\partial}{\partial x_i}(\epsilon v_i'') = 0, \tag{25}$$

subject to the boundary conditions

$$v_i'' = 0, \text{ on } \Sigma_\beta, i = 1, 2, 3 \text{ and } \beta = 1, 2, \ldots, N, \tag{26}$$

$$\iint_{S_\alpha} v_i'' n_i dS_\alpha = 0, \qquad \alpha = 1, 2, \ldots, M, \tag{27}$$

$$\iint_{A_\gamma} v_i'' n_i dA_\gamma = 0, \qquad \gamma = 1, 2, \ldots, P. \tag{28}$$

It can now be shown that ($v_i = v_i' - \epsilon v_i''$, $p = p' - \epsilon p''$) satisfies the equations of motion (17) with neglect only of terms of the second degree or higher in ϵ, and satisfies exactly the continuity equation (14) and the boundary conditions (10)–(12).

Introducing ($v_i = v'_i - \epsilon v''_i, p = p' - \epsilon p''$) into equations (17) yields

$$\mu \frac{\partial^2 v'_i}{\partial x_j \partial x_j} - \epsilon\mu \frac{\partial^2 v''_i}{\partial x_j \partial x_j} + \epsilon \frac{\partial P_{ij}}{\partial x_j} + \rho f_i = \frac{\partial p'}{\partial x_i} - \epsilon \frac{\partial p''}{\partial x_i}, \qquad i = 1, 2, 3. \tag{29}$$

However, (v'_i, p') satisfies equations (18), and (v''_i, p'') satisfies equations (24), so that equations (29) reduce to

$$\epsilon \frac{\partial P_{ij}}{\partial x_j} + \rho(f'_i - f_i) = 0, \qquad i = 1, 2, 3. \tag{30}$$

However, from the definitions of P_{ij}, P'_{ij}, and the kinematic tensors $A^{(\gamma)}_{ij}$, $\gamma = 1, 2, \ldots, R$, we have

$$P_{ij} = P'_{ij} + E_{ij}, \qquad i, j = 1, 2, 3, \tag{31}$$

where each of the quantities E_{ij}, $i, j = 1, 2, 3$, is of order ϵ. It is seen, from (23), that equations (30) are identically satisfied with neglect only of terms of the second degree or higher in ϵ.

It is clear that the velocity components v_i, $i = 1, 2, 3$, satisfy the continuity equation (14); indeed, with equations (19) and (25), we have

$$\frac{\partial v_i}{\partial x_j} = \frac{\partial}{\partial x_i}(v'_i - \epsilon v''_i) = \frac{\partial v'_i}{\partial x_i} - \epsilon \frac{\partial v''_i}{\partial x_i} = 0. \tag{32}$$

The boundary conditions (10)–(12) are satisfied, for on Σ_β we have, from equations (20) and (26),

$$v_i = v'_i - \epsilon v''_i = V_i, \qquad i = 1, 2, 3, \tag{33}$$

from equations (21) and (27),

$$\begin{aligned} \iint_{S_\alpha} v_i n_i \, dS_\alpha &= \iint_{S_\alpha} (v'_i - \epsilon v''_i) n_i \, dS_\alpha \\ &= \iint_{S_\alpha} v'_i n_i \, dS_\alpha - \epsilon \iint_{S_\alpha} v''_i n_i \, dS_\alpha \\ &= F_\alpha, \qquad \alpha = 1, 2, \ldots, M, \end{aligned} \tag{34}$$

and from equations (22) and (27),

$$\iint_{A_\gamma} v_i n_i \, dA_\gamma = \iint_{A_\gamma} (v_i' - \epsilon v_i'') n_i \, dA_\gamma$$

$$= \iint_{A_\gamma} v_i' n_i \, dA_\gamma - \epsilon \iint_{A_\gamma} v_i'' n_i \, dA_\gamma$$

$$= Q_\gamma, \qquad \gamma = 1, 2, \ldots, P. \tag{35}$$

We have then shown that flows of a fluid possessing the stress-deformation relation (16) subject to the boundary conditions (10)–(12) and to the incompressibility condition (14) can be constructed by the following procedure:

Step 1. A solution (v_i', p') to the equations of motion for an incompressible Newtonian fluid with coefficient of viscosity μ is calculated neglecting inertia subject to the boundary conditions (10)–(12);

Step 2. The solution (v_i', p') calculated in Step 1 is introduced into the equations of motion (17) and the additional body forces required to support this solution are calculated;

Step 3. A solution $(\epsilon v_i'', \epsilon p'')$ to the equations of motion for an incompressible Newtonian fluid with coefficient of viscosity μ under the action of the body forces calculated in Step 2 is calculated neglecting inertia subject to the boundary conditions of zero velocity on Σ_β, $\beta = 1, 2, \ldots, N$, zero flux across S_α, $\alpha = 1, 2, \ldots, M$, and zero efflux from the point-sources σ_γ, $\gamma = 1, 2, \ldots, P$;

Step 4. Then $(v_i = v_i' - \epsilon v_i'', p = p' - \epsilon p'')$ is an approximate solution to the equations of motion (17), the error involving terms of order ϵ^2, and satisfies exactly the boundary conditions (10)–(12) and the incompressibility condition (14).

It is clear that an analogous theorem can be proven for two-dimensional flows. In fact, the theorem proven above applies to two-dimensional flows without sources. For flows with sources, slight modification must be made, for two-dimensional point-sources are the equivalent of three-dimensional line-sources, which have not been considered explicitly in the above analysis.

A somewhat analogous theorem has been proven for the second order elasticity theory of isotropic materials by Rivlin (Ref. 6) and for that of anisotropic materials by Rivlin and Topakaglu (Ref. 7).

3. Some General Remarks on Flows with Axial Symmetry

Since the remainder of the paper will be concerned with a flow which is symmetric about an axis, it will be helpful to discuss some general aspects of flows of this type before considering the specific problem.

If a spherical polar coordinate system (r, θ, φ) is defined by

$$r = x^1 = \sqrt{\bar{x}_1^2 + \bar{x}_2^2 + \bar{x}_3^2}, \tag{36a}$$

$$\theta = x^2 = \arctan\left(\frac{\sqrt{\bar{x}_1^2 + \bar{x}_2^2}}{\bar{x}_3}\right), \tag{36b}$$

$$\varphi = x^3 = \arctan(\bar{x}_2/\bar{x}_1), \tag{36c}$$

where $(\bar{x}_1, \bar{x}_2, \bar{x}_3)$ is a Cartesian system, then the covariant metric tensor g_{ij} for this polar system, defined by

$$g_{ij} = \frac{\partial \bar{x}_k}{\partial x^i}\frac{\partial \bar{x}_k}{\partial x^j}, \qquad i, j = 1, 2, 3, \tag{37}$$

is given by

$$[G_{ij}] = \begin{bmatrix} 1 & 0 & 0 \\ 0 & r^2 & 0 \\ 0 & 0 & r^2 \sin^2\theta \end{bmatrix}, \tag{38}$$

whence the contravariant metric tensor g^{ij} is given by

$$[G_{ij}] = \begin{bmatrix} 1 & 0 & 0 \\ 0 & 1/r^2 & 0 \\ 0 & 0 & 1/(r^2 \sin^2\theta) \end{bmatrix}. \tag{39}$$

The Christoffel symbols of the first kind for this system, defined by

$$\Gamma_{i;jk} = \frac{1}{2}\left(\frac{\partial g_{ji}}{\partial x^k} + \frac{\partial g_{ik}}{\partial x^j} - \frac{\partial g_{jk}}{\partial x^i}\right), \qquad i, j, k = 1, 2, 3, \tag{40}$$

vanish except for

$$\Gamma_{1;22} = -r, \qquad \Gamma_{1;33} = -r\sin^2\theta, \tag{41a}$$

$$\Gamma_{2;12} = \Gamma_{2;21} = r, \qquad \Gamma_{2;33} = -r\sin^2\theta\cos\theta, \tag{41b}$$

$$\Gamma_{3;\,13} = \Gamma_{3;\,31} = r\sin^2\theta, \qquad \Gamma_{3;\,23} = \Gamma_{3;\,32} = r^2\sin\theta\cos\theta. \tag{41c}$$

The Christoffel symbols of the second kind are defined by

$$\Gamma^i_{jk} = g^{i\ell}\Gamma_{\ell;\,jk}, \qquad i, j, k = 1, 2, 3. \tag{42}$$

The nonvanishing components are

$$\Gamma^1_{22} = -r, \qquad \Gamma^1_{33} = -r\sin^2\theta, \tag{43a}$$

$$\Gamma^2_{12} = \Gamma^2_{21} = \frac{1}{r}, \qquad \Gamma^2_{33} = -\sin\theta, \cos\theta, \tag{43b}$$

$$\Gamma^3_{13} = \Gamma^3_{31} = \frac{1}{r}, \qquad \Gamma^3_{23} = \Gamma^3_{32} = \cot\theta. \tag{43c}$$

If the physical components of velocity are denoted by V_r, V_θ, V_φ, then the corresponding covariant components, defined by

$$v_1 = \sqrt{g_{11}}\,V_r, \qquad v_2 = \sqrt{g_{22}}\,V_\theta, \qquad v_3 = \sqrt{g_{33}}\,V_\varphi, \tag{44}$$

are, from equation (38), given by

$$v_1 = V_r, \qquad v_2 = V_\theta, \qquad v_3 = r\sin\theta\,V_\varphi, \tag{45}$$

and the corresponding contravariant components, defined by

$$v^1 = V_r/\sqrt{g_{11}}, \qquad v^2 = V_\theta/\sqrt{g_{22}}, \qquad v^3 = V_\varphi/\sqrt{g_{33}}, \tag{46}$$

are given by

$$v^1 = V_r, \qquad v^2 = V_\theta/r, \qquad v^3 = V_\varphi/r\sin\theta. \tag{47}$$

Since the covariant derivative of the contravariant vector v^i is defined by

$$v^i_{,j} = \frac{\partial v^i}{\partial x^j} + \Gamma^i_{kj}\,v^k, \tag{48}$$

the incompressibility condition

$$v^i_{,i} = 0 \tag{49}$$

becomes, with equations (43) and (47),

$$r\frac{\partial V_r}{\partial r} + 2V_r + \frac{\partial V_\theta}{\partial^\theta} + \cot\theta\, V_\theta + \frac{1}{\sin\theta}\frac{\partial V_\varphi}{\partial\varphi} = 0, \tag{50}$$

or for flows which are symmetric about the $\theta = 0$ axis,

$$r\frac{\partial V_r}{\partial r} + 2V_r + \frac{\partial V_\theta}{\partial^\theta} + \cot\theta\, V_\theta = 0. \tag{51}$$

Since the covariant derivative of the covariant vector v_i is given by

$$v_{i,j} = \frac{\partial v_i}{\partial x^j} - \Gamma^k_{ij}\, v_k, \tag{52}$$

the components of the covariant kinematic tensor $A^{(1)}_{ij}$, defined by

$$A^{(1)}_{ij} = v_{i,j} = v_{j,i}, \qquad i,j = 1,2,3, \tag{53}$$

are given by

$$A^{(1)}_{11} = 2\frac{\partial V_r}{\partial r}, \tag{54a}$$

$$A^{(1)}_{22} = 2r\frac{\partial V_\theta}{\partial\theta} + V_r), \tag{54b}$$

$$A^{(1)}_{33} = 2r\sin\theta\left(\frac{\partial V_\varphi}{\partial\varphi} + \sin\theta\, V_r + \cos\theta\, V_\theta\right), \tag{54c}$$

$$A^{(1)}_{23} = r\left(\sin\theta\frac{\partial V_\varphi}{\partial\theta} + \frac{\partial V_\theta}{\partial\varphi} - \cos\theta\, V_\varphi\right), \tag{54d}$$

$$A^{(1)}_{31} = \frac{\partial V_r}{\partial\varphi} + r\sin\theta\frac{\partial V_\varphi}{\partial r} - \sin\theta\, V_\varphi, \tag{54e}$$

$$A^{(1)}_{12} = r\frac{\partial V_\theta}{\partial r} + \frac{\partial V_r}{\partial\theta} - V_\theta. \tag{54f}$$

The recurrence relations

$$A^{(\gamma)}_{ij} = v^k\frac{\partial A^{(\gamma-1)}_{ij}}{\partial x^k} + A^{(\gamma-1)}_{ik}\frac{\partial v^k}{\partial x^j} + A^{(\gamma-1)}_{kj}\frac{\partial v^k}{\partial x^i},$$
$$i,j = 1,2,3 \quad \text{and} \quad \gamma = 2,3,\ldots,R, \tag{55}$$

for the components of the covariant kinematic tensors $A^{(\gamma)}_{ij}$, $\gamma = 2, 3, \ldots, R$, when the fluid is in a steady state of flow, yield

$$A_{11}^{(\gamma)} = V_r \frac{\partial A_{11}^{(\gamma-1)}}{\partial r} + \frac{V_\theta}{r}\frac{\partial A_{11}^{(\gamma-1)}}{\partial \theta} + \frac{V_\varphi}{r\sin\theta}\frac{\partial A_{11}^{(\gamma-1)}}{\partial \varphi}$$

$$+ 2A_{11}^{(\gamma-1)}\frac{\partial V_r}{\partial r} + 2A_{12}^{(\gamma-1)}\frac{\partial}{\partial r}\left(\frac{V_\theta}{r}\right) + \frac{2}{\sin\theta}A_{13}^{(\gamma-1)}\frac{\partial}{\partial r}\left(\frac{V_\varphi}{r}\right), \tag{56a}$$

$$A_{22}^{(\gamma)} = V_r \frac{\partial A_{22}^{(\gamma-1)}}{\partial r} + \frac{V_\theta}{r}\frac{\partial A_{22}^{(\gamma-1)}}{\partial \theta} + \frac{V_\varphi}{r\sin\theta}\frac{\partial A_{22}^{(\gamma-1)}}{\partial \varphi}$$

$$+ 2A_{21}^{(\gamma-1)}\frac{\partial V_r}{\partial \theta} + \frac{2}{r}A_{22}^{(\gamma-1)}\frac{\partial V_\theta}{\partial \theta} + \frac{2}{r}A_{23}^{(\gamma-1)}\frac{\partial}{\partial \theta}\left(\frac{V_\varphi}{\sin\theta}\right), \tag{56b}$$

$$A_{33}^{(\gamma)} = V_r \frac{\partial A_{33}^{(\gamma-1)}}{\partial r} + \frac{V_\theta}{r}\frac{\partial A_{33}^{(\gamma-1)}}{\partial \theta} + \frac{V_\varphi}{r\sin\theta}\frac{\partial A_{33}^{(\gamma-1)}}{\partial \varphi}$$

$$+ 2A_{31}^{(\gamma-1)}\frac{\partial V_r}{\partial \varphi} + \frac{2}{r}A_{32}^{(\gamma-1)}\frac{\partial V_\theta}{\partial \varphi} + \frac{2}{r\sin\theta}A_{33}^{(\gamma-1)}\frac{\partial V_\varphi}{\partial \varphi}, \tag{56c}$$

$$A_{23}^{(\gamma)} = V_r \frac{\partial A_{23}^{(\gamma-1)}}{\partial r} + \frac{V_\theta}{r}\frac{\partial A_{23}^{(\gamma-1)}}{\partial \theta} + \frac{V_\varphi}{r\sin\theta}\frac{\partial A_{23}^{(\gamma-1)}}{\partial \varphi}$$

$$+ A_{23}^{(\gamma-1)}\left(\frac{1}{r}\frac{\partial V_\theta}{\partial \theta} + \frac{1}{r\sin\theta}\frac{\partial V_\varphi}{\partial \varphi}\right) + A_{31}^{(\gamma-1)}\frac{\partial V_r}{\partial \theta}$$

$$+ A_{12}^{(\gamma-1)}\frac{\partial V_r}{\partial \varphi} + \frac{1}{r}A_{22}^{(\gamma-1)}\frac{\partial V_\theta}{\partial \varphi} + \frac{1}{r}A_{33}^{(\gamma-1)}\frac{\partial}{\partial \theta}\left(\frac{V_\varphi}{\sin\theta}\right), \tag{56d}$$

$$A_{31}^{(\gamma)} = V_r \frac{\partial A_{31}^{(\gamma-1)}}{\partial r} + \frac{V_\theta}{r}\frac{\partial A_{31}^{(\gamma-1)}}{\partial \theta} + \frac{V_\varphi}{r\sin\theta}\frac{\partial A_{31}^{(\gamma-1)}}{\partial \varphi}$$

$$+ A_{31}^{(\gamma-1)}\left(\frac{\partial V_r}{\partial r} + \frac{1}{r\sin\theta}\frac{\partial V_\varphi}{\partial \varphi}\right) + \frac{1}{r}A_{12}^{(\gamma-1)}\frac{\partial V_\theta}{\partial \varphi}A_{23}^{(\gamma-1)}\frac{\partial}{\partial r}\left(\frac{V_\theta}{r}\right)$$

$$+ \frac{1}{\sin\theta}A_{33}^{(\gamma-1)}\frac{\partial}{\partial r}\left(\frac{V_\varphi}{r}\right) + A_{11}^{(\gamma-1)}\frac{\partial V_r}{\partial \varphi}, \tag{56e}$$

$$A_{12}^{(\gamma)} = V_r \frac{\partial A_{12}^{(\gamma-1)}}{\partial r} + \frac{V_\theta}{r}\frac{\partial A_{12}^{(\gamma-1)}}{\partial \theta} + \frac{V_\varphi}{r\sin\theta}\frac{\partial A_{12}^{(\gamma-1)}}{\partial \varphi}$$

$$+ A_{12}^{(\gamma-1)}\left(\frac{\partial V_r}{\partial r} + \frac{1}{r}\frac{\partial V_\theta}{\partial \theta}\right) + \frac{1}{\sin\theta}A_{23}^{(\gamma-1)}\frac{\partial}{\partial r}\left(\frac{V_\varphi}{r}\right)$$

$$+ \frac{1}{r}A_{31}^{(\gamma-1)}\frac{\partial}{\partial \theta}\left(\frac{V_\varphi}{\sin\theta}\right) + A_{11}^{(\gamma-1)}\frac{\partial V_r}{\partial \theta} + A_{22}^{(\gamma-1)}\frac{\partial}{\partial r}\left(\frac{V_\theta}{r}\right). \tag{56f}$$

For flows with axial symmetry, equations (54) for the components of the covariant kinematic tensor $A_{ij}^{(1)}$ reduce to

$$A_{11}^{(1)} = 2\frac{\partial V_r}{\partial r}, \tag{57a}$$

$$A_{22}^{(1)} = 2r\left(\frac{\partial V_\theta}{\partial \theta} + V_r\right), \tag{57b}$$

$$A_{33}^{(1)} = 2r \sin\theta\,(\sin\theta\, V_r + \cos\theta\, V_\theta), \tag{57c}$$

$$A_{23}^{(1)} = r\left(\sin\theta \frac{\partial V_\varphi}{\partial \theta} - \cos\theta\, V_\varphi\right), \tag{57d}$$

$$A_{31}^{(1)} = \sin\theta\left(r\frac{\partial V_\varphi}{\partial r} - V_\varphi\right), \tag{57e}$$

$$A_{12}^{(1)} = r\frac{\partial V_\theta}{\partial r} + \frac{\partial V_r}{\partial \theta} - V_\theta, \tag{57f}$$

and the recurrence relations (56) for the components of the covariant kinematic tensors $A_{ij}^{(1)}$, $\gamma = 2, 3, \ldots, R$, become

$$A_{11}^{(\gamma)} = V_r \frac{\partial A_{11}^{(\gamma-1)}}{\partial r} + \frac{V_\theta}{r}\frac{\partial A_{11}^{(\gamma-1)}}{\partial \theta}$$

$$+ 2A_{11}^{(\gamma-1)}\frac{\partial V_r}{\partial r} + 2A_{12}^{(\gamma-1)}\frac{\partial}{\partial r}\left(\frac{V_\theta}{r}\right) + \frac{2}{\sin\theta}A_{13}^{(\gamma-1)}\frac{\partial}{\partial r}\left(\frac{V_\varphi}{r}\right), \tag{58a}$$

$$A_{22}^{(\gamma)} = V_r \frac{\partial A_{22}^{(\gamma-1)}}{\partial r} + \frac{V_\theta}{r}\frac{\partial A_{22}^{(\gamma-1)}}{\partial \theta}$$

$$+ 2A_{21}^{(\gamma-1)}\frac{\partial V_r}{\partial \theta} + \frac{2}{r}A_{22}^{(\gamma-1)}\frac{\partial V_\theta}{\partial \theta} + \frac{2}{r}A_{23}^{(\gamma-1)}\frac{\partial}{\partial \theta}\left(\frac{V_\varphi}{\sin\theta}\right), \tag{58b}$$

$$A_{33}^{(\gamma)} = V_r \frac{\partial A_{33}^{(\gamma-1)}}{\partial r} + \frac{V_\theta}{r}\frac{\partial A_{33}^{(\gamma-1)}}{\partial \theta}, \tag{58c}$$

$$A_{23}^{(\gamma)} = V_r \frac{\partial A_{23}^{(\gamma-1)}}{\partial r} + \frac{V_\theta}{r}\frac{\partial A_{23}^{(\gamma-1)}}{\partial \theta}$$

$$+ \frac{1}{r}A_{23}^{(\gamma-1)}\frac{\partial V_\theta}{\partial \theta} + A_{31}^{(\gamma-1)}\frac{\partial V_r}{\partial \theta} + \frac{1}{r}A_{33}^{(\gamma-1)}\frac{\partial}{\partial \theta}\left(\frac{V_\varphi}{\sin\theta}\right), \tag{58d}$$

$$A_{31}^{(\gamma)} = V_r \frac{\partial A_{31}^{(\gamma-1)}}{\partial r} + \frac{V_\theta}{r} \frac{\partial A_{31}^{(\gamma-1)}}{\partial \theta}$$

$$+ A_{31}^{(\gamma-1)} \frac{\partial V_r}{\partial r} + A_{23}^{(\gamma-1)} \frac{\partial}{\partial r}\left(\frac{V_\theta}{r}\right) + \frac{1}{\sin\theta} A_{33}^{(\gamma-1)} \frac{\partial}{\partial r}\left(\frac{V_\varphi}{r}\right), \tag{58e}$$

$$A_{12}^{(\gamma)} = V_r \frac{\partial A_{12}^{(\gamma-1)}}{\partial r} + \frac{V_\theta}{r} \frac{\partial A_{12}^{(\gamma-1)}}{\partial \theta}$$

$$+ A_{12}^{(\gamma-1)} \left(\frac{\partial V_r}{\partial r} + \frac{1}{r}\frac{\partial V_\theta}{\partial \theta}\right) + \frac{1}{\sin\theta} A_{23}^{(\gamma-1)} \frac{\partial}{\partial r}\left(\frac{V_\varphi}{r}\right)$$

$$+ \frac{1}{r} A_{31}^{(\gamma-1)} \frac{\partial}{\partial \theta}\left(\frac{V_\varphi}{\sin\theta}\right) + A_{11}^{(\gamma-1)} \frac{\partial V_r}{\partial \theta} + A_{22}^{(\gamma-1)} \frac{\partial}{\partial r}\left(\frac{V_\theta}{r}\right). \tag{58f}$$

If, in addition to possessing axial symmetry, the flow is such that the component of velocity V_φ is zero, then equations (57) for the components of the covariant kinematic tensor $A_{ij}^{(1)}$ reduce to

$$A_{11}^{(1)} = 2\frac{\partial V_r}{\partial r}, \tag{59a}$$

$$A_{22}^{(1)} = 2r\left(\frac{\partial V_\theta}{\partial \theta} + V_r\right), \tag{59b}$$

$$A_{33}^{(1)} = 2r \sin\theta\, (\sin\theta\, V_r + \cos\theta\, V_\theta), \tag{59c}$$

$$A_{23}^{(1)} = 0, \tag{59d}$$

$$A_{31}^{(1)} = 0, \tag{59e}$$

$$A_{12}^{(1)} = r\frac{\partial V_\theta}{\partial r} + \frac{\partial V_r}{\partial \theta} - V_\theta, \tag{59f}$$

and the recurrence relations (58) for the components of the covariant kinematic tensors $A_{ij}^{(\gamma)}$, $\gamma = 2, 3, \ldots, R$, become

$$A_{11}^{(\gamma)} = V_r \frac{\partial A_{11}^{(\gamma-1)}}{\partial r} + \frac{V_\theta}{r}\frac{\partial A_{11}^{(\gamma-1)}}{\partial \theta} + 2A_{11}^{(\gamma-1)}\frac{\partial V_r}{\partial r} + 2A_{12}^{(\gamma-1)}\frac{\partial}{\partial r}\left(\frac{V_0}{r}\right), \tag{60a}$$

$$A_{22}^{(\gamma)} = V_r \frac{\partial A_{22}^{(\gamma-1)}}{\partial r} + \frac{V_\theta}{r}\frac{\partial A_{22}^{(\gamma-1)}}{\partial \theta} + 2A_{12}^{(\gamma-1)}\frac{\partial V_r}{\partial \theta} + \frac{2}{r}A_{22}^{(\gamma-1)}\frac{\partial V_\theta}{\partial \theta}, \tag{60b}$$

$$A_{33}^{(\gamma)} = V_r \frac{\partial A_{33}^{(\gamma-1)}}{\partial r} + \frac{V_\theta}{r} \frac{\partial A_{33}^{(\gamma-1)}}{\partial \theta}, \tag{60c}$$

$$A_{23}^{(1)} = 0, \tag{60d}$$

$$A_{31}^{(1)} = 0, \tag{60e}$$

$$A_{12}^{(\gamma)} = V_r \frac{\partial A_{12}^{(\gamma-1)}}{\partial r} + \frac{V_\theta}{r} \frac{\partial A_{12}^{(\gamma-1)}}{\partial \theta} + A_{12}^{(\gamma-1)} \left(\frac{\partial V_r}{\partial r} + \frac{1}{r} \frac{\partial V_\theta}{\partial \theta} \right)$$

$$+ A_{11}^{(\gamma-1)} \frac{\partial V_r}{\partial \theta} + A_{22}^{(\gamma-1)} \frac{\partial}{\partial r}\left(\frac{V_\theta}{r} \right). \tag{60f}$$

If inertia terms are neglected, it is seen from Love (Ref. 8) that, for the spherical polar system (r, θ, φ), the equations of steady motion are

$$-\rho f_r = \frac{\partial \tau_{11}}{\partial r} + \frac{1}{r}\frac{\partial \tau_{12}}{\partial \theta} + \frac{1}{r \sin\theta}\frac{\partial \tau_{13}}{\partial \varphi} + \frac{1}{r}(2\tau_{11} - \tau_{22} - \tau_{33} + \cot\theta\, \tau_{12}), \tag{61a}$$

$$-\rho f_\theta = \frac{\partial \tau_{12}}{\partial r} + \frac{1}{r}\frac{\partial \tau_{22}}{\partial \theta} + \frac{1}{r \sin\theta}\frac{\partial \tau_{23}}{\partial \varphi} + \frac{1}{r}[(\tau_{22} - \tau_{33})\cot\theta - 3\tau_{12}], \tag{61b}$$

$$-\rho f_\varphi = \frac{\partial \tau_{13}}{\partial r} + \frac{1}{r}\frac{\partial \tau_{23}}{\partial \theta} + \frac{1}{r \sin\theta}\frac{\partial \tau_{33}}{\partial \varphi} + \frac{1}{r}(3\tau_{13} + 2\tau_{23}\cot\theta), \tag{61c}$$

where f_r, f_θ, f_φ are the physical components of body force per unit mass of the fluid and τ_{ij}, $i, j = 1, 2, 3$, are the physical components of stress defined by

$$\tau_{ij} = \sqrt{\frac{g_{ij}}{g_{jj}}}\, t_j^i, \qquad \text{not summed.} \tag{62}$$

For flows with axial symmetry the equations of motion (61) become

$$-\rho f_r = \frac{\partial \tau_{11}}{\partial r} + \frac{1}{r}\frac{\partial \tau_{12}}{\partial \theta} + \frac{1}{r}(2\tau_{11} - \tau_{22} - \tau_{33} + \cot\theta\, \tau_{12}), \tag{63a}$$

$$-\rho f_\theta = \frac{\partial \tau_{12}}{\partial r} + \frac{1}{r}\frac{\partial \tau_{22}}{\partial \theta} + \frac{1}{r}[(\tau_{22} - \tau_{33})\cot\theta + 3\tau_{12}], \tag{63b}$$

$$-\rho f_\varphi = \frac{\partial \tau_{13}}{\partial r} + \frac{1}{r}\frac{\partial \tau_{23}}{\partial \theta} + \frac{1}{r}(3\tau_{13} + 2\tau_{23}\cot\theta). \tag{63c}$$

If, in addition to possessing axial symmetry, the flow is such that V_φ is zero, then $\tau_{13} = \tau_{23} = 0$ and equation (63c) becomes

$$f_\varphi = 0. \tag{64}$$

For an incompressible Newtonian fluid with coefficient of viscosity μ,

$$t^i_j = -p\delta^i_j + \mu A^{(1)i}_j, \tag{65}$$

so that, with equations (54), the definition (62) and the relation

$$A^{(1)i}_j = g^{ik}A^{(1)}_{kj}, \tag{66}$$

the equations of motion (63) become, for this case,

$$\frac{\partial p}{\partial r} - \rho f_r = \mu\left(\nabla^2 V_r - \frac{2}{r^2}V_r - \frac{2}{r^2}\frac{\partial V_\theta}{\partial \theta} - \frac{2V_\theta \cot \partial}{r^2} - \frac{2}{r^2 \sin\theta}\frac{\partial V_\varphi}{\partial \varphi}\right), \tag{67a}$$

$$\frac{1}{r}\frac{\partial p}{\partial \theta} - \rho f_\theta = \mu\left(\nabla^2 V_\theta - \frac{2}{r^2}\frac{\partial V_r}{\partial \theta} - \frac{V_\theta}{r^2 \sin^2\theta} - \frac{2\cos\theta}{r^2\sin^2\theta}\frac{\partial V_\varphi}{\partial \varphi}\right), \tag{67b}$$

$$\frac{1}{r\sin\theta}\frac{\partial p}{\partial \varphi} - \rho f_\varphi = \mu\left(\nabla^2 V_\varphi - \frac{V_\varphi}{r^2\sin^2\theta} + \frac{2}{r^2\sin\theta}\frac{\partial V_r}{\partial \varphi} + \frac{2\cos\theta}{r^2\sin^2\theta}\frac{\partial V_\theta}{\partial \varphi}\right), \tag{67c}$$

where

$$\nabla^2 = \frac{\partial^2}{\partial r^2} + \frac{2}{r}\frac{\partial}{\partial r} + \frac{1}{r^2}\frac{\partial^2}{\partial \theta^2} + \frac{\cot\theta}{r^2}\frac{\partial}{\partial \theta} + \frac{1}{r^2\sin^2\theta}\frac{\partial^2}{\partial \varphi^2}. \tag{68}$$

For flows with axial symmetry,

$$\frac{\partial p}{\partial r} - \rho f_r = \mu\left(\bar{\nabla}^2 V_r - \frac{2}{r^2}V_r - \frac{2}{r^2}\frac{\partial V_\theta}{\partial \theta} - \frac{2V_\theta\cot\theta}{r^2}\right), \tag{69a}$$

$$\frac{1}{r}\frac{\partial p}{\partial \theta} - \rho f_\theta = \mu\left(\bar{\nabla}^2 V_\theta - \frac{2}{r^2}\frac{\partial V_r}{\partial \theta} - \frac{V_0}{r^2\sin^2\theta}\right), \tag{69b}$$

$$-\rho f_\varphi = \mu\left(\bar{\nabla}^2 V_\varphi - \frac{V_\varphi}{r^2\sin^2\theta}\right), \tag{69c}$$

where

$$\overline{\nabla}^2 = \frac{\partial^2}{\partial r^2} + \frac{2}{r}\frac{\partial}{\partial r} + \frac{1}{r^2}\frac{\partial^2}{\partial \theta^2} + \frac{\cot\theta}{r^2}\frac{\partial}{\partial \theta}. \tag{70}$$

If a stream function ψ is introduced such that

$$V_r = \frac{1}{r^2 \sin\theta}\frac{\partial \psi}{\partial \theta}, \tag{71a}$$

$$V_\theta = \frac{1}{r \sin\theta}\frac{\partial \psi}{\partial r}, \tag{71b}$$

then the continuity equation (51) is identically satisfied and p can be eliminated from equations (69a, b), yielding the partial differential equation

$$E^4\psi = \frac{\rho}{\mu}\left[\frac{\partial}{\partial r}(rf_\theta) - \frac{\partial f_r}{\partial \theta}\right]\sin\theta, \tag{72}$$

where

$$E^4 = \left(\frac{\partial^2}{\partial r^2} - \frac{\cot\theta}{r^2}\frac{\partial}{\partial \theta} + \frac{1}{r^2}\frac{\partial^2}{\partial \theta^2}\right)^2. \tag{73}$$

It is readily seen that

$$E^4\left(\frac{F(\theta)}{r^n}\right) = \frac{1}{r^{n+4}}\left[\frac{d^2}{d\theta^2} - cot\,\theta\frac{d}{d\theta} + (n+2)(n+3)\right] \times \left[\frac{d^2}{d\theta^2} - \cot\theta\frac{d}{d\theta} + n(n+1)\right]F(\theta), \tag{74}$$

where F(θ) is any function of θ with four derivatives.

4. Convergent Flow in a Right-Circular Cone

We shall now consider a mass of the incompressible, isotropic, slightly viscoelastic material having the stress-deformation relation (16) to be contained within a rigid right-circular cone of semivertical angle α. We shall consider that fluid is being annihilated at the vertex at a steady rate Q.

Let (r, θ, φ) be a spherical polar system with origin at the vertex of the cone, where r is the distance measured from the vertex, θ is the colatitude measured from the cone axis, and φ is any longitude. As in Section 3, we will sometimes use the general tensor notation

$$x^1 = r, \qquad x^2 = \theta, \qquad x^3 = \varphi. \tag{75}$$

The velocity must be zero at each point of the cone, so that

$$V_r(r, \alpha, \varphi) = V_\theta(r, \alpha, \varphi) = V_\varphi(r, \alpha, \varphi) = 0. \tag{76}$$

The velocity component V_r must also satisfy the volume flow condition

$$2\pi \int_0^\alpha r^2 V_r \sin\theta d\theta = -Q. \tag{77}$$

If an incompressible, Newtonian fluid with coefficient of viscosity μ is contained within the cone, a solution to the equations of motion (69) with no body forces subject to the incompressibility condition (50) and the subsidiary conditions (76), (77) is

$$V'_r = \frac{1}{r^2} f(\theta), \tag{78a}$$

$$V'_\theta = 0, \tag{78b}$$

$$V'_\varphi = 0, \tag{78c}$$

with

$$p' = \frac{2\mu}{r^3} f(\theta) + \frac{K\mu}{3r^3} + \text{const}, \tag{79}$$

where

$$K = \frac{3Q(1 - 3\cos^2\alpha)}{\pi(1 - 3\cos^2\alpha + 2\cos^3\alpha)}, \tag{80}$$

and

$$f(\theta) = \frac{3Q}{2\pi} \frac{\cos^2\alpha - \cos^2\theta}{1 - 3\cos^2\alpha + 2\cos^3\alpha}. \tag{81}$$

We shall now calculate the body forces which must be applied in order to maintain the flow described by equations (78) in a fluid having the stress-deformation relation (16).

From equations (59), it is seen that the components of the covariant kinematic tensor $A_{ij}^{(1)}$ for the flow described by equations (78) are given by

$$A_{11}^{(1)} = -\frac{4}{r^3} f(\theta), \tag{82a}$$

$$A_{22}^{(1)} = \frac{2}{r} f(\theta), \tag{82b}$$

$$A_{33}^{(1)} = \frac{2}{r} \sin^2\theta\, f(\theta), \tag{82c}$$

$$A_{23}^{(1)} = 0, \tag{82d}$$

$$A_{31}^{(1)} = 0, \tag{82e}$$

$$A_{12}^{(1)} = \frac{1}{2} f'(\theta), \tag{82f}$$

and the recurrence relations (60) for the components of the covariant kinematic tensors $A_{ij}^{(\gamma)}$, $\gamma = 2, 3, \ldots, R$, become

$$A_{11}^{(\gamma+1)} = \frac{f}{r^2} \frac{\partial A_{11}^{(\gamma)}}{\partial r} - \frac{4f}{r^3} A_{11}^{(\gamma)}, \qquad \gamma = 1, 2, \ldots, R-1, \tag{83a}$$

$$A_{22}^{(\gamma+1)} = \frac{f}{r^2} \frac{\partial A_{22}^{(\gamma)}}{\partial r} + \frac{2f'}{r^2} A_{12}^{(\gamma)}, \qquad \gamma = 1, 2, \ldots, R-1, \tag{83b}$$

$$A_{33}^{(\gamma+1)} = \frac{f}{r^2} \frac{\partial A_{33}^{(\gamma)}}{\partial r}, \qquad \gamma = 1, 2, \ldots, R-1, \tag{83c}$$

$$A_{23}^{(\gamma+1)} = 0, \qquad \gamma = 1, 2, \ldots, R-1, \tag{83d}$$

$$A_{31}^{(\gamma+1)} = 0, \qquad \gamma = 1, 2, \ldots, R-1, \tag{83e}$$

$$A_{12}^{(\gamma+1)} = \frac{f}{r^2} \frac{\partial A_{12}^{(\gamma)}}{\partial r} + \frac{f'}{r^2} A_{11}^{(\gamma)} - \frac{2f}{r^3} A_{12}^{(\gamma)}, \qquad \gamma = 1, 2, \ldots, R-1. \tag{83f}$$

Explicitly, the components of the covariant kinematic tensors $A_{ij}^{(\gamma)}$, $\gamma = 1, 2, \ldots, R$, are given by

$$A_{11}^{(\gamma)} = \frac{\Gamma\left(\gamma + \frac{4}{3}\right)}{\Gamma\left(\frac{4}{3}\right)} \left(-\frac{3f}{r^3}\right)^{\gamma}, \tag{84a}$$

$$A_{22}^{(\gamma)} = \frac{1}{r^4} \frac{\Gamma\left(\gamma - \frac{2}{3}\right)}{\Gamma\left(\frac{4}{3}\right)} \left(-\frac{3f}{r^3}\right)^{\gamma - 2} [\gamma(\gamma - 1)f'^2 - 2f^2], \tag{84b}$$

$$A_{33}^{(\gamma)} = \frac{2}{9} r^2 \sin^2\theta \frac{\Gamma\left(\gamma - \frac{2}{3}\right)}{\Gamma\left(\frac{4}{3}\right)} \left(-\frac{3f}{r^3}\right)^{\gamma}, \tag{84c}$$

$$A_{23}^{(\gamma)} = 0, \tag{84d}$$

$$A_{31}^{(\gamma)} = 0, \tag{84e}$$

$$A_{12}^{(\gamma)} = \frac{f'}{r^2} \left(-\frac{3f}{r^3}\right)^{\gamma - 1} \frac{\Gamma\left(\gamma + \frac{1}{3}\right)}{\Gamma\left(\frac{4}{3}\right)} \gamma, \tag{84f}$$

where $\Gamma(x)$ denotes the gamma function.

Using equation (39) and the relation

$$A_j^{(\gamma)i} = g^{ik} A_{kj}^{(\gamma)}, \qquad i, j = 1, 2, 3 \text{ and } \gamma = 1, 2, \ldots, R,$$

it is seen that the components of the corresponding mixed kinematic tensors $A_j^{(\gamma)i}$, $\gamma = 1, 2, \ldots, R$, are then given by

$$A_1^{(\gamma)1} = \frac{\Gamma\left(\gamma + \frac{4}{3}\right)}{\Gamma\left(\frac{4}{3}\right)} \left(-\frac{3f}{r^3}\right)^{\gamma}, \tag{85a}$$

$$A_2^{(\gamma)2} = \frac{1}{r^6}\frac{\Gamma\left(\gamma - \frac{2}{3}\right)}{\Gamma\left(\frac{4}{3}\right)}\left(-\frac{3f}{r^3}\right)^{\gamma-2}[\gamma(\gamma - 1)f'^2 - 2f^2)], \tag{85b}$$

$$A_3^{(\gamma)3} = \frac{2}{9}\frac{\Gamma\left(\gamma - \frac{2}{3}\right)}{\Gamma\left(\frac{4}{3}\right)}\left(-\frac{3f}{r^3}\right)^{\gamma}, \tag{85c}$$

$$A_3^{(\gamma)2} = A_2^{(\gamma)3} = 0, \tag{85d}$$

$$A_1^{(\gamma)3} = A_3^{(\gamma)1} = 0, \tag{85e}$$

$$A_2^{(\gamma)1} = \frac{f'}{r^2}\left(-\frac{3f}{r^3}\right)^{\gamma-1}\frac{\Gamma\left(\gamma + \frac{1}{3}\right)}{\Gamma\left(\frac{4}{3}\right)}\gamma, \tag{85f}$$

$$A_1^{(\gamma)2} = \frac{f'}{r^4}\left(-\frac{3f}{r^3}\right)^{\gamma-1}\frac{\Gamma\left(\gamma + \frac{1}{3}\right)}{\Gamma\left(\frac{4}{3}\right)}\gamma. \tag{85g}$$

Substituting the expression (81) for $f(\theta)$ into equations (85), we then have, for $\gamma = 1, 2, \ldots, R$,

$$A_1^{(\gamma)1} = \frac{\Gamma\left(\gamma + \frac{4}{3}\right)}{\Gamma\left(\frac{4}{3}\right)}\left(-\frac{3C}{r^3}\right)^{\gamma}(\cos^2\alpha - \cos^2\theta)^{\gamma}, \tag{86a}$$

$$A_2^{(\gamma)2} = \frac{2}{9}\frac{\Gamma\left(\gamma - \frac{2}{3}\right)}{\Gamma\left(\frac{4}{3}\right)}\left(-\frac{3C}{r^3}\right)^{\gamma}(\cos^2\alpha - \cos^2\theta)^{\gamma-2},$$

$$\times\ [2\gamma(\gamma - 1)(\cos^2\theta - \cos^4\theta) - (\cos^2\alpha - \cos^2\theta)^2], \tag{86b}$$

$$A_3^{(\gamma)3} = -\frac{2}{9}\frac{\Gamma\left(\gamma - \frac{2}{3}\right)}{\Gamma\left(\frac{4}{3}\right)}\left(-\frac{3C}{r^3}\right)^{\gamma}(\cos^2\alpha - \cos^2\theta)^{\gamma}, \tag{86c}$$

$$A_3^{(\gamma)2} = A_2^{(\gamma)3} = 0, \tag{86d}$$

$$A_1^{(\gamma)3} = A_3^{(\gamma)1} = 0, \tag{86e}$$

$$A_2^{(\gamma)1} = -\frac{2}{3}\gamma r \frac{\Gamma\left(\gamma + \frac{1}{3}\right)}{\Gamma\left(\frac{4}{3}\right)} \left(-\frac{3C}{r^3}\right)^{\gamma} (\cos^2\alpha - \cos^2\theta)^{\gamma-1} \cos\theta \sin\theta, \tag{86f}$$

$$A_1^{(\gamma)2} = -\frac{2}{3}\frac{\gamma}{r} \frac{\Gamma\left(\gamma + \frac{1}{3}\right)}{\Gamma\left(\frac{4}{3}\right)} \left(-\frac{3C}{r^3}\right)^{\gamma} (\cos^2\alpha - \cos^2\theta)^{\gamma-1} \cos\theta \sin\theta, \tag{86g}$$

where

$$C = \frac{3Q}{2\pi\,(1 - 3\cos^2\alpha + 2\cos^3\alpha)}. \tag{87}$$

It is clear from equations (86) that the scalar invariants of the R kinematic tensors $A_j^{(\gamma)i}$, $\gamma = 1, 2, \ldots, R$, are expressible as polynomials in $(1/r^3)$ with coefficients which are polynomials in $\cos^2\theta$. None of these invariants has a term in $(1/r^3)^0$ or in $(1/r^3)^1$.

The stress-deformation relation (16) becomes, in general tensor notation,

$$t_j^i = -p\delta_j^i + \mu A_j^{(1)i} + \epsilon p_j^i, \qquad i, j = 1, 2, 3, \tag{88}$$

where p_j^i is the same tensor polynomial function of the mixed tensors $A_j^{(\gamma)i}$ $\gamma = 1, 2, \ldots, R$, that P_{ij} is of the $A_{ij}^{(\gamma)}$, $\gamma = 1, 2, \ldots, R$, in a Cartesian system, for both sides of equations (88) transform as components of mixed tensors of rank two, and, when the coordinate system is Cartesian, equations (88) reduce to equations (16). It is then clear from equations (86) that the mixed components of stress for the flow described by equations (78) with $f(\theta)$ given by equation (81) are expressible in the forms

$$t_1^1 = -p - \frac{4\mu C}{r^3}(\cos^2\alpha - \cos^2\theta) + \epsilon \sum_{\beta=2}^{L} \frac{1}{r^{3\beta}} \sum_{\gamma=0}^{\beta} a_{\gamma\beta}\cos^{2\gamma}\theta, \tag{89a}$$

$$t_2^2 = -p + \frac{2\mu C}{r^3}(\cos^2\alpha - \cos^2\theta) + \epsilon \sum_{\beta=2}^{L} \frac{1}{r^{3\beta}} \sum_{\gamma=0}^{\beta} b_{\gamma\beta}\cos^{2\gamma}\theta, \tag{89b}$$

$$t_3^3 = -p + \frac{2\mu C}{r^3}(\cos^2\alpha - \cos^2\theta) + \epsilon \sum_{\beta=2}^{L} \frac{1}{r^{3\beta}} \sum_{\gamma=0}^{\beta} c_{\gamma\beta}\cos^{2\gamma}\theta, \tag{89c}$$

$$t_3^2 = t_2^3 = 0, \tag{89d}$$

$$t_1^3 = t_3^1 = 0, \tag{89e}$$

$$t_2^1 = \frac{2\mu C}{r^2}\cos\theta\sin\theta + \epsilon\sin\theta\sum_{\beta=2}^{L}\frac{1}{r^{3\beta-1}}\sum_{\gamma=0}^{\beta-1} d_{\gamma\beta}\cos^{2\gamma+1}\theta, \tag{89f}$$

$$t_1^2 = \frac{2\mu C}{r^4}\cos\theta\sin\theta - \epsilon\sin\theta\sum_{\beta=2}^{L}\frac{1}{r^{3\beta+1}}\sum_{\gamma=0}^{\beta-1} d_{\gamma\beta}\cos^{2\gamma+1}\theta, \tag{89g}$$

where the value of the integer L and the values of the constants $a_{\gamma\beta}$, $b_{\gamma\beta}$, $c_{\gamma\beta}$, $d_{\gamma\beta}$, $\gamma = 0, 1, \ldots, \beta$, and $\beta = 2, 3, \ldots, L$, depend on the form of the tensor polynomial p_j^i.

Using equation (38) and the relation

$$\tau_{ij} = \sqrt{\frac{g_{ii}}{g_{jj}}}\, t_j^i, \qquad \text{not summed}, \qquad i, j = 1, 2, 3,$$

the physical components of stress τ_{ij}, $i, j = 1, 2, 3$, are then given by

$$\tau_{11} = -p - \frac{4\mu C}{r^3}(\cos^2\alpha - \cos^2\theta) + \epsilon\sum_{\beta=2}^{L}\frac{1}{r^{3\beta}}\sum_{\gamma=0}^{\beta} a_{\gamma\beta}\cos^{2\gamma}\theta, \tag{90a}$$

$$\tau_{22} = -p + \frac{2\mu C}{r^3}(\cos^2\alpha - \cos^2\theta) + \epsilon\sum_{\beta=2}^{L}\frac{1}{r^{3\beta}}\sum_{\gamma=0}^{\beta} b_{\gamma\beta}\cos^{2\gamma}\theta, \tag{90b}$$

$$\tau_{33} = -p + \frac{2\mu C}{r^3}(\cos^2\alpha - \cos^2\theta) + \epsilon\sum_{\beta=2}^{L}\frac{1}{r^{3\beta}}\sum_{\gamma=0}^{\beta} c_{\gamma\beta}\cos^{2\gamma}\theta, \tag{90c}$$

$$\tau_{23} = 0, \tag{90d}$$

$$\tau_{31} = 0, \tag{90e}$$

$$\tau_{12} = \frac{2\mu C}{r^3}\cos\theta\sin\theta + \epsilon\sin\theta\sum_{\beta=2}^{L}\frac{1}{r^{3\beta}}\sum_{\gamma=0}^{\beta-1} d_{\gamma\beta}\cos^{2\gamma+1}\theta. \tag{90f}$$

Unless the quantity p in equations (90) is specified, the equations of motion (63a, b) do not uniquely determine the components of body force f_r', f_θ' required to maintain the flow described by equations (78). If we also require that $p = p'$, where p' is given by equation (79), then substituting the results (90) into the equations of motion (63a, b) yields

$$-\rho f_r' = \epsilon\sum_{\beta=2}^{L}\frac{1}{r^{3\beta+1}}\Bigg\{[(2 - 3\beta)a_{0\beta} - b_{0\beta} - c_{0\beta} - d_{0\beta}]$$

$$+ \sum_{\gamma=1}^{\beta-1}[(2 - 3\beta)a_{\gamma\beta} - b_{\gamma\beta} - c_{\gamma\beta} - (2\gamma + 1)d_{\gamma\beta} + (2\gamma + 1)d_{(\gamma-1)\beta}]\cos^{2\gamma}\theta$$

$$+[(2-3\beta)a_{\beta\beta} - b_{\beta\beta} - c_{\beta\beta} + (2\beta+1)d_{(\beta-1)\beta}]\cos^{2\beta}\theta\Big\}, \tag{91a}$$

$$-\rho f_\theta' = \epsilon \sum_{\beta=2}^{L} \frac{1}{r^{3\beta+1}} \Big\{ \sin\theta \sum_{\gamma=1}^{\beta} [3(1-\beta)d_{(\gamma-1)\beta} - 2\gamma b_{\gamma\beta}]\cos^{2\gamma-1}\theta$$
$$+ \frac{1}{\sin\theta} \sum_{\gamma=0}^{\beta} (b_{\gamma\beta} - c_{\gamma\beta})\cos^{2\gamma+1}\theta \Big\}. \tag{91b}$$

It is clear[2] that $\sin^2\theta$ is a factor of $\tau_{22} - \tau_{33}$, so that we may let

$$\sum_{\gamma=0}^{\beta} (b_{\gamma\beta} - c_{\gamma\beta})\cos^{2\gamma+1}\theta = \sin^2\theta \sum_{\gamma=1}^{\beta} e_{\gamma\beta}\cos^{2\gamma-1}\theta, \tag{92}$$

---

[2]From equations (86), we have that each of the matrices $(A_j^{(\gamma)i})$ is expressible in the form

$$[A_j^{(\gamma)i}] = \begin{bmatrix} A_\gamma & C_\gamma \sin\theta & 0 \\ B_\gamma \sin\theta & D_\gamma \sin^2\theta + E_\gamma & 0 \\ 0 & 0 & E_\gamma \end{bmatrix},$$

where $A_\gamma, B_\gamma, C_\gamma, D_\gamma, E_\gamma$ are polynomials in $1/r$ and $\cos\theta$. We then have

$$[A_k^{(\gamma)i} A_j^{(\delta)k}] = \begin{bmatrix} A_{\gamma\delta} & C_{\gamma\delta} \sin\theta & 0 \\ B_{\gamma\delta} \sin\theta & D_{\gamma\delta} \sin^2\theta + E_{\gamma\delta} & 0 \\ 0 & 0 & E_{\gamma\delta} \end{bmatrix},$$

where

$$A_{\gamma\delta} = A_\gamma A_\delta + C_\gamma B_\delta \sin^2\theta, \qquad B_{\gamma\delta} = A_\delta B_\gamma + B_\delta (D_\gamma \sin^2\theta + E_\gamma),$$
$$C_{\gamma\delta} = A_\gamma C_\delta + C_\gamma (D_\delta \sin^2\theta + E_\delta), \quad D_{\gamma\delta} = B_\gamma C_\delta + D_\gamma D_\delta \sin^2\theta + E_\gamma D_\delta + E_\delta D_\gamma,$$
$$E_{\gamma\delta} = E_\gamma E_\delta.$$

Proceeding in this manner, it is evident that $[P_j^i]$ is of the form

$$[P_j^i] = \begin{bmatrix} A & C\sin\theta & 0 \\ B\sin\theta & D\sin^2\theta + E & 0 \\ 0 & 0 & E \end{bmatrix},$$

where A, B, C, D, E are polynomials in $1/r$ and $\cos\theta$. We then have

$$\tau_{22} - \tau_{33} = \mu\,(A_2^{(1)2} - A_3^{(1)3}) + \dot{\epsilon}(P_2^2 - P_3^3) = \sin^2\theta\,(\mu D_1 + \epsilon D).$$

where $e_{\gamma\beta}$, $\gamma = 1,2,\ldots,\beta$ and $\beta = 2, 3, \ldots, L$, are constants. Equation (91b) becomes

$$-\rho f'_\theta = \epsilon \sin\theta \sum_{\beta=2}^{L} \frac{1}{r^{3\beta+1}} \sum_{\gamma=1}^{\beta} [3(1-\beta)d_{(\gamma-1)\beta} - 2\gamma b_{\gamma\beta} + e_{\gamma\beta}] \cos^{2\gamma-1}\theta. \quad (93)$$

From equations (93) and (91a), we then see that

$$\rho\left[\frac{\partial}{\partial r}(rf'_\theta) - \frac{\partial f'_r}{\partial\theta}\right] = \epsilon \sin\theta \sum_{\beta=2}^{L} \frac{1}{r^{3\beta+1}} \sum_{\gamma=1}^{\beta} C_{\beta\gamma}\cos^{2\gamma-1}\theta, \quad (94)$$

where

$$\begin{aligned} C_{\beta\gamma} = {} & 3\beta[3(1-\beta)d_{(\gamma-1)\beta} - 2\gamma b_{\gamma\beta} + e_{\gamma\beta}] - 2\gamma[(2-3\beta)a_{\gamma\beta} - b_{\gamma\beta} - c_{\gamma\beta} \\ & -(2\gamma+1)d_{\gamma\beta} + (2\gamma+1)d_{(\gamma-1)\beta}], \\ & \gamma = 1, 2, \ldots, \beta \text{ and } \beta = 2, 3, \ldots, L, \end{aligned} \quad (95a)$$

$$\begin{aligned} C_{\beta\beta} = {} & 3\beta[3(1-\beta)d_{(\beta-1)\beta} - 2\beta b_{\beta\beta} + e_{\beta\beta}] - 2\beta[(2-3\beta)a_{\beta\beta} - b_{\beta\beta} - c_{\beta\beta} \\ & +(2\beta+1)d_{(\beta-1)\beta}], \qquad \beta = 2, 3, \ldots, L. \end{aligned} \quad (95b)$$

If no body forces are applied to the fluid, the velocity componets V'_r, V'_θ given by equations (78) with $p = p'$ as given by equation (79) do not satisfy the equations of motion (63a, b). As was shown in Section 2, if we calculate a solution ($\epsilon V''_r$, $\epsilon V''_\theta$, $\epsilon p''$) to the equations of motion (63a, b) for an incompressible, isotropic, Newtonian fluid with coefficient of viscosity μ, under the action of the body forces f'_r, f'_θ , subject to the incompressibility condition (51) and to the boundary conditions

$$V''_r(r,\alpha) = V''_\theta(r,\alpha) = \lim_{r\to\infty} V''_r(r,\theta) = \lim_{r\to\infty} V''_\theta(r,\theta) = 0, \quad (96a)$$

$$\int_0^\alpha r^2 V''_r \sin\theta d\theta = 0, \quad (96b)$$

then ($V_r = V'_r - \epsilon V''_r$, $V_\theta = V'_\theta - \epsilon V''_\theta$, $p = p' - \epsilon p''$) satisfies the equations of motion (63a, b) with neglect only of terms of the second degree or higher in ϵ and satisfies exactly the incompressibility condition (51) and the boundary conditions (76), (77).

If we introduce a stream function $\psi''(r, \theta)$ such that

$$V_r'' = \frac{1}{r^2 \sin\theta} \frac{\partial \psi''}{\partial \theta}, \tag{97a}$$

$$V_\theta'' = -\frac{1}{r \sin\theta} \frac{\partial \psi''}{\partial r}, \tag{97b}$$

it follows from equations (72) and (94) that

$$E^4 \psi'' = \sum_{\beta=2}^{L} \frac{1}{r^{3\beta+1}} \sum_{\gamma=0}^{\beta} X_{\beta\gamma} \cos^{2\gamma+1}\theta, \tag{98}$$

where

$$X_{\beta 0} = \frac{1}{\mu} C_{\beta 1}, \qquad \beta = 2, 3, \ldots, L, \tag{99a}$$

$$X_{\beta\gamma} = \frac{1}{\mu}(C_{\beta(\gamma+1)} - C_{\beta\gamma}),$$
$$\gamma = 1, 2, \ldots, \beta - 1 \quad \text{and} \quad \beta = 2, 3, \ldots, L, \tag{99b}$$

$$X_{\beta\beta} = -\frac{1}{\mu} C_{\beta\beta}, \qquad \beta = 2, 3, \ldots, L. \tag{99c}$$

An immediate consequence of equations (99) is that

$$\sum_{\gamma=0}^{\beta} X_{\beta\gamma} = 0, \qquad \beta = 2, 3, \ldots, L. \tag{100}$$

Since both V_r'' and V_θ'' must vanish on the cone $\theta = \alpha$, ψ'' must be constant and $\partial\psi''/\partial\theta$ must vanish on this cone. Since the flow is symmetric about the $\theta = 0$ axis, V_θ'' must vanish on this axis, so that from equation (97b) ψ'' must be constant along $\theta = 0$. If V_r'' is to be finite except at the origin, we see from equation (97a) that $\partial\psi''/\partial\theta$ must vanish on $\theta = 0$. The boundary condition (96b) implies that ψ'' must assume the same value on $0 = \alpha$ as on $\theta = 0$, which value may, without loss of generality, be taken to be zero. Since both V_r'' and V_θ'' must approach zero as r approaches infinity, both $(1/r^2)(\partial\psi''/\partial\theta)$ and $(1/r)(\partial\psi''/\partial r)$ must approach zero as r approaches infinity. The boundary conditions to be imposed on ψ'' are, then,

$$\psi''(r,0) = \psi''(r,\alpha) = \frac{\partial\psi''(r,0)}{\partial\theta} = \frac{\partial\psi''(r,\alpha)}{\partial\theta}$$

$$= \lim_{r\to\infty}\frac{1}{r^2}\frac{\partial\psi''(r,\theta)}{\partial\theta} = \lim_{r\to\infty}\frac{1}{r}\frac{\partial\psi''(r,\theta)}{\partial r} = 0. \qquad (101)$$

After a solution ψ'' to the partial differential equation (98) is calculated subject to the boundary conditions (101), the resultant velocity components, given by equations (97), can be substituted into the equations of motion (69a, b), which can then be integrated, determining p'' up to an additive constant.

Using equation (74), it is seen that a solution to the partial differential equation (98), subject to the boundary conditions (101), is

$$\psi'' = \sum_{\beta=2}^{L}\frac{\varphi_\beta(\theta)}{r^{3\beta-3}}, \qquad (102)$$

where the functions $\varphi_\beta(\theta)$, $\beta = 2, 3, \ldots, L$, satisfy the ordinary differential equations

$$\left[\frac{d^2}{d\theta^2} - \cot\theta\frac{d}{d^\theta} + 3\beta(3\beta-1)\right]\left[\frac{d^2}{d\theta^2} - \cot\theta\frac{d}{d^\theta} + (3\beta-2)(3\beta-3)\right]\varphi_\beta(\theta)$$

$$= \sum_{\gamma=0}^{\beta} X_{\beta\gamma}\cos^{2\gamma+1}\theta, \qquad \beta = 2, 3, \ldots, L, \quad (103)$$

subject to the boundary conditions

$$\varphi_\beta(0) = \varphi_\beta(\alpha) = \frac{d\varphi_\beta(0)}{d\theta} = \frac{d\varphi_\beta(\alpha)}{d\theta} = 0, \qquad \beta = 2, 3, \ldots, L. \quad (104)$$

Particular integrals of the equations (103) can be constructed with the aid of the identity

$$\left[\frac{d^2}{d\theta^2} - \cot\theta\frac{d}{d\theta} + 3\beta(3\beta-1)\right]\left[\frac{d^2}{d\theta^2} - \cot\theta\frac{d}{d^\theta} + (3\beta-2)(3\beta-3)\right]\cos^\delta\theta$$
$$= [3\beta(3\beta-1) + \delta(1-\delta)][(3\beta-2)(3\beta-3) + \delta(1-\delta)]\cos^\delta\theta$$
$$+ \delta(\delta-1)[3\beta(3\beta-1) + (3\beta-2)(3\beta-3) + 6\delta - 2\delta^2 - 6)]\cos^{\delta-2}\theta$$
$$+ \delta(\delta-1)(\delta-2)(\delta-3)\cos^{\delta-4}\theta, \ \beta = 2, 3, \ldots, L \text{ and all } \delta. \qquad (105)$$

If we let

$$\varphi_\beta^{(p)}(\theta) = \sum_{\gamma=0}^{\beta} Y_{\beta\gamma} \cos^{2\gamma+1}\theta, \qquad \beta = 2, 3, \ldots, L, \tag{106}$$

then $\varphi_\beta^{(p)}(\theta)$, $\beta = 2, 3, \ldots, L$, are particular integrals of the equations (103), provided the coefficients $Y_{\beta\gamma}$, $\gamma = 0, 1, \ldots, \beta$ and $\beta = 2, 3, \ldots, L$, are chosen to satisfy the linear equations

$$\begin{aligned}[3\beta(3\beta-1) - 2\beta(2\beta+1)][(3\beta-2)(3\beta-3)\\ - 2\beta(2\beta+1)]\, Y_{\beta\beta} = X_{\beta\beta}, \qquad \beta = 2, 3, \ldots, L,\end{aligned} \tag{107a}$$

$$\begin{aligned}&2\beta(2\beta+1)[3\beta(3\beta-1) + (3\beta-2)(3\beta-3) + 6(2\beta+1)\\ &-2(2\beta+1)^2 - 6]\, Y_{\beta\beta} + [3\beta(3\beta-1) - 2(2\beta-2)(2\beta-1)][(3\beta-2)(3\beta-3)\\ &- (2\beta-2)(2\beta-1)]\, Y_{\beta(\beta-1)} = X_{\beta(\beta-1)}, \qquad \beta = 2, 3, \ldots, L,\end{aligned} \tag{107b}$$

$$\begin{aligned}&(2\gamma+5)(2\gamma+4)(2\gamma+3)(2\gamma+2)\, Y_{\beta(\gamma+2)} + (2\gamma+3)(2\gamma+2)[3\beta(3\beta-1)\\ &+ (3\beta-2)(3\beta-3) + 6(2\gamma+3) - 2(2\gamma+3)^2 - 6]\, Y_{\beta(\gamma+1)} + [3\beta(3\beta-1)\\ &- 2(2\gamma+1)][(3\beta-2)(3\beta-3) - 2\gamma(2\gamma+1)]\, Y_{\beta\gamma} = X_{\beta\gamma},\\ &\gamma = 0, 1, \ldots, \beta-2 \text{ and } \beta = 2, 3, \ldots, L.\end{aligned} \tag{107c}$$

If, for a fixed value of β, we add the $(\beta + 1)$ equations represented by (107), we see that

$$3\beta(3\beta-1)(3\beta-2)(3\beta-3) \sum_{\gamma=0}^{\beta} Y_{\beta\gamma} = \sum_{\gamma=0}^{\beta} X_{\beta\gamma}, \qquad \beta = 2, 3, \ldots, L, \tag{108}$$

so that, with equations (100),

$$\sum_{\gamma=0}^{\beta} Y_{\beta\gamma} = 0, \qquad \beta = 2, 3, \ldots, L. \tag{109}$$

If we set $\theta = 0$ in equations (106), equations (109) then imply

$$\varphi_\beta^{(p)}(0) = 0, \qquad \beta = 2, 3, \ldots, L. \tag{110}$$

It is clear from equations (106) that we also have

$$\frac{d\varphi_\beta^{(p)}(0)}{d\theta} = 0, \qquad \beta = 2, 3, \ldots, L,$$

so that if we can find two independent solutions $\varphi_\beta^{(1)}$, $\varphi_\beta^{(2)}$ to each of the homogeneous differential equations

$$\left[\frac{d^2}{d\theta^2} - \cot\theta\,\frac{d}{d\theta} + 3\beta(3\beta - 1)\right]$$
$$\times\left[\frac{d^2}{d\theta^2} - \cot\theta\,\frac{d}{d^{\,\theta}} + (3\beta - 2)(3\beta - 3)\right]\varphi_\beta(\theta) = 0, \qquad \beta = 2, 3, \ldots, L, \quad (111)$$

such that

$$\varphi_\beta^{(1)}(0) = \varphi_\beta^{(2)}(0) = \frac{d\varphi_\beta^{(1)}(0)}{d\theta} = \frac{d\varphi_\beta^{(2)}(0)}{d\theta} = 0, \qquad \beta = 2, 3, \ldots, L, \quad (112)$$

then

$$\varphi_\beta(\theta) = a_\beta\varphi_\beta^{(1)}(\theta) + b_\beta\varphi_\beta^{(2)}(\theta) + \varphi_\beta^{(p)}(\theta), \qquad \beta = 2, 3, \ldots, L \quad (113)$$

are the solutions to the differential equations (103) subject to the boundary conditions (104), provided we choose the constants a_β, b_β, $\beta = 2, 3, \ldots, L$, so that

$$a_\beta\varphi_\beta^{(1)}(\alpha) = b_\beta\varphi_\beta^{(2)}(\alpha) + \varphi_\beta^{(p)}(\alpha) = 0, \qquad \beta = 2, 3, \ldots, L, \quad (114a)$$

$$a_\beta\frac{d\varphi_\beta^{(1)}(\alpha)}{d\theta} + b_\beta\frac{d\varphi_\beta^{(2)}(\alpha)}{d\theta} + \frac{d\varphi_\beta^{(p)}(\alpha)}{d\theta} = 0, \qquad \beta = 2, 3, \ldots, L, \quad (114b)$$

i.e.,

$$a_\beta = \frac{\varphi_\beta^{(2)}(\alpha)\dfrac{d\varphi_\beta^{(p)}(\alpha)}{d\theta} - \varphi_\beta^{(p)}(\alpha)\dfrac{d\varphi_\beta^{(2)}(\alpha)}{d\theta}}{\varphi_\beta^{(1)}(\alpha)\dfrac{d\varphi_\beta^{(2)}(\alpha)}{d\theta} - \varphi_\beta^{(2)}(\alpha)\dfrac{d\varphi_\beta^{(1)}(\alpha)}{d\theta}}, \quad (115a)$$

$$b_\beta = \frac{\varphi_\beta^{(p)}(\alpha)\dfrac{d\varphi_\beta^{(1)}(\alpha)}{d\theta} - \varphi_\beta^{(1)}(\alpha)\dfrac{d\varphi_\beta^{(p)}(\alpha)}{d\theta}}{\varphi_\beta^{(1)}(\alpha)\dfrac{d\varphi_\beta^{(2)}(\alpha)}{d\theta} - \varphi_\beta^{(2)}(\alpha)\dfrac{d\varphi_\beta^{(1)}(\alpha)}{d\theta}}. \quad (115b)$$

The problem of finding the functions $\varphi_\beta^{(1)}$, $\varphi_\beta^{(2)}$, $\beta = 2, 3, \ldots, L$, is simplified by the fact that the operators $[d^2/d\theta^2 - \cot\theta\; d/d\theta + 3\beta(3\beta - 1)]$ and $[d^2/d\theta^2 - \cot\theta\; d/d\theta + 3\beta(3\beta - 2)(3\beta - 3)]$ commute, so that for fixed β any solution to the homogeneous differential equation represented by (111) is either a solution to

$$\left[\frac{d^2}{d\theta^2} - \cot\theta \frac{d}{d\theta} + 3\beta(3\beta - 1)\right]\varphi_\beta(\theta) = 0, \tag{116}$$

or to

$$\left[\frac{d^2}{d\theta^2} - \cot\theta \frac{d}{d\theta} + (3\beta - 2)(3\beta - 3)\right]\varphi_\beta(\theta) = 0. \tag{117}$$

With the changes of variable

$$z = \cos^2\frac{\theta}{2} = \frac{1}{2}(1 + \cos\theta), \tag{118a}$$

$$\varphi_\beta(\theta) = \sin^2\theta\, g_\beta(z), \tag{118b}$$

equation (116) becomes the hypergeometric equation

$$\left[z(1 - z)\frac{d^2}{dz^2} + (2 - 4z)\frac{d}{dz} + (3\beta - 2)(3\beta + 1)\right]g_\beta(z) = 0, \tag{119}$$

an entire solution of which is the Jacobi polynomial $G_{(3\beta-2)}(3, 2, z)$, where the notation of Courant and Hilbert (Ref. 9) has been used. The Jacobi polynomial $G_{(3\beta-2)}(2, 3, z)$ is defined by

$$G_{(3\beta-2)}(3, 2, z) = \frac{1}{(3\beta - 1)!}\frac{1}{z(1 - z)}\frac{d^{(3\beta-2)}}{dz^{(3\beta-2)}}[z^{3\beta-1}(1 - z)^{3\beta-1}]. \tag{120}$$

Similarly, with the changes of variable (118), equation (117) becomes the hypergeometric equation

$$\left[z(1 - z)\frac{d^2}{dz^2} + (2 - 4z)\frac{d}{dz} + (3\beta - 4)(3\beta - 1)\right]g_\beta(z) = 0, \tag{121}$$

an entire solution of which is the Jacobi polynomial $G_{(3\beta-4)}(3, 2, z)$.

Since any entire function of θ multiplied by $\sin^2\theta$ vanishes together with its first derivative at $\theta = 0$, the required solutions $\varphi_\beta^{(1)}$ and $\varphi_\beta^{(2)}$ are given by

$$\varphi_\beta^{(1)} = \sin^2\theta\, G_{(3\beta-2)}\left(3, 2, \cos^2\left(\frac{\theta}{2}\right)\right), \qquad \beta = 2, 3, \ldots, L, \tag{122a}$$

$$\varphi_\beta^{(2)} = \sin^2\theta\, G_{(3\beta-4)}\left(3, 2, \cos^2\left(\frac{\theta}{2}\right)\right), \qquad \beta = 2, 3, \ldots, L. \quad (122b)$$

For illustration, let us consider the case

$$p_j^i = \nu A_j^{(2)i}, \qquad i, j = 1, 2, 3, \quad (123)$$

where ν is a constant. Equation (88) then reduces to

$$t_j^i = -p\delta_j^i + \mu A_j^{(1)i} + \epsilon\nu A_j^{(2)i}, \qquad i, j = 1, 2, 3. \quad (124)$$

With the results (86), a comparison of equation (124) with the equations (89) and (92) reveals that $L = 2$ and that

$$a_{02} = 28\nu C^2\cos^4\alpha, \quad a_{12} = -56\nu C^2\cos^2\alpha, \quad a_{22} = 28\nu C^2, \quad (125a)$$

$$b_{02} = -2\nu C^2\cos^4\alpha, \quad b_{12} = 4\nu C^2(\cos^2\alpha + 2), \quad b_{22} = -10\nu C^2, \quad (125b)$$

$$c_{02} = -2\nu C^2\cos^4\alpha, \quad c_{12} = 4\nu C^2\cos^2\alpha, \quad c_{22} = -2\nu C^2, \quad (125c)$$

$$d_{02} = -16\nu C^2\cos^2\alpha, \quad d_{12} = 16\nu C^2, \quad (125d)$$

$$e_{12} = 0, \quad e_{22} = 8\nu C^2. \quad (125e)$$

From equations (95), we then have

$$C_{21} = 16(1 - 6\cos^2\alpha)\nu C^2, \quad (126a)$$

$$C_{22} = 80\,\nu C^2. \quad (126b)$$

Substituting the results (126) into equations (99) yields

$$X_{20} = 16(1 - 6\cos^2\alpha)\nu C^2/\mu, \quad (127a)$$

$$X_{21} = 32(2 + 3\cos^2\alpha)\nu C^2/\mu, \quad (127b)$$

$$X_{22} = -80\,\nu C^{22}/\mu. \quad (127c)$$

Equations (107) for the constants $Y_{2\gamma}$, $\gamma = 0, 1, 2$, become

$$-80\,Y_{22} = X_{22}, \quad (128a)$$

$$144\,Y_{21} + 320\,Y_{22} = X_{21}, \quad (128b)$$

$$360\,Y_{20} + 216\,Y_{21} + 120\,Y_{22} = X_{20}. \tag{128c}$$

Solving this system with $X_{2\gamma}$, $\gamma = 0, 1, 2$, given by equations (127) yields

$$Y_{20} = (7 - 6\cos^2\alpha)N, \tag{129a}$$

$$Y_{21} = (6\cos^2\alpha - 16)N, \tag{129b}$$

$$Y_{22} = 9N, \tag{129c}$$

where

$$N = \frac{\nu C^2}{9\mu}. \tag{130}$$

It is seen from equation (120), or from Courant and Hilbert (Ref. 9), that

$$G_2(3, 2, z) = 1 - 5z + 5z^2, \tag{131a}$$

$$G_4(3, 2, z) = 1 - 14z + 56z^2 - 84z^3 + 42z^4. \tag{131b}$$

Setting $\beta = 2$ in equations (122), we then have

$$\varphi_1^{(1)} = \sin^2\theta\left(1 - 14\cos^2\left(\frac{\theta}{2}\right) + 56\cos^4\left(\frac{\theta}{2}\right) - 84\cos^6\left(\frac{\theta}{2}\right) + 42\cos^8\left(\frac{\theta}{2}\right)\right)$$

$$= \frac{1}{8}\sin^2\theta\,(1 - 14\cos^2\theta + 21\cos^4\theta), \tag{132a}$$

$$\varphi_2^{(2)} = \sin^2\theta\left(1 - 5\cos^2\left(\frac{\theta}{2}\right) + 5\cos^4\left(\frac{\theta}{2}\right)\right) = \frac{1}{4}\sin^2\theta\,(5\cos^2\theta - 1). \tag{132b}$$

With equations (102), (106), (113), (129), and (132) we then have

$$\begin{aligned}\frac{r^3\psi''}{N} &= a\sin^2\theta\,(1 - 14\cos^2\theta + 21\cos^4\theta)\\ &+ b\sin^2\theta\,(1 - 5\cos^2\theta) + (7 - 6\cos^2\alpha)\cos\theta\\ &+ (6\cos^2\alpha - 16)\cos^3\theta + 9\cos^5\theta,\end{aligned} \tag{133}$$

where, in accordance with equations (114), the constants a and b are chosen so that

$$\psi''(r, \alpha) = \frac{\partial \psi''(r, \alpha)}{\partial \theta} = 0. \tag{134}$$

If we define a stream function ψ' such that V'_r and V'_θ, the velocity components of the basic flow, are given by

$$V'_r = \frac{1}{r^2 \sin \theta} \frac{\partial \psi'}{\partial \theta}, \tag{135a}$$

$$V'_\theta = \frac{1}{r \sin \theta} \frac{\partial \psi'}{\partial r}, \tag{135b}$$

then, from equations (78), (81), and (87), we must see that

$$\psi'(r, \theta) = (C/3)(\cos^3 \theta - 3 \cos^2 \alpha \cos \theta) + \text{const.} \tag{136}$$

If we select the constant in equation (136) so that

$$\psi'(r, 0) = 0, \tag{137}$$

we then have

$$\psi'(r, \theta) = (C/3)(3 \cos^3 \alpha - 3 \cos^2 \alpha \cos \theta + \cos^3 \theta - 1). \tag{138}$$

With equations (133) and (138), the stream function $\psi = \psi' - \epsilon\psi''$ for the resultant flow is given by

$$\begin{aligned} \psi = {} & (C/3)\,[3 \cos^2 \alpha - 3 \cos^2 \alpha \cos \theta + \cos^3 \theta - 1)] \\ & -(\epsilon N/r^3)\,[a \sin^2 \theta\,(1 - 14 \cos^2 \theta + 21 \cos^4 \theta) + b \sin^2 \theta\,(1 - 5 \cos^2 \theta) \\ & + (7 - 6 \cos^2 \alpha) \cos \theta + (6 \cos^2 \alpha - 16) \cos^3 \theta + 9 \cos^5 \theta]. \end{aligned} \tag{139}$$

It is thus seen that $V''_\theta = -(1/r \sin \theta)(\partial \psi'/\partial r)$ is zero only if $N = 0$, so that otherwise the flow is not radial. However, ψ'' behaves as $1/r^3$, whereas ψ' is independent of r. Far from the cone vertex the flow is, then, essentially that of a Newtonian fluid. As we approch the vertex, ψ'' begins to become important, so that V_θ is appreciably different from zero and the streamlines curve. Near the vertex, the assumption that the resultant flow differs only slightly from the basic flow is not valid, so that this region must be excluded from our consideration. It is seen from equation (139) that we may consider only the region such that

$$r >> \sqrt[3]{|\epsilon N/C|}\,. \tag{140}$$

Some of the streamlines for the case $\alpha = 60°$ are illustrated in Fig. 1 for $\epsilon N/C$ positive and in Fig. 2 for $\epsilon N/C$ negative. The unit of length in both figures is $\sqrt[3]{|\epsilon N/C|}$.

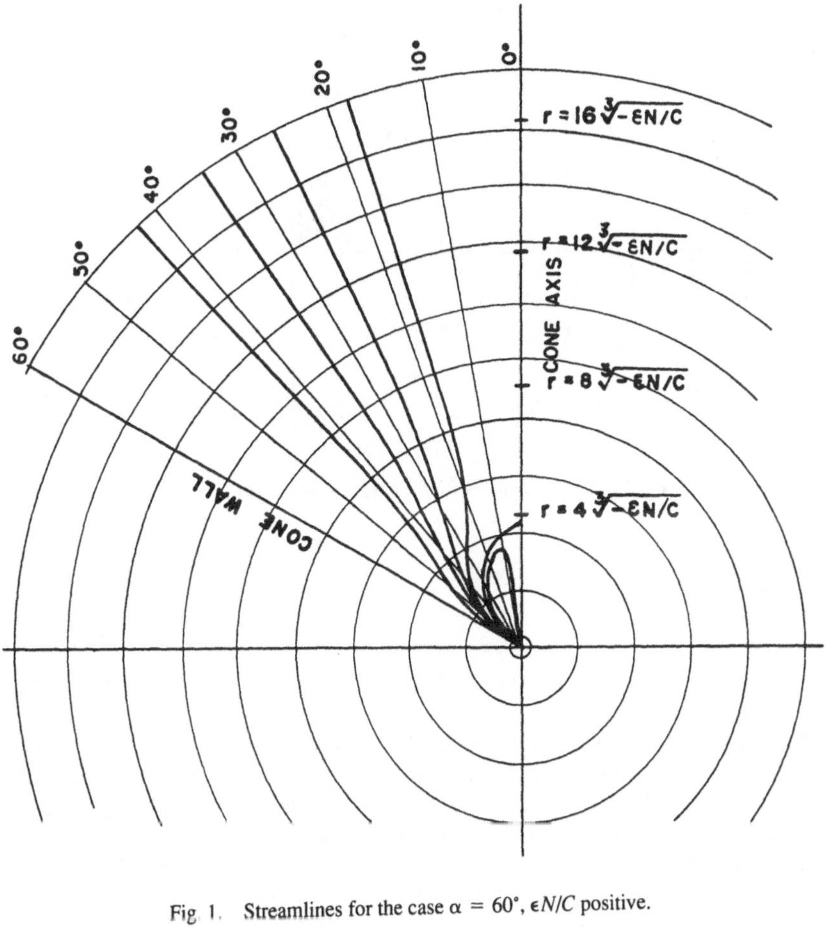

Fig. 1. Streamlines for the case $\alpha = 60°$, $\epsilon N/C$ positive.

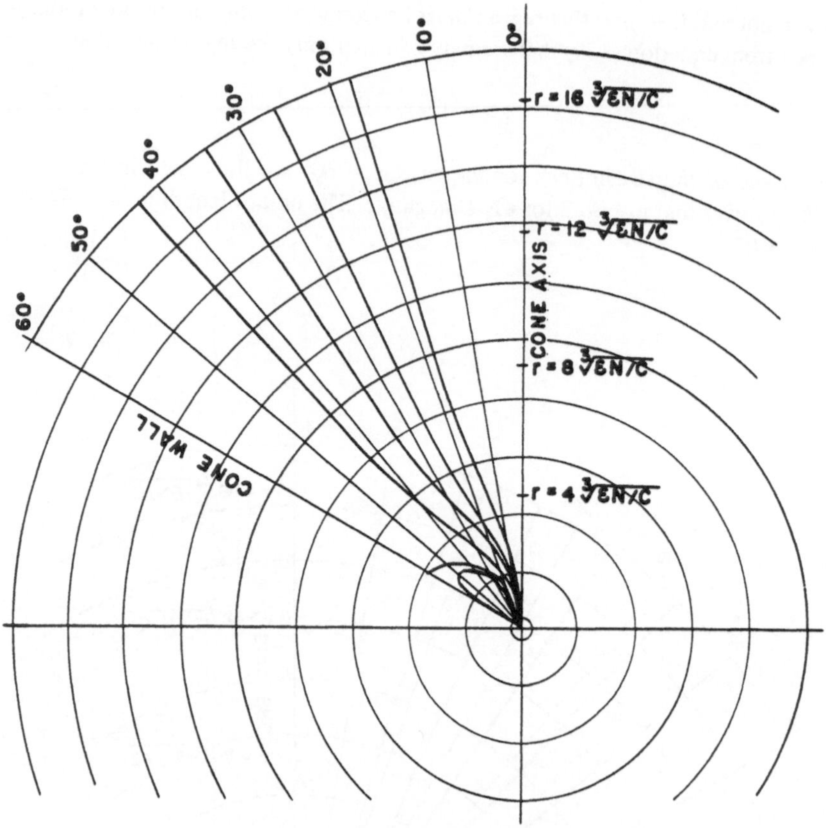

Fig. 2. Streamlines for the case $\alpha = 60°$, $\epsilon N/C$ negative.

References

1. Langlois, W. E., *Steady Flow of Slightly Visco-elastic Fluids*, PhD Thesis, Brown University, 1957.
2. Rivlin, R. S., and Ericksen, J. L., *Stress-Deformation Relations for Istropic Materials*, Journal of Rational Mechanics and Analysis, Vol. 4, pp. 323–354, 1955.
3. Rivlin, R.S., *Further Remarks on the Stress-Deformation Relations for Isotropic Materials*, Journal of Rational Mechanics and Analysis, Vol. 4, pp. 681–702, 1955.
4. Green, A. E., and Rivlin, R. S., *Steady Flow of Non-Newtonian Fluids Through Tubes*, Quarterly of Applied Mathematics, Vol. 14, pp. 299–308, 1956.
5. Ericksen, J. L., *Overdetermination of the Speed in Rectilinear Motion of Non-Newtonian Fluids*, Quarterly of Applied Mathematics, Vol. 14, pp. 318–321, 1956.

6. RIVLIN, R. S., *The Solution of Problems in Second Order Elasticity Theory*, Journal of Rational Mechanics and Analysis, Vol. 2, pp. 53–81, 1953.
7. RIVLIN, R. S., and TOPAKAGLU, C., *A Theorem in the Theory of Finite Elastic Deformations*, Journal of Rational Mechanics and Analysis, Vol. 3, pp. 581–589, 1954.
8. LOVE, A. E. H., *A Treatise on the Mathematical Theory of Elasticity*, Dover Publications, New York, New York, 1945.
9. COURANT, R., and HILBERT, D., *Methods of Mathematical Physics*, Interscience Publishers, New York, New York, 1953.

Supplementary References

10. LANGLOIS, W. E., *Steady Flow of Slightly Viscoelastic Fluids Between Rotating Spheres*, Quarterly of Applied Mathematics, Vol. 21, pp. 61–71, 1963.
11. GIESEKUS, H., *Some Secondary Flow Phenomena in General Visco-Elastic Fluids*, Proceedings of the Fourth International Conference on Rheology, Part I, Edited by E. H. Lee, Brown University, Providence, Rhode Island, p. 249, 1963.
12. LANGLOIS, W. E., and RIVLIN, R. S., *Slow Steady-State Flow of Visco-Elastic Fluids Through Non-Circular Tubes*, Rendiconti di Matematica e delle Sue Applicazioni, Vol. 22, p. 169, 1963.
13. LANGLOIS, W. E., *A Recursive Approach to the Theory of Slow, Steady-State Visco-Elastic Flow*, Transactions of the Society of Rheology, Vol. 7, pp. 75–99, 1963.
14. LANGLOIS, W. E., *The Recursive Theory of Slow Viscoelastic Flow Applied to Three Basic Problems of Hydrodynamics*, Transactions of the Society of Rheology, Vol. 8, pp. 33–60, 1964.
15. KAPUR, J. N., and SRIVASTAVA, R. C., *A Note on the Impossibility of Some Flows for General Reiner–Rivlin Fluids*, The Mathematics Seminar, Delhi, India, Vol. 4, p. 38, 1966.
16. LANGLOIS, W. E., *Slow Viscoelastic Flow at Large Distance from an Axis of Symmetry*, Journal of Applied Mechanics, Vol. 39, pp. 78–82, 1972.

9

Some Anomalies in the Structure of Elastic–Plastic Theory at Finite Strain

E. H. LEE

Abstract. Plastic flow at finite deformation has turned out to be a challenging phenomenon to model in analytical form. It involves two quite disparate behaviors, elastic and plastic, the former solid and the latter of fluid type. The combination demands modification of the usual continuum-mechanics theories of the components in intricate ways. For example, the unstressed, and hence elastically unstrained, configuration which forms a basic fixed reference configuration in elasticity theory, is in continuous motion in the present case because of the simultaneous plastic deformation progress. This paper aims to clarify some seemingly still open questions concerning these matters.

Because of the widely different stiffnesses associated with the interacting components, nonlinear geometrical effects can become significant at unexpectedly small strains. Because of the function nature of the elasticity law and the functional characteristics of plasticity relations, uncoupling the properties of each component from experimental measurements can pose problems. The selection of appropriate variables with which to formulate the theory has led to anomalies. The study of objectivity under superimposed rigid body rotation poses problems concerning the assignment of certain terms to one or other of the behavioral components and whether independent superimposed spins are required for the separate components. An example of a completed theory for a combined isotropic–kinematic hardening plasticity model is discussed.

Key Words. Finite deformation, plastic deformation, ductile metals, elasticity, plasticity, plastic strains, incremental variables, elastic–plastic tan-

---

E. H. Lee • Professor Emeritus, Division of Applied Mechanics, Stanford University, Stanford, California 94305-4040 and Department of Mechanical Engineering, Aeronautical Engineering and Mechanics, Rensselaer Polytechnic Institute, Troy, New York 12180.

Nonlinear Effects in Fluids and Solids, edited by M. M. Carroll and M. Hayes, Plenum Press, New York, 1996.

gent modulus, nonlinear effects, destressing, rate of deformation, prior plastic flow, tension test, induced anisotropy and spin effects, back stress in kinematic hardening, objectivity of the kinematics, Bauschinger effect, negative plastic flow, full invariance condition, Jaumann derivative, Kirchhoff stress, isotropic–kinematic hardening, variable plastic strain, plastic Lagrange strain, yield surface, hardening coefficients, plastic stretching tensor.

1. Introduction

The structure of nonlinear continuum mechanics, which involves stress analysis in materials subjected to finite deformation, incorporates much more of the essence of the underlying physics and geometry than do the classical linearized approximations restricted to infinitesimal-strain applications. It provides a powerful tool for generating the structure of constitutive relations for materials when the main features of the physical deformation mechanisms are known. Rivlin (Ref. 1) pioneered this approach in establishing the general law for the hydrodynamics of non-Newtonian fluids and provided solutions to a number of basic boundary-value problems. He also pioneered a similar program on large elastic deformation of isotropic materials (Refs. 2–5), and with Saunders (Ref. 6) reported extensive experiments to establish the constitutive relation for a vulcanized rubber.

The analogous problem for plastic deformation of ductile metals is still being investigated and what has been done has been subjected to considerable misunderstanding and controversy. This in part arises because plastic deformation always occurs in conjunction with elastic deformation and the coupling of the two in the same material particle is quite complex. Moreover, the constitutive relations for the two phenomena are quite disparate: elasticity requiring a function relation between stress and deformation, and plasticity, a functional law involving stress rate and rate of deformation. The present paper attempts to elucidate some of the issues and point out some of the anomalies evident in the literature.

Plastic stress and deformation analysis of polycrystalline metals arise in the design of processes for the manufacture of structural components by metalforming, and also in the design of structures. Analyses involving finite strain are needed in the former, and also in the latter for structures which must withstand severe loads or impacts such as in the crashworthy design of transportation vehicles and the design of structures for survival in earthquakes.

Deformation of structural metals at ambient temperatures is considered so that plasticity theory of the classical time-rate-independent type is appropriate. Commonly termed incremental or of flow type, it is expressed as a functional relation of the histories of stress and deformation in which time derivatives of stress and strain appear homogeneously of first order so that any variable which increases monotonically with time can replace time in the analysis without affecting the stress variation deduced. Such a variable could be the displacement of a piston driving an extrusion process. The stress–strain relations for such materials are insensitive to the magnitude of the rate of strain, and this is the case for structural metals at ambient temperatures for which the tensile stress–strain curve is effectively independent of the rate of strain. In contrast, at higher temperatures (greater than 0.2 of the absolute temperature of melting) strain rate-dependent laws termed viscoplastic are needed.

In problems of the type already specified, involving finite deformation, elastic strain components in general do not exceed the order 10^{-3} (yield stress divided by elastic modulus magnitudes) so that plastic strains will generally dominate elastic strains. However, in spite of this, it is usually necessary to adopt elastic–plastic theory in order to satisfy boundary conditions on the whole boundary and determine in which regions of the body plastic flow is taking place and which are deforming purely elastically. In most cases this can only be carried out by using elastic–plastic theory and finding that in certain regions the stress falls inside the yield surface and the plastic components of strain rate are zero. Only in special circumstances can plastic–rigid theory be utilized because in rigid regions the stress field cannot in principle be determined and no satisfactory method of applying the lower-bound limit-analysis theorem to determine the elastic regions precisely has been devised.

In incremental or flow type theory, the elastic–plastic moduli required relate increments of stress and strain, or equivalently, rates of stress and strain. In simple shearing or tensile tests they correspond to the tangent moduli given by the gradients of the corresponding stress–strain curves and are known as the shearing or tensile strain-hardening moduli. When plastic flow sets in, the stress–strain curves flatten sharply and the tangent modulus can fall rapidly to the order 10^{-3} times the elastic modulus.

This drastic reduction in the magnitude of the controlling moduli has a marked influence on the mathematical structure of the stress-analysis equations. Hill (Ref. 7, p. 54) demonstrated that infinitesimal-displacement theory, used in classical elasticity, may no longer be valid in elastic–plastic analysis because the convected terms in the rate of change of the stress acting on a material particle may then not be negligible.

Since flow or incremental type theory is to be considered, it is insightful to utilize incremental variables. Following Hill, let $\sigma_{ij}(x_k, t)$ be the Cauchy, or true, stress tensor acting at the position x_k at time t. Then, when t increases by dt, the increment of stress at the point x_k fixed in space will be

$$\delta\sigma_{ij} = \frac{\partial\sigma_{ij}}{\partial t}dt. \tag{1}$$

The corresponding increment of stress in the material particle at x_k at time t is denoted by $d\sigma_{ij}$,

$$d\sigma_{ij} = \frac{\partial\sigma_{ij}}{\partial t}dt + du_k\frac{\partial\sigma_{ij}}{\partial x_k}, \tag{2}$$

in which du_k is the displacement of the particle during the time increment dt. With no body forces acting, the equilibrium of the initial stress requires

$$\partial\sigma_{ij}/\partial x_j = 0. \tag{3}$$

The incremental stress must also satisfy equilibrium, giving

$$\partial(\delta\sigma_{ij})/\partial x_j = 0. \tag{4}$$

Substituting (3) and (4) into (2) differentiated with respect to x_j yields the equilibrium condition in terms of $d\sigma_{ij}$, the stress increment moving with a material particle which arises directly in the constitutive relation,

$$\frac{\partial(d\sigma_{ij})}{\partial x_j} = \frac{\partial(du_k)}{\partial x_j}\frac{\partial\sigma_{ij}}{\partial x_k}. \tag{5}$$

Classical elasticity is based appropriately on infinitesimal-displacement theory which neglects the difference between $\delta(\sigma_{ij})$ and $d(\sigma_{ij})$ and thus yields the equilibrium equations

$$\frac{\partial(d\sigma_{ij})}{\partial x_j} = \frac{\partial(\delta\sigma_{ij})}{\partial x_j} = 0. \tag{6}$$

This implies that the right-hand side of (5) is negligible.

In order to assess the significance of the terms involved, a heuristic argument based on scalar variables expressing the magnitudes of the quantities will be adopted. The symmetric part of the displacement gradient term in (5), $(\partial(du_k)/\partial x_j)_S$, comprises the increment of strain in the material particle. For purely elastic deformation its magnitude is of the order $d\sigma/E$ where $d\sigma$ is the stress increment magnitude and E a representative elastic modulus. The contribution of this term to the magnitude of the right-hand side of (5) can be estimated as

$$\frac{d\sigma}{E}\,(\text{stress gradient}) = \frac{d\sigma}{\sigma}\frac{\sigma}{E}\,(\text{stress gradient}). \tag{7}$$

Since the stress σ will be less than the yield stress Y, which, as has already been stated, is of the order $E \times 10^{-3}$, the magnitude (7) will be of the order of 10^{-3} times the gradient of the stress increment, because of the factor $d\sigma/\sigma$, and thus can correctly be neglected in (5). The term arising from the skew-symmetric part of $\partial(du_k)/\partial x_j$ can be expected to make a comparable contribution. Thus, either of the equilibrium equations (6) is adequate for elastic analysis of metals.

When plastic flow sets in, as has already been discussed, the elastic–plastic tangent modulus, E^t, can rapidly fall to the order of magnitude $E \times 10^{-3}$ and thus be comparable to the yield stress Y. Substituting this condition into the previous assessment of (7) determines the right-hand side to have the estimated contribution from the strain-rate component to be of the order

$$\frac{d\sigma}{E^t}\,(\text{stress gradient}) \sim \frac{d\sigma}{\sigma}\frac{Y}{E^t}\,(\text{stress gradient}), \tag{8}$$

which, with $Y/E^t \sim 1$, will be comparable with the stress increment gradient terms. Thus, with plasticity the nonlinear convected terms are comparable to the incremental stress gradient terms. The spin terms emanating from the skew-symmetric component of the displacement gradient will have a comparable influence so that it will be necessary to adopt nonlinear kinematics usually only considered necessary in the case of finite deformation. Related concepts were discussed by Rice (Ref. 8). Note that the reduction in the tangent modulus can occur at strains less than 1%, so that, in the presence of plastic flow, nonlinear effects can play an important role even at small strains. This requirement calls for a precise analysis of the coupling of elastic and plastic deformations to generate the total deformation.

2. Elastic–Plastic Coupling

In order to measure elastic–plastic material properties it is clearly necessary to be able to uncouple the elastic and plastic components of strain generated in a loading test. This is achievable because of the quite disparate characteristics of elastic and plastic deformation. Elasticity is due to the variation with the separation of the atoms of the interatomic forces which hold the crystal lattice together, so that elastic strain is directly related to distortion of the crystal lattice. Zero elastic strain arises when there are no external tractions applied to the crystal and the lattice is in equilibrium. Such tractions are essentially limited by the yield stress and this in general limits the elastic strain to the order 10^{-3}. The elastic constitutive relation is thus effectively a linear reversible function relation between stress and elastic strain—Hooke's law. In contrast, plastic deformation is associated with the migration through the lattice of defects, such as dislocations, and this leave the crystallites and the specimen permanently deformed when the external traction has been removed. As discussed in the previous section, the macroscopic deformation is governed by a functional law in which the stress rate is a function of the plastic strain rate. The deformation is dominated by shearing processes and is thus deviatoric in nature.

These two deformation mechanisms interact in the tension test measurements, depicted in Fig. 1, by means of which determination of the evolution of both the elastic and the plastic strains can be attempted. In one approach the test specimen in the undisturbed initial state X is deformed into the current state x. Plastic flow commences at the initial yield point A beyond which elastic and plastic increments of strain occur simultaneously. Having reached x the specimen is unloaded along xp, usually elastically, unless the plastic strain has been large, which is considered here not to have been the case. The trajectory xp will then be linear as depicted. Since only elastic changes of strain occur during unloading along xp, the plastic strain ϵ^P expresses the plastic deformation in both the configurations x and p, which is consistent with the physical circumstance that no dislocation migration has occurred during the elastic unloading along xp.

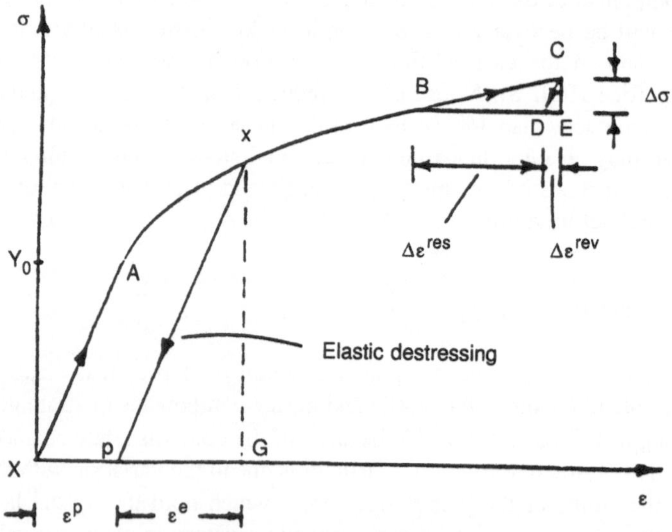

Fig. 1. Elastic and plastic strains in a tension test.

When the stress is zero the elastic strain is zero, so that the tensile test followed by sequential destressing to zero stress precisely separates out the plastic strain. That the theory can be formulated to achieve this even when some plastic flow occurs on destressing has been pointed out (Ref. 9). Use of this technique of isolating the plastic strain by reducing the stress, and hence the elastic strain, to zero was not utilized historically because with the older design of testing machines it was difficult to make precise measurements of the strain when the stress was reduced to zero and the specimen became loose in the fixtures holding it.

In practice, elastic and plastic strains have been determined experimentally by stepping up the stress–strain curve in increments of stress $\Delta\sigma$ as illustrated in Fig. 1. At each step $\Delta\sigma$ is applied and then removed to determine a residual strain increment $\Delta\epsilon^{res}$ and a reversible one $\Delta\epsilon^{rev}$. The increment $\Delta\epsilon^{res}$ is considered to be the plastic strain increment $\Delta\epsilon^{p}$ and $\Delta\epsilon^{rev}$ the elastic one $\Delta\epsilon^{e}$. In the case of finite deformation they are defined as increments of natural strain, the increment of extension divided by the current length. The plastic and elastic strains are then obtained by summing these increments over the whole straining history.

As discussed by Lee (Ref. 9), this procedure does not achieve precise separation of the elastic and plastic strains since the increment of stress $\Delta\sigma$ produces deformation in material already stressed to yield and thus strained elastically. Elastic and plastic components of strain increments will be mutually interacting there such that $\Delta\epsilon^{res}$ and $\Delta\epsilon^{rev}$ will not be exactly the increments of plastic and elastic strain, respectively. In order to analyze the situation precisely, the kinematics must be based on strict nonlinear finite-deformation-valid continuum theory. This reinforces the previous deduction that nonlinear convected terms in the incremental equilibrium equation must be retained.

In the case of three-dimensional finite strain, Fig. 2 illustrates the kinematics. The initial configuration of the body is expressed by the coordinates of ma-

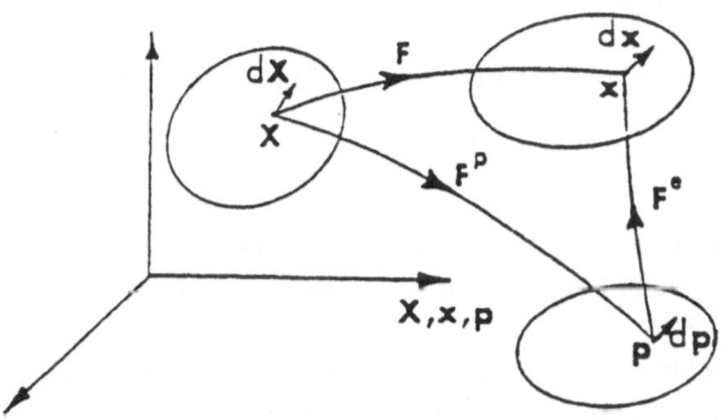

Fig. 2. Elastic–plastic deformation.

terial points X in that configuration. The positions of these material points after elastic–plastic deformation are expressed by the motion x,

$$x = x(X, t), \tag{9}$$

as a function of time t. It is convenient to express the deformation in terms of the deformation gradient F,

$$F_{ij} = \partial x_i/\partial X_j, \qquad F = \partial x/dX, \tag{10}$$

the latter in matrix notation with position vectors expressed as column matrices. Just as illustrated in Fig. 1, following elastic–plastic deformation from the undisturbed configuration X to x, destressing to zero stress occurs from x to p expressed by the superimposed deformation gradient $(F^e)^{-1}$, the inverse of the elastic deformation gradient, F^e, accumulated in the deformation from X to x. Since the configuration p is unstressed, the elastic strain there is zero and the strain in p is therefore totally plastic. Again the destressing procedure has uncoupled the plastic strain from the elastic. From the geometry of sequentially imposed deformations, the plastic deformation gradient constitutes the initial total deformation followed by elastic destressing which is expressed by the matrix product relation

$$F^p = (F^{e-1})F. \tag{11}$$

The purely plastically deformed configuration $p(x, t)$ is introduced in order to simplify the formulation of the elastic–plastic constitutive relation. It is a conceptual local configuration associated with each material neighborhood which places no restriction on the material continuity of the elastically plastically deformed physical state. It contrasts with the situation when surface tractions are removed from a nonhomogeneously plastically deformed body which is generally left in a state of residual stress and hence of retained elastic strains.

Premultiplying both sides of (11) by F^e gives (Ref. 10)

$$F = F^e F^p. \tag{12}$$

This matrix product relation is in general noncommutative and so cannot result in an additive uncoupling of elastic and plastic variables as occurs in the infinitesimal strain case.

The plasticity law is of incremental or flow type which can be expressed in terms of the rate of deformation, the symmetric part of the velocity gradient. The particle velocity in the state x is given by

$$v = \partial x/\partial t|_X \tag{13}$$

and so the velocity gradient L in the configuration x is

$$L = \frac{\partial v}{\partial x} = \frac{\partial v}{\partial X}\frac{\partial X}{\partial x} = \dot{F}F^{-1}, \tag{14}$$

where the superposed dot denotes the material derivative, time differentiation at a fixed material particle X. Substituting the expression (12) for F then gives

$$L = \dot{F}^{e}F^{e^{-1}} + F^{e}\dot{F}^{p}F^{p^{-1}}F^{e^{-1}}. \tag{15}$$

The unstressed configuration p is not uniquely determined since arbitrary rotations of material neighborhoods will leave them unstressed. On the basis of the polar decomposition theorem such rotations can be chosen so that the elastic deformation gradient expresses pure strain without rotation, so that without loss of generality F^e can be selected to be symmetric,

$$F^{e} = V^{e}, \qquad V^{e} = V^{eT}. \tag{16}$$

The rate of deformation D in the current configuration x, the symmetric part of L (15) then takes the form

$$D = D^{e} + (V^{e}D^{p}V^{e^{-1}})_{S} + (V^{e}W^{p}V^{e^{-1}})_{S}, \tag{17}$$

where D^e, W^e, D^p and W^p are the symmetric and antisymmetric parts of $\dot{V}^{e}V^{e^{-1}}$ and $\dot{F}^{p}F^{p^{-1}}$, respectively, and the subscript S denotes the symmetric part.

In general, all three terms on the right-hand side of (17) are nonzero which indicates a very involved coupling of the elastic deformation V^e, the plastic deformation rate D^p, and spin W^p in combining to generate the total rate for deformation D. It is clearly impossible to express the rate of total deformation in terms of elastic and plastic rates of deformation only in order to combine the elastic and plastic laws to generate a constitutive relation involving rate of stress and rate of total strain. Many authors have simply defined the sum of the last two terms in (17) to constitute the rate of plastic strain and hence obtained the total rate of deformation as the sum of elastic and plastic components. However, this clearly hides a much more involved coupling between elastic and plastic variables including the plastic spin W^p.

The deformation of a material element dX in the straining–destressing experiment depicted in Fig. 2 is given by

$$dx = FdX, \tag{18a}$$

$$dp = F^{\mathrm{P}}dX, \tag{18b}$$

$$dx = F^{\mathrm{e}}dp, \tag{18c}$$

so that the element lengths in X, x, and p, dS, ds, and ds^{P}, respectively, are given by

$$dS^2 = dX^{\mathrm{T}}dX, \tag{19}$$

$$ds^2 = dx^{\mathrm{T}}dx = dX^{\mathrm{T}}F^{\mathrm{T}}FdX = dp^{\mathrm{T}}F^{\mathrm{e}^{\mathrm{T}}}F^{\mathrm{e}}dp, \tag{20}$$

$$(ds^{\mathrm{P}})^2 = dX^{\mathrm{T}}F^{\mathrm{p}^{\mathrm{T}}}F^{\mathrm{p}}dX. \tag{21}$$

Thus, the total strain, E, from the initial configuration X to the deformed configuration x is given by

$$ds^2 - dS^2 = dX^{\mathrm{T}}(2E)dX = dX^{\mathrm{T}}(F^{\mathrm{T}}F - I)dX. \tag{22}$$

Similarly the plastic strain E^{p} from X to p satisfies

$$(ds^{\mathrm{p}})^2 - dS^2 = dX^{\mathrm{T}}(2E^{\mathrm{p}})dX = dX^{\mathrm{T}}(F^{\mathrm{p}^{\mathrm{T}}}F^{\mathrm{p}} - I)dX. \tag{23}$$

The elastic strain, E^{e}, associated with the deformation from the destressed, and hence elastically unstrained, configuration p to the current configuration x is determined by

$$ds^2 - (ds^{\mathrm{p}})^2 = dp^{\mathrm{T}}(2E^{\mathrm{e}})dp = dp^{\mathrm{T}}(V^{\mathrm{e}}V^{\mathrm{e}} - I)dp, \tag{24}$$

since $F^{\mathrm{e}} = V^{\mathrm{e}}$ by (16). Equations (22), (23), and (24) determine

$$2E = F^{\mathrm{T}}F - I, \tag{25a}$$

$$2E^{\mathrm{p}} = F^{\mathrm{p}^{\mathrm{T}}}F^{\mathrm{p}} - \mathrm{I}, \tag{25b}$$

$$2E^{\mathrm{e}} = V^{\mathrm{e}}V^{\mathrm{e}} - I. \tag{25c}$$

Using (18b) to change the reference configuration p in (24) to X gives

$$ds^2 - (ds^{\mathrm{p}})^2 = dx^{\mathrm{T}}F^{\mathrm{p}^{\mathrm{T}}}(2E^{\mathrm{e}})F^{\mathrm{p}}dX = dX^{\mathrm{T}}(2\bar{E}^{\mathrm{e}})dX, \tag{26}$$

in which $\bar{E}^{\mathrm{e}}$ defines the elastic strain expressed from the basis of the initial undeformed configuration X. Subtracting (23) from (22) gives

$$\begin{aligned} ds^2 - (ds^p)^2 &= dX^T(F^TF - F^{p^T}F^p)dX \\ &= dX^T2(E - E^p)dX = dX^T2\bar{E}^e dX, \end{aligned} \tag{27}$$

showing that the strain components $\bar{E}^e$ and E^p are additive,

$$E = \bar{E}^e + E^p. \tag{28}$$

Transforming the basis to the elastically unstrained configuration p through (18b) gives

$$ds^2 - (ds^p)^2 = dp^T F^{p^{-T}} 2\bar{E}^e F^{p^{-1}} dp. \tag{29}$$

Comparison of (24) and (29) determines

$$\bar{E}^e = F^{p^T} E^e F^p. \tag{30}$$

In view of (28) and (30), E^e and E^p are far removed from additivity to give the total strain E in the presence of finite plastic strain.

Equation (26) can be interpreted as a definition of elastic strain, $\bar{E}^e$, in terms of convected corrdinates X deforming with the unstressed, and hence elastically unstrained, configuration p, thus deforming with the purely plastic flow. This approach was adopted by Sedov (Ref. 11), Skanchenko and Sporykhin (Ref. 12) and pointed out by N. Fox (see Green and Naghdi, Ref. 13). As expressed by (28), this approach achieves additivity of elastic and plastic strains, but at the expense of involving the plastic deformation in the elastic constitutive relation. This is evident in (30) which expresses $\bar{E}^e$ in terms of E^e the elastic strain based on the elastically unstrained reference configuration p, the simplest and standard choice in elasticity theory.

Since the range of the elastic strain is small, of the order 10^{-3}, Hooke's law provides an appropriate form for experimentally checking the elastic law in the presence of plasticity. The elastic moduli of polycrystalline metals, measured as tangent moduli in the current elastic region, are known to be insensitive to prior plastic flow, which is consistent with the physical theory that distortion of the crystal lattice, which does not change with plastic flow, constitutes the main contribution to elastic strain. In the presence of very large plastic strain, preferred orientation of the crystallographic planes in the crystallites and oriented distortion of the shapes of the crystallites can produce some elastic anisotropy but this is usually negligible compared with the strain-induced anisotropy of the plastic characteristics. Howard and Smith (Ref. 14) showed that on unloading following plastic flow in a tension test, the elastic modulus measured by increments of stress and strain was not influenced by prior plastic flow. This circumstance is also evident in that, for elastic structural analysis, handbook values of elastic constants, depending only on the material involved, are adequate, essentially independent of the forming processes used in the manufacture of the structural elements. This is not so for plastic properties. The elastic moduli measured in terms of E^e, the elastic strain from the elastically unstrained reference configuration p, conforms with this circumstance. Thus, adoption of the strain variable E^e uncouples the elastic law from the influence of plastic flow. Equation (30) then shows that an elastic law expressed in terms of $\bar{E}^e$ would involve a strong coupling with the changing plastic deformation at finite strain. This result quantifies earlier discussions of this question (Refs. 9 and 15).

Equation (17) demonstrates that elastic and plastic rate of deformation components are approximately additive to yield the resultant deformation rate D, although there remains some coupling with the elastic deformation gradient V^e. As mentioned in Section 1, the elastic strain is of the order 10^{-3} so that from (25c) in the form

$$V^e V^e = I + 2E^e, \tag{31}$$

it follows that

$$V^e = \mathrm{I} + \delta, \qquad \delta \sim 10^{-3}. \tag{32}$$

Thus, in (17), $V^e D^p V^{e^{-1}}$ is approximately equal to D^p and $V^e W^p V^{e^{-1}}$ approximately equal to W^p, and taking the symmetric parts,

$$D \simeq D^e + D^p. \tag{33}$$

In many cases this can supply a satisfactory approximation for stress-analysis evaluations.

In the special case when principal axes of the applied strain, rate of strain or stress are fixed in a body homogeneously loaded, and the material has initially isotropic elastic and plastic properties, the elastic and plastic rates of deformation and the stress generated will have the same principal axes. The elastic and plastic deformation gradient matrix product in (12) will then be commutative and summation of elastic and plastic strains and strain rates to give the resultant values will be valid in terms of logarithmic or natural strain as observed by Lee and Wierzbicki (Ref. 16). This accounts for the summability suggested in Fig. 1. However, such straining, for example the tension test, does not provide much of the information needed for general stress analysis, such as plastic strain-induced anisotropy and spin effects, a particular example being data needed to analyze the evolution, including spin, of the back stress in kinematic hardening.

The elastic strain, E^e, based on deformation from the unstressed, and hence elastically unstrained, configuration and the associated deformation gradient V^e (or F^e) variables are commonly used in expressing the finite-deformation-valid elastic constitutive relation. In spite of the small elastic strain magnitudes usual in the presence of plasticity, the nonlinear formulation is needed in elastic–plastic theory as discussed in Section 1 and illustrated in Section 4.

In the following section, examination of the mathematical structure of elastic–plastic continuum theory will elucidate some anomalies which have arisen in attempts to interpret the elastic–plastic coupling relations.

3. Objectivity

In view of the somewhat complicated coupling of the elastic and plastic straining variables evident in (17), it is important to consider objectivity of the kinematics to establish which terms influence the elastic and plastic components of the analysis, respectively. Transformation characteristics are therefore examined of the quantities involved in superimposed rigid-body rotation of the deformed configuration, x, by the proper orthogonal transformation $Q(t)$. The modified values of the variables are indicated by an asterisk. Thus, following Lubarda and Lee (Ref. 17), $X^* = X$ and

$$x^* = Qx. \tag{34}$$

The deformation gradient changes to

$$F^* = QF. \tag{35}$$

Since elastic destressing is considered to occur without rotation, each element of the unstressed configuration must be subjected to the same rotation, $Q(t)$, as the current configuration and this constraint must be introduced into the objectivity requirements. Thus,

$$F^{p*} = QF^p, \tag{36}$$

$$V^{e*} = QV^eQ^T. \tag{37}$$

It is clear that (35), (36), and (37) are consistent with (12) and (16).

The velocity gradient L [see (15)], in the form

$$L = \dot{V}^e V^{e-1} + V^e(D^p + W^p)V^{e-1}, \tag{38}$$

becomes

$$L^* = \dot{Q}Q^T + Q\dot{V}^e V^{e-1}Q^T + QV^e\dot{Q}^TQV^{e-1}Q^T$$
$$+ QV^eQ^T\dot{Q}V^{e-1}QT + QV^e\dot{F}^pF^{p-1}V^{e-1}Q^T, \tag{39}$$

where the first line on the right-hand side comes from the elastic rate component term in (38) and the second line from the plastic rate components. Substituting the elastic and plastic rate of deformation and spin terms D^e, D^p, W^e, and W^p used in (17) gives (39) in the form

$$L^* = \dot{Q}Q^T + Q(D^e + \mathrm{W}^e)Q^T + QV^e\dot{Q}^TQV^{e-1}Q^T$$
$$+ QV^eQ^T\dot{Q}V^{e-1}Q^T + QV^e(D^p + W^p)V^{e-1}Q^T. \tag{40}$$

Since Q is an orthogonal tensor, differentiation of $Q^TQ = I$ gives

$$\dot{Q}^TQ + Q^T\dot{Q} = 0, \tag{41}$$

so that two terms in (40) cancel. Taking the symmetric and antisymmetric parts of (40) then gives

$$\mathrm{D}^* = Q[D^e + (V^eD^pV^{e-1})_S + (V^eW^pV^{e-1})_S]Q^T, \tag{42a}$$

$$W^* = \dot{Q}Q^T + Q[W^e + (V^eD^pV^{e-1})_A + (V^eW^pV^{e-1})_A]Q^T, \tag{42b}$$

where the subscript A denotes the antisymmetric part. Thus, the rate of total deformation is objective.

In order to check objectivity of an elastic–plastic stress-analysis problem it is of course necessary to include the influence of stress on the deformation. In the previous section it was explained that for polycrystalline metals the influence of plastic flow on elastic properties was in general negligible. Even at strains of the order 1%, the plastic yield contribution can exhibit severe anisotropy as shown, for example, by Helling, Miller, and Stout (Ref. 18) for several ductile metals. As observed in Section 2, with essentially no loss of applicability, linear isotropic elasticity with constant moduli can be adopted in the presence of anisotropic plastic response; the latter here in the form of combined isotropic–kinematic hardening. The restriction of isotropic elasticity implies that rotation of the unstressed state does not affect the elastic strain generated by an applied stress. This permits the choice defined in (16) that $F^e = V^e = V^{eT}$ without modifying the elastic law.

Since a nonlinear finite-deformation-valid law is needed, the form

$$\sigma = \frac{2}{\det(V^e)} V^e \frac{\partial\psi}{\partial C^e} V^{eT} \tag{43}$$

is adopted in which σ is the Cauchy or true stress, $\psi(C^e)$ is the isotropic strain-energy function, and $C^e = V^e V^e$ is the Cauchy–Green tensor. Since in this case the three tensors on the right-hand side have the same principal directions, they combine to form a function of C^e only so that the simpler form,

$$\sigma = \frac{2}{\det(V^e)} C^e \frac{\partial \psi}{\partial C^e}, \tag{44}$$

can be utilized with $(\det V^e)^2 = (\det C^e)$. Note that because of (16), the right Cauchy–Green tensor,

$$C^e = V^{eT} V^e = V^e V^e = V^e V^{eT} = B^e, \tag{45}$$

the left Cauchy–Green tensor.

As has been observed (Ref. 19), the last term in the square brackets in (42) comprises an elastic component of the rate of strain and should therefore be combined with D^e. It is the symmetric part of a sequence of three motions. The final postmultiplier $V^{e^{-1}}$, which operates first, represents the elastic deformation associated with destressing. Then follows a rotation after which the elastic strain is reapplied. This sequence results in a change in the elastic deformation equal to the difference of elastic deformations due to the stress applied in incrementally rotated directions relative to the body. This will appear as a residual strain but is part of the elastic strain increments and should be so allocated. However, many authors have erroneously simply allocated the term D^e alone to the elastic deformation rate and the remainder involving $V^e L^p V^{e^{-1}}$ to plasticity. A similar situation arises with regard to the two self-canceling nonobjective terms in (40). That the second one expresses an elastic phenomenon, although emanating from the plastic part of L in (38), provides a reason for self-canceling these terms before identifying the elastic and plastic contributions.

In assessing objectivity, arbitrary rigid body rotation of the deformed configuration x has been considered. The configuration x could comprise any physically achievable state such as the unstressed state p following plastic flow for materials which remain elastic on destressing to zero stress. As plastic strains increase, the Bauschinger effect, the lowering of the magnitude of the yield stress in reverse straining, can grow so that the yield stress on reversed loading is greatly reduced. This trend can continue to the extent that negative plastic flow can occur before the body is unloaded. The experiments by Helling et al. (Ref. 18) illustrate this condition. The inability to unload elastically to zero stress prevents direct measurement of the plastic strain in a stressed condition by simply reducing the stress elastically to zero stress, and determining the plastic strain since the elastic strain is zero at zero stress. If no plastic flow occurs on unloading, the plastic strain measured at zero stress applies throughout the unloading process.

When reversed or negative plastic flow occurs on unloading before zero stress is reached, a formal measure of plastic strain achieved before the unloading commenced can be determined by extrapolating the strain to zero stress on the basis of the known elastic law. This is equivalent to assuming that all plastic-flow mechanisms, such as dislocations, are locked in place. Although this provides only a formal measure of plastic strain, it will give the correct value for the strain at any stress which can be reached by means of elastic straining only, because of the reversibility of the elastic law. This includes all stresses within and on the yield surface. Thus, in a stress-analysis computation, the application of the known elastic law will determine correct stress–strain trajectories as long as they lie within or on the yield surface. Used in an elastic–plastic code, such stress–strain points will be valid until the yield surface is reached when the elastic–plastic routine will come into play involving changing configuration of the yield surface (hardening) and of the plastic strain tensor. Whereas with a less severe Bauschinger effect the unstressed configuration can be reached physically without change in the plastic strain, such a point at zero stress on the stress–strain trajectory cannot be reached by means of elastic strains only in the presence of a more severe Bauschinger effect. If the plasticity hardening laws which determine the changing shape of the yield surface are known, the elastic–plastic code will operate with the formal definition of plastic-strain just as it does when the unstressed condition can be reached physically.

In an elastic–plastic theory of the type discussed in this paper, the need for a full invariance condition has been stated (Ref. 13) which requires that: if two physically possible configurations arise in the theory, independent checks of objectivity are needed by considering independent rigid-body rotations superimposed simultaneously on these configurations. In the development of the theory discussed in this paper the two configurations involved are the finally deformed state x in Fig. 2 subjected to combined elastic–plastic deformation and a destressed configuration p subjected only to the plastic strain, since the stress and hence the elastic component of strain are zero. Thus, rigid-body rotation of the destressed configuration leaves it unstressed; this provides alternative unstressed states on the basis of which to develop the theory. Since, as already discussed, the consideration of isotropic elasticity is appropriate, the basic theory is not modified by the choice of this rotation. Thus, without essential loss of applicability, the rotation was selected such that destressing occurred without rotation which simplified the analysis. These configurations being thus linked, objectivity was checked only for rotation of the configuration x, and p was subjected to the same rotation because of the de-stressing choice made. If destressing to zero stress is achievable, x can be varied to pass through this state simply by destressing the body, so that the objectivity of p is already established through that of x. On this basis it was suggested (Ref. 9) that the separate independent check is unnecessary.

Since the initial development of the theory, it has been realized, as already discussed in this section, that the theory can be applied when, because of a strong Bauschinger effect, destressing to zero stress without generating further plastic flow is not possible. This involves utilizing the known elastic law extrapolated to zero stress involving the formal plastic-strain concept already discussed. Destressing to zero stress is then strictly a thought experiment, but does permit general applicaton of the theory. In analyzing an actual stress-analysis problem all particles of the body will pass only through physically achievable configurations x. Only the associated configurations p are formal. They are not in fact explicitly involved in the usual elastic–plastic computer program which, at each step, evaluates only increments of stress and strain from the current state. In this case it seems that according to the discussion by Casey and Naghdi (Ref. 20), since p is not physically achievable, the independent objectivity check proposed of the configuration p is not needed and the approach presented by Lee (Ref. 9) and in the present paper fulfills the suggested necessary requirements. This appears to lead to an inconsistency in the suggested requirement of full invariance when the configuration p is achievable, but not in the case of a strong Bauschinger effect for exactly the same analytical structure.

If an independent superimposed spin $\bar{Q}$ of the destressed configuration p is adopted, it will be necessary to replace V^e by a nonsymmetric tensor F^e since the destressing deformation may then involve a rotational component. It turns out that applying the full invariance condition only modifies the terms which cancel in (39) and (40) and the objective relations (42a) and (42b) are recovered with F^e replacing V^e, Q being the rotation superimposed on the current configuration x only.

4. Elastic–Plastic Constitutive Relations

In order to avoid complications in the deformation rate equation (17) because of the coupling with the plastic spin W^p, elimination of this in favor of deformation-rate type terms was achieved by eliminating W^p from the velocity gradient equation (38) (Ref. 21), as described below. In order to be able to separate elastic and plastic terms, the contribution of these processes to the rate of deformation D in the configuration x is deduced. $\hat{D}^e$ expresses the contribution to the symmetric part of the velocity gradient in x due to the changing elastic deformation V^e from the configuration p. $\hat{D}^p$ expresses the contribution to the symmetric part of the velocity gradient in x due to the velocity field in p caused by plastic flow modified by the elastic straining from p to x. Thus, the total rate of deformation in x is expressed in the form

$$D = \hat{D}^e + \hat{D}^p. \tag{46}$$

These components of rate of deformation are additive since they are symmetrical parts of the gradients of velocity contributions in the same configuration x. Such velocity vectors are additive to give the resultant velocity field.

Equation (38) can be modified by premultiplying it by $V^{e^{-1}}$ and postmultiplying by V^e to give

$$D^p + W^p = -V^{e^{-1}}\dot{V}^e + V^{e^{-1}}[\hat{D}^e + \hat{D}^p + W]C^eV^{e^{-1}}. \tag{47}$$

Taking the symmetric part eliminates W^p to give

$$D^p = (-V^{e^{-1}}\dot{V}^e)_S + V^{e^{-1}}[(\hat{D}^e + \hat{D}^p + W)C^e]_S V^{e^{-1}}. \tag{48}$$

Equating the plasticity terms containing D^p and $\hat{D}^p$ gives

$$D^p = V^{e^{-1}}(\hat{D}^pC^e)_S V^{e^{-1}}, \tag{49}$$

hence

$$C^e\hat{D}^p + \hat{D}^pC^e = 2V^eD^pV^e. \tag{50}$$

Thus, $2V^eD^pV^e$ is given by a linear elastic operator Λ on $\hat{D}^p$,

$$\Lambda : \hat{D}^p = 2V^eD^pV^e, \tag{51}$$

where Λ is a fourth-order tensor with components

$$\Lambda_{ijk\ell} = C^e_{ik}\delta_{j\ell} = C^e_{j\ell}\delta_{ik}. \tag{52}$$

Having dealt with the plastic part of (48), the remainder takes the following form after pre- and postmultiplying by V^e and utilizing $\dot{C}^e = \dot{V}^eV^e + V^e\dot{V}^e$:

$$\dot{C}^e - WC^e + C^eW = \overset{\circ}{C}{}^e = C^e\hat{D}^e + \hat{D}^eC^e = \Lambda : \hat{D}^e, \tag{53}$$

where $\overset{\circ}{C}{}^e$ is the Jaumann derivative of C^e. These terms are all associated with elastic deformation. Thus, the choice of the plastic relation (49) has produced a separation of elastic and plastic rate terms, $\hat{D}^e$ and $\hat{D}^p$, with coupling only between $\hat{D}^p$ and elastic strain because this modifies the plastic contribution to the strain rate in the configuration x. The spin W^p does not appear at all and the spin W plays an independent role in the constitutive relation through the Jaumann derivative and also in the variational principle for stress analysis.

Thus, making use of the inverse operator Λ^{-1}, the kinematic relations take the form: additivity of deformation rates $\hat{D}^e$ and $\hat{D}^p$ [see (46)] and the expressions for these in terms of familiar deformation variables,

$$\hat{D}^e = \Lambda^{-1} : \overset{\circ}{C}{}^e, \tag{54}$$

$$\hat{D}^p = \Lambda^{-1} : (2V^e D^p V^e), \tag{55}$$

which display a structure analogous to classical infinitesimal strain theory: $\hat{D}^e$ comprising the rate of elastic strain, but modified by an operator depending on the elastic strain, and $\hat{D}^p$ comprising the plastic rate of deformation or stretching tensor similarly modified by the elastic strain. The modification arises because motion in the configuration x involves the plastic deformation on which is superimposed elastic strain. No restrictions with regard to anisotropic plastic properties have been assumed.

In view of the fact that plastic flow is incompressible, hence $\det(F^p) = 1$, it is advantageous to introduce the Kirchhoff stress, $\tau = \sigma \det(F)$, into the elastic constitutive relation (45), which then takes the form

$$\tau = 2C^e(\partial\psi/\partial C^e). \tag{56}$$

Moreover, τ is appropriate for the flow or incremental type plastic constitutive relation since it involves power dissipation per unit initial volume of material, and this generates symmetric elastic–plastic stiffness matrices, important for stress evaluations.

Since the kinematic relation (53) involves the Jaumann derivative of C^e, the appropriate rate form of (56), needed for combining with the plastic flow type relation, is the Jaumann rate which determines

$$\overset{\circ}{\tau} = L : \overset{\circ}{C}{}^e, \tag{57}$$

where

$$L_{ijk\ell} = 2\delta_{ik}\left(\frac{\partial\psi}{\partial C^e}\right)_{ij} + C^e_{im}\left(\frac{\partial^2\psi}{\partial C^{e^2}}\right)_{mjk\ell}. \tag{58}$$

Substituting for $\overset{\circ}{C}{}^e$ from (53) into (57) and incorporating (46) gives

$$\overset{\circ}{\tau} = (L : \Lambda) : (D - \hat{D}^p). \tag{59}$$

Inverting (49) and rearranging the terms yields

$$\hat{D}^p = D - (L : \Lambda)^{-1} : \overset{\circ}{\tau}, \tag{60}$$

with the symmetries

$$(L:\Lambda)_{ijk\ell} = (L:\Lambda)_{ij\ell k} = (L:\Lambda)_{jik\ell} = (L:\Lambda)_{k\ell ij}, \tag{61}$$

so that, as shown by Hill (Ref. 22), $(L:\Lambda)^{-1} : \overset{\circ}{\tau}$ is derivable from an elastic rate potential which means that $(\hat{D}^e\, dt)$ will be recovered on unloading of the Jaumann stress increment $(\overset{\circ}{\tau}\, dt)$. The Jaumann stress increment is used since it eliminates the effect on the stress increment of a possible spin difference in the loading and unloading processes. Since the unloading is elastic, $(\hat{D}^p\, dt)$ is the residual increment of the strain $(D\, dt)$ after a loading/unloading cycle of the stress increment. This conclusion provides a precise interpretation of the reversible and residual strain increments referred to in Section 2.

Equation (60) contains an expression for the part of the rate of deformation or stretching tensor, D, due to the elastic straining which takes the form of an elastic operator on the Jaumann derivative of the Kirchhoff stress. The formulation places no constraint on the constitutive relation for the plastic component. An analogous relation is needed for the plastic-straining contribution to D in terms of $\overset{\circ}{\tau}$, so that an elastic–plastic rate operator Π can be deduced which determines the stress rate from the deformation rate D,

$$\overset{\circ}{\tau} = \Pi : D. \tag{62}$$

This can then be introduced into computer codes for elastic–plastic stress and deformation analysis. In Agah-Tehrani *et al.* (Ref. 21) such a finite-deformation-valid plastic analysis involving strain-induced anisotropy in the form of combined isotropic–kinematic hardening is presented. It includes a general constitutive relation for the evolution of the kinematic-hardening back stress.

An aspect of plasticity theory which has generated considerable discussion over the years has been the role played in plasticity theory by the variable plastic strain.

Elasticity at finite strain, involving a function relation between stress and deformation, is most simply expressed in terms of strain from the unstressed reference state. Whatever the deformation, the latter can always be retrieved by reducing the stress to zero and remains as a permanent unstrained basis for expressing the kinematic structure. The stress arises from stretching the material elements due to the deformation from the undeformed configuration to the deformed configuration so that the strain incorporates information concerning both of these states. In contrast, plastic flow involves permanent deformation caused by motion of dislocations and other imperfections through the crystal lattice which produces transport of deformation through the lattice. This causes a permanent macroscopic strain to spread such that on removal of the driving stress the initial unstrained configuration is not retrieved. The plastically deformed configuration then comprises the new unstressed configuratioan to be further deformed by the next increment of plastic strain. Thus, at each instant the stress drives the motion of the imperfections through the lattice, maintaining the rate of deformation. The process is thus associated with rate of strain about the current state expressed mathematically by the rate of deformation or stretching tensor, the symmetric part of the velocity gradient, or the rate of strain at the instant the strain is zero. There is an influence of the previous deformation in that flow of the imperfections generates greater disturbance of the crystal lattice which can inhibit continued flow thus causing strain hardening. When the displacement is infinitesimal the usual various definitions of strain coincide because the displacement variable expresses the deformation and the basic configuration is considered not to have changed. Thus, the rate of strain and the symmetrical part of the velocity gradient coincide and classical elastic–plastic theory simply introduces additivity of elastic and plastic strains and a rate of plastic strain into the flow law. In the case of finite deformation, the rate of change of the various strain definitions and the stretching tensor do not coincide and care must be exercised in formulating the plastic flow rule. Also the influence of the superimposed usually-small elastic strain must be included, as, for example, expressed by the pre- and postmultiplication terms involving V^e in (17).

Differentiation of the plastic Lagrange strain (25b) yields

$$\dot{E}^p = F^{p^T} D^p F^p, \tag{63}$$

as shown by Green and Naghdi (Ref. 13). At finite plastic strain this indicates a major disparity between the strain rate and the stretching tensor which expresses the influence of the initial configuration on the strain rate definition. The description of the physical mechanism of plastic flow shows that D^p and not $\dot{E}^p$ expresses the appropriate rate of plastic flow and hence of strain hardening. Thus, based on $\dot{E}^p$, in tension the predicted hardening would increase more rapidly ($F^p_{11} > 1$) than with natural strain, while in compression more slowly ($F^p_{11} < 1$), in disagreement with experiment. Tensile and compressive true stress–natural strain curves for ductile metals are found to be quite similar.

The difficulty associated with the use of the variable, plastic strain, and hence plastic strain rate in applications of the theory of plasticity is compounded since the structural elements or workpieces concerning which stress analyses are sought have universally been subjected to plastic flow during their manufacture by forming or forging and such histories of deformation are not available. Thus, plasticity theory must permit carrying out stress and deformation analysis of a structural component without knowing its previous strain history. Measuring the current yield surface and hardening coefficients must permit consistent determination of the future stress and deformation variations. The use of the plastic stretching tensor in the flow law permits this. The need to know deformation from some virgin state, which cannot be reconstructed from the current state as is the unstressed reference state in elasticity, is clearly inappropriate for the application of plasticity theory.

In plasticity literature the term "strain rate" is commonly used to signify rate effects in general which is appropriate in infinitesimal-displacement theory. However, more precision is now called for, but traditional usage inhibits this as is even evident in the introductory discussion in the present paper. Such general interpretation of strain rate probably arose since initially such considerations were limited to the analysis of tension tests expressed in terms of logarithmic strain for which strain rate and the stretching tensor are identical and additivity of elastic and plastic components is valid. This only applies in general when principal directions are fixed in the body.

Acknowledgement

This work was supported in part by the U.S. Army Research Office through a contract with RPI. The author expresses his appreciation for this support.

References

1. Rivlin, R. S., *The Hydrodynamics of Non-Newtonian Fluids, I*, Proceedings of the Royal Society of London, Vol. 193, pp. 260–281, 1948.
2. Rivlin, R. S. *Large Elastic Deformations of Isotropic Materials, I, Fundamental Concepts*, Philosophical Transactions of the Royal Society of London, Vol. A240, pp. 459–490, 1948.
3. Rivlin, R. S. *Large Elastic Deformations of Isotropic Materials II, Uniqueness Theorems for Pure Homogeneous Deformation*, Philosophical Transactions of the Royal Society of London, Vol. A240, pp. 491–508, 1948.
4. Rivlin, R. S. *Large Elastic Deformations of Isotropic Materials, III, Some Simple Problems in Cylindrical Polar Coordinates*, Philosophical Transactions of the Royal Society of London, Vol. A240 pp. 509–525, 1948.
5. Rivlin, R. S., *Large Elastic Deformations of Isotropic Materials, IV, Further Developments of the General Theory*, Philosophical Transactions of the Royal Society of London, Vol. A241, pp. 379–397, 1948.
6. Rivlin, R. S., and Saunders, D. W., *Large Elastic Deformations of Isotropic Materials, VII, Experiments on the Deformation of Rubber*, Philosophical Transactions of the Royal Society of London, Vol. A243, pp. 251–288, 1951.
7. Hill, R., *The Mathematical Theory of Plasticity*, Clarendon Press, Oxford, England, 1950.
8. Rice, J. R., *A Note on the 'Small Strain' Formulation for Elastic–Plastic Problems*, Technical Report N00014-67-A-000318, Division of Engineering, Brown University, 1970.

9. LEE, E. H., *Some Comments on Elastic–Plastic Analysis*, International Journal of Solids and Structures, Vol. 17, pp. 859–872, 1981.
10. LEE, E. H., *Elastic–Plastic Deformation at Finite Strains*, Journal of Applied Mechanics, Vol. 36, pp. 1–6, 1969.
11. SEDOV, L. I., *Introduction to Mechanics of Continua*, M. Fizmatgiz, Moscow, Russia, 1962.
12 SKANCHENKO, A. V., and SPORYKHIN, A. N., *On the Addivity of Tensors of Strains and Displacements for Finite Elastoplastic Deformations*, Prikladnaia Matematika i Mekhanika (PMM), Vol. 44, pp. 1145–1146, 1947.
13. GREEN, A. E., and NAGHDI, P. M., *Some Remarks on Elastic–Plastic Deformation at Finite Strain*, International Journal of Engineering Science, Vol. 9, pp. 1219–1229, 1971.
14. HOWARD, J. V., and SMITH, S. L. *Recent Developments in Tensile Testing*, Proceedings of the Royal Society of London, Vol. A107, pp. 113–125, 1925.
15. LEE, E. H., and LIU, D. T., *Finite-Strain Elastic–Plastic Theory with Application to Plane-Wave Analysis*, Journal of Applied Physics, Vol. 38, pp. 19–27, 1967.
16. LEE, E. H., and WIERZBICKI, T., *Analyses of the Propagation of Plane Elastic–Plastic Waves of Finite Strain*, Journal of Applied Mechanics, Vol. 34, pp. 931–936, 1967.
17. LUBARDA, V. A. and LEE, E. H., *A Correct Definition of Elastic and Plastic Deformation and Its Computational Significance*, Journal of Applied Mechanics, Vol. 48, pp. 35–40, 1981.
18. HELLING, D. E., MILLER, A. K., and STOUT, M. G., *An Experimental Investigation of the Yield Loci of 1100-0 Aluminum, 70 : 30 Brass, and an Overaged 2024 Aluminum Alloy after Various Prestrains*, Journal of Engineering Materials and Technology, Vol. 108, pp. 313–320, 1986.
19. LEE, E. H., and MCMEEKING, R. M., *Concerning Elastic and Plastic Components of Deformation*, International Journal of Solids and Structures, Vol. 16, pp. 715–721, 1980.
20. CASEY, J., and NAGHDI, P. M., *Discussion of : A Correct Definition of Elastic and Plastic Deformation and Its Computational Significance*, Journal of Applied Mechanics, Vol. 48, pp. 983–984, 1981.
21. AGAH-TEHRANI, A., LEE, E. H., MALLETT, R. L., and ONAT, E. T., *The Theory of Elastic–Plastic Deformation at Finite Strain with Induced Anisotropy Modeled as Combined Isotropic–Kinematic Hardening*, Journal of the Mechanics and Physics of Solids, Vol. 35, pp. 519–539, 1987.
22. HILL, R., *Some Basic Principles in the Mechanics of Solids without a Natural Time*, Journal of the Mechanics and Physics of Solids, Vol. 7, pp. 209–225, 1959.

Supplementary Reference

23. CASEY, J., and NAGHDI, P. M., *A Remark on the Use of the Decomposition* $F = F_e F_p$ *in Plasticity*, Journal of Applied Mechanics, Vol. 47, pp. 672–675, 1980.

10

Nonlinear Thermoelastic Constitutive Equations for Transversely Isotropic and Orthotropic Materials

J. M. O'NEILL AND A. J. M. SPENCER

Abstract. We consider thermoelastic materials in which the Helmholtz free energy is a function of the deformation and the temperature and the heat flux vector is a function of the deformation, the temperature gradient, and the temperature. Material symmetries restrict the manner in which the stress and the heat flux depend on the deformation and temperature gradient. Explicit results are known for isotropic materials. Here we derive corresponding results for materials whose symmetry groups are: (1) one of the five symmetry groups which include the set of rotations about an axis (which can collectively be described as the transverse isotropy groups) or (2) the symmetry group which corresponds to orthotropic symmetry. These results are derived and simplified by application of representation theorems for vector and tensor polynomials which were established by Rivlin and others. The distinctions between the various material symmetries are illustrated by applying the results to several problems of material in simple shear, in which the planes and directions of shear are either normal or parallel to the material symmetry axes, and subject to uniform temperature gradients, in directions which are also either normal or parallel to the material symmetry axes.

Key Words. Finite thermoelastic materials, Helmholtz free energy, canonical forms, heat flux vector, transversely isotropic materials, orthotropic symmetry, deformation gradient tensor, Cauchy stress, first Piola–Kirchhoff stress, invariance under rigid rotations, dissipation inequality, symmetry group, scalar in-

J. M. O'Neill • Research Assistant, Department of Theoretical Mechanics, University of Nottingham, Nottingham NG7 2RD, England. **A. J. M. Spencer** • Professor Emeritus, Department of Theoretical Mechanics, University of Nottingham, Nottingham NG7 2RD, England.

Nonlinear Effects in Fluids and Solids, edited by M. M. Carroll and M. Hayes, Plenum Press, New York, 1996.

variant, form-invariant vector function, continuous groups, minimal integrity bases, Rivlin–Spencer identity, simple shearing deformations.

1. Introduction

The general theory of finite thermoelasticity is well established; two of the many available accounts are those in the first few sections of Carlson (Ref. 1) and in Chadwick and Seet (Ref. 2).

In the standard formulation the Helmholtz free energy A is a function of the deformation and the temperature, and serves as a potential function for the stress and the entropy, and the heat flux vector $\mathbf{Q}$ is a vector function of the deformation, the temperature, and the temperature gradient.

Any material symmetry restricts the manner in which A and $\mathbf{Q}$ may depend on their arguments. In the case of isotropic materials (with or without a center of symmetry), explicit representations for A and $\mathbf{Q}$ were derived by Pipkin and Rivlin (Ref. 3). These results are also given in Green and Adkins (Ref. 4) and elsewhere. Chadwick and Seet (Ref. 2) gave corresponding results for certain classes of transversely isotropic materials, but did not consider all of the five symmetry groups which include among their symmetry transformations the set of rotations about a given axis. Further results for all five of these symmetry groups were obtained by Spencer (Ref. 5) but these results are not optimal in the sense that the canonical forms given there contain some redundant terms. Some preliminary results were also given in Ref. 5 for materials with orthotropic symmetry.

In this paper we investigate in more detail the canonical forms of the Helmholtz free energy function and the heat flux vector for two classes of anisotropic thermoelastic materials. The first of these classes comprises materials whose symmetry group is one of the five symmetry groups which contain all rotations about a given axis as symmetry transformations. These we describe loosely as transversely isotropic materials. The second class consists of materials with orthotropic symmetry. For each of these symmetries we obtain explicit representations for A and $\mathbf{Q}$ which, in most cases, are sharper than those which have been recorded previously.

We also describe a number of illustrative examples for each of these material symmetries, in which the material undergoes a variety of simple shearing deformations (with the planes and directions of shear either normal or parallel to the material symmetry axes) and is subject to uniform temperature gradient, in directions which are also either normal or parallel to the symmetry axes. For each of these deformations and temperature fields the heat flux response is determined. These examples illustrate clearly the restrictions which the various material symmetries impose on the heat flux response.

2. Notation and General Theory

We adopt a conventional formulation of thermoelasticity, and in the main follow the notation adopted by Chadwick and Seet (Ref. 2).

All vector and tensor components are components referred to a fixed rectangular coordinate system. Let **X** denote the position vector of a typical particle in a stress-free reference configuration at a uniform reference temperature, and **x** the position vector in a subsequent deformed configuration. The deformation is described by the equations

$$\mathbf{x} = \mathbf{x}(\mathbf{X}), \tag{1}$$

or in components,

$$x_i = x_i(X_R), \qquad i, R = 1, 2, 3.$$

The deformation gradient tensor **F** has components F_{iR}, where

$$F_{iR} = \partial x_i/\partial X_R = x_{i,R}, \tag{2}$$

and the deformation tensors **B**, **C** and their components B_{ij}, C_{RS} are defined as

$$\mathbf{B} = \mathbf{F}\mathbf{F}^{\mathrm{T}}, \qquad B_{ij} = F_{iR}F_{jR} = x_{i,R}\,x_{j,R}, \tag{3a}$$

$$\mathbf{C} = \mathbf{F}^{\mathrm{T}}\mathbf{F}, \qquad C_{RS} = F_{iR}F_{iS} = x_{i,R}\,x_{i,S}. \tag{3b}$$

The Cauchy stress tensor and the first Piola–Kirchhoff stress tensor are denoted by $\boldsymbol{\sigma}$ and $\boldsymbol{\Sigma}$, respectively, where

$$\boldsymbol{\Sigma}^{\mathrm{T}} = (\rho_0/\rho)\mathbf{F}^{-1}\boldsymbol{\sigma}, \tag{4}$$

and ρ_0 and ρ are densities in the reference and deformed configurations, respectively.

The excess temperature above the reference temperature θ_0 is denoted by θ. The spatial temperature gradient **g** has components g_i given by

$$g_i = \partial\theta/\partial x_i = \theta_{,i}, \tag{5}$$

and the spatial heat flux vector $\mathbf{q}$ has components q_i. We also introduce the referential temperature gradient $\mathbf{G}$ and referential heat flux $\mathbf{Q}$, with components G_R and Q_R, respectively, where

$$G_R = \partial\theta/\partial X_R = \theta_{,R}, \qquad \mathbf{G} = \mathbf{F}^{\mathrm{T}}\mathbf{g}, \tag{6a}$$

$$\mathbf{Q} = (\rho_0/\rho)\mathbf{F}^{-1}\mathbf{q}, \qquad \mathbf{q} = (\rho/\rho_0)\mathbf{F}\mathbf{Q}. \tag{6b}$$

With the usual constitutive assumptions, requirements of invariance under rigid rotations of the reference configuration determine that A and $\mathbf{Q}$ have the forms

$$A = A(\mathbf{C}, \theta), \qquad \mathbf{Q} = \mathbf{Q}(\mathbf{C}, \mathbf{G}, \theta), \tag{7}$$

and by standard arguments we derive the constitutive equations for $\boldsymbol{\sigma}$ (components σ_{ij}) and entropy S as

$$\sigma_{ij} = \rho F_{iR} F_{jS}\left[\frac{\partial A}{\partial C_{RS}} + \frac{\partial A}{\partial C_{SR}}\right], \qquad S = \frac{\partial A}{\partial \theta}. \tag{8}$$

The dissipation inequality takes either of the forms

$$\mathbf{Q}\cdot\mathbf{G} \leqslant 0 \quad \text{or} \quad \mathbf{q}\cdot\mathbf{g} \leqslant 0. \tag{9}$$

By using (4) it is possible to work, if desired, in terms of Piola–Kirchhoff rather than Cauchy stress. The relations (6) enable us, if we so wish, to employ $\mathbf{g}$ rather than $\mathbf{G}$ as an independent variable and $\mathbf{q}$ rather than $\mathbf{Q}$ as a dependent variable. In some cases, it is convenient to introduce the deformation tensor $\mathbf{B}$. For definiteness, we shall formulate constitutive equations for A and $\mathbf{Q}$ of the form (7).

If the symmetry group of the material contains a symmetry transformation $\mathbf{M}$, where $\mathbf{M}$ is an orthogonal tensor, then by the standard arguments it follows that the forms of $\mathbf{Q}$ and A are subject to the requirement that

$$\mathbf{M}\mathbf{Q}(\mathbf{C}, \mathbf{G}, \theta) = \mathbf{Q}(\mathbf{M}\mathbf{C}\mathbf{M}^{\mathrm{T}}, \mathbf{M}\mathbf{G}, \theta), \qquad A(\mathbf{C}, \theta) = A(\mathbf{M}\mathbf{C}\mathbf{M}^{\mathrm{T}}, \theta). \tag{10}$$

Thus, under the appropriate transformation group, A is a scalar invariant of $\mathbf{C}$ and θ and $\mathbf{Q}$ is a form-invariant vector function of $\mathbf{C}$, $\mathbf{G}$, and θ. Consequently, the problem of determining explicit expressions for A and $\mathbf{Q}$ reduces to the algebraic problem of determining the invariants of $\mathbf{C}$ and the form-invariant vector functions of $\mathbf{C}$ and $\mathbf{G}$ for each symmetry group considered. Since the scalar variable θ does not enter into the algebraic arguments, henceforth it will be suppressed, but of course it remains present in the physical problem. It is assumed that the same symmetry group governs the forms of both A and $\mathbf{Q}$.

For isotropic materials the symmetry group consists of all orthogonal transformations, if the material has a center of symmetry, and of all rotations (proper orthogonal transformations) if there is no reflectional symmetry. For these cases the reduced forms for A, $\mathbf{Q}$, and $\mathbf{q}$ are given in Refs. 2–5. For other types of material symmetry, including transverse isotropy and the various crystal symmetries, the necessary invariance results are available, and are mainly due to J. E. Adkins (for transverse isotropy) and G. F. Smith and R. S. Rivlin (for the crystal classes). Many of these results are summarized in Ref. 6, with reference to the original works (see also Ref. 7 for recent results relating to transverse isotropy). However, we shall proceed in a slightly different manner which does not make use of these results, and which was also used in Ref. 5.

3. Constitutive Equations for Transversely Isotropic Materials

In this section we extend and sharpen some results given in Ref. 5. We consider materials which are characterized by a single preferred direction, which is defined by a unit vector $\mathbf{a}_0$ in the reference configuration. For convenience we adopt terminology associated with fiber-reinforced materials and call the direction of $\mathbf{a}_0$ the fiber direction and its trajectories fibers; however, it is not necessary for $\mathbf{a}_0$ to represent physical fibers, but merely to determine a preferred direction in the material. If the fibers convect with the material, then in the deformed configuration the new fiber direction $\mathbf{a}$ and the fiber stretch λ_a are given by

$$\lambda_a \mathbf{a} = \mathbf{F}\mathbf{a}_0, \qquad \lambda_a^2 = \mathbf{a}_0\mathbf{C}\mathbf{a}_0. \tag{11}$$

We consider materials with at least symmetry with respect to all rotations which have $\mathbf{a}_0$ as rotation axis. The rotation tensor for a rotation through α about $\mathbf{a}_0$ is denoted $\mathbf{M}(\alpha)$. We also employ the following reflection and rotation tensors:

$\mathbf{R}_T$ = reflection in a plane containing $\mathbf{a}_0$,
$\mathbf{R}_L$ = reflection in the plane perpendicular to $\mathbf{a}_0$,
$\mathbf{D}$ = rotation through π about an axis perpendicular to $\mathbf{a}_0$.

Thus, if we choose rectangular Cartesian axes so that $\mathbf{a}_0 = (0, 0, 1)$, the axis of rotation for $\mathbf{D}$ is $(0, 1, 0)$ and the normal to the reflection plane for $\mathbf{R}_T$ is $(1, 0, 0)$, then the matrices of components of $\mathbf{M}(\alpha)$, $\mathbf{R}_T$, $\mathbf{R}_L$, $\mathbf{D}$ are

$$\mathbf{M}_\alpha = \begin{bmatrix} \cos\alpha & -\sin\alpha & 0 \\ \sin\alpha & \cos\alpha & 0 \\ 0 & 0 & 1 \end{bmatrix}, \quad \mathbf{D} = \begin{bmatrix} -1 & 0 & 0 \\ 0 & 1 & 0 \\ 0 & 0 & -1 \end{bmatrix}, \tag{12a}$$

$$\mathbf{R}_T = \begin{bmatrix} -1 & 0 & 0 \\ 0 & 1 & 0 \\ 0 & 0 & 1 \end{bmatrix}, \quad \mathbf{R}_L = \begin{bmatrix} 1 & 0 & 0 \\ 0 & 1 & 0 \\ 0 & 0 & -1 \end{bmatrix}, \tag{12b}$$

but a special choice of coordinate system is not essential, and $\mathbf{a}_0$ is not necessarily constant, but may depend on $\mathbf{X}$.

The transformations $\mathbf{M}_\alpha$, $\mathbf{D}$, $\mathbf{R}_T$, $\mathbf{R}_L$ generate five distinct continuous groups, as follows:

T_1	:	$\mathbf{M}(\alpha)$	(rotational symmetry),
T_2	:	$\mathbf{M}(\alpha)$, $\mathbf{R}_T$	(usual definition of transverse isotropy),
T_3	:	$\mathbf{M}(\alpha)$, $\mathbf{R}_L$,	
T_4	:	$\mathbf{M}(\alpha)$, $\mathbf{R}_T$, $\mathbf{R}_L$	(also includes $\mathbf{D} = \mathbf{R}_T\mathbf{R}_L$),
T_5	:	$\mathbf{M}(\alpha)$, $\mathbf{D}$.	

It has been shown (Ref. 8) that invariants of $\mathbf{C}$ and $\mathbf{G}$ under any of these groups are isotropic invariants of $\mathbf{C}$, $\mathbf{G}$, and $\mathbf{a}_0$ (although the converse is not always true). Since tables of isotropic invariants of vectors and second-order tensors are available (Ref. 6) we may read off from these tables the set of isotropic invariants of $\mathbf{C}$, $\mathbf{G}$, and $\mathbf{a}_0$. This was done in Ref. 5 to produce the list shown in Table 1, where square brackets denote scalar triple products. (Note: The invariant K_8 was inadvertently omitted from Table 1 in Ref. 5) This set is complete in the sense that any invariant of $\mathbf{C}$, $\mathbf{G}$, and $\mathbf{a}_0$ under any of the groups T_α can be expressed as a function of the listed set. However, certain of the scalar quantities of Table 1 change sign under $\mathbf{R}_T$, $\mathbf{R}_L$, or $\mathbf{D}$; these sign changes are indicated by a $(-)$ sign in the appropriate column of Table 1. Such scalars are not invariant under symmetry groups which include the relevant transformation.

To obtain sets of invariants for the various symmetry groups, we proceed as follows. For the group T_1, every listed scalar in Table 1 is an invariant and this set is, we believe, complete and irreducible. For the group T_2, we see that I_6, K_6, K_7, and K_8 change sign under $\mathbf{R}_T$. Nevertheless, their squares and products in pairs are invariant, so we must give consideration to the inclusion in the list of

Table 1. Isotropic scalar invariants of **C**, **G**, $\mathbf{a}_0$

	Invariant	M(α)	$\mathbf{R}_T$	$\mathbf{R}_L$	D
C	$I_1 = \mathrm{tr}\,\mathbf{C}$	+	+	+	+
	$I_2 = \frac{1}{2}\{(\mathrm{tr}\,\mathbf{C})^2 - \mathrm{tr}\,\mathbf{C}^2\}$	+	+	+	+
	$I_3 = \det \mathbf{C}$	+	+	+	+
G	$K_1 = \mathbf{G} \cdot \mathbf{G}$	+	+	+	+
C, **G**	$K_2 = \mathbf{GCG}$	+	+	+	+
C, $\mathbf{a}_0$	$I_4 = \mathbf{a}_0\mathbf{C}\mathbf{a}_0$	+	+	+	+
	$I_5 = \mathbf{a}_0\mathbf{C}^2\mathbf{a}_0$	+	+	+	+
	$I_6 = [\mathbf{a}_0, \mathbf{C}\mathbf{a}_0, \mathbf{C}^2\mathbf{a}_0]$	+	−	+	−
G, $\mathbf{a}_0$	$K_3 = \mathbf{a}_0 \cdot \mathbf{G}$	+	+	−	−
C, **G**, $\mathbf{a}_0$	$K_4 = \mathbf{a}_0\mathbf{CG}$	+	+	−	−
	$K_5 = \mathbf{a}_0\mathbf{C}^2\mathbf{G}$	+	+	−	−
	$K_6 = [\mathbf{a}_0, \mathbf{G}, \mathbf{C}\mathbf{a}_0]$	+	−	−	+
	$K_7 = [\mathbf{a}_0, \mathbf{G}, \mathbf{C}^2\mathbf{a}_0]$	+	−	−	+
	$K_8 = [\mathbf{a}_0, \mathbf{G}, \mathbf{CG}]$	+	−	+	−

invariants of

$$I_6^2, I_6K_6, I_6K_7, I_6K_8, K_6^2, K_6K_7, K_6K_8, K_7^2, K_7K_8, K_8^2. \tag{13}$$

However, it can be shown (the methods employed, with some examples, are outlined below) that each of (13) can be expressed as a polynomial in I_1 to I_5 and K_1 to K_5, and so need not be included. Thus, I_1 to I_5 and K_1 to K_5 form a complete basis for the group T_2.

For the group T_3, I_1 to I_6, K_1, K_2, and K_8 are invariant, but K_3 to K_7 change sign under $\mathbf{R}_L$. Thus, we must consider squares and products in pairs of K_3 to K_7. Many of these squares and products are reducible (that is, expressible as polynomials in I_1 to I_6, K_1, K_2, and K_8) but reductions have not been obtained for those listed under T_3 in Table 2, and so these are retained. Similarly, for the groups T_4 and T_5, we retain the squares and products listed in the appropriate columns of Table 2. In the case of T_4 it is necessary to consider some products of three scalars (for example, none of K_3, K_6, and I_6, nor any of their products in pairs, is an invariant under T_4, but $K_3K_6I_6$ is invariant). However, all such invariant products of three factors are easily shown to be reducible. Thus, Table 2 gives complete sets of invariants for all five symmetry groups.

It now follows from (10) that A can be expressed as a function of θ and I_1 to I_5 for the groups T_2, T_4, and T_5, and of θ and I_1 to I_6 in the case of T_1 and T_3. The corresponding expressions for $\boldsymbol{\sigma}$ were given in Ref. 5. We therefore concentrate

Table 2. Invariants of **C** and **G** under transformation groups T_1 to T_5.

T_1	T_2	T_3	T_4	T_5
I_1	I_1	I_1	I_1	I_1
I_2	I_2	I_2	I_2	I_2
I_3	I_3	I_3	I_3	I_3
I_4	I_4	I_4	I_4	I_4
I_5	I_5	I_5	I_5	I_5
I_6		I_6		
K_1	K_1	K_1	K_1	K_1
K_2	K_2	K_2	K_2	K_2
K_3	K_3			
K_4	K_4			
K_5	K_5			
K_6				K_6
K_7				K_7
K_8		K_8		
				I_6K_3
		K_3^2	K_3^2	K_3^2
		K_3K_4	K_3K_4	K_3K_4
		K_3K_5	K_3K_5	K_3K_5
		K_3K_6		
		K_3K_7		
				K_3K_8
		K_4^2	K_4^2	K_4^2
		K_4K_6		
		K_4K_7		
		K_5K_6		

on the heat flux vector. Following a procedure due to Pipkin and Rivlin (Ref. 3) we note that, for an arbitrary vector **u**, **Q.u** is an invariant of **C**, **G**, and **u** under the relevant transformation group, and linear in **u**. Hence (Ref. 8) **Q.u** is an isotropic invariant of **C**, **G**, $\mathbf{a}_0$, and **u**, and linear in **u**. Therefore, from tables of isotropic invariants we find, as in Ref. 5, that **Q** can be expressed in the form

$$\mathbf{Q} = \sum_{\alpha} \psi_\alpha \mathbf{Q}_\alpha, \tag{14}$$

where ψ_α are scalar functions of the invariants listed in Table 1, and $\mathbf{Q}_\alpha$ are the vectors listed in Table 3.

The vectors $\mathbf{Q}_\alpha$ all have the required invariance property (10) under the transformations $\mathbf{M}(\alpha)$, but some of them change sign under one or more of the transformations $\mathbf{R}_T$, $\mathbf{R}_L$, and **D**; these sign changes are recorded in Table 3. However, the products $I_\beta \mathbf{Q}_\alpha$, $K_\beta \mathbf{Q}_\alpha$ will have the required invariance properties if I_β, K_β, and $\mathbf{Q}_\alpha$ individually undergo changes of sign under transformation. For ex-

Table 3. Vector functions of **C**, **G**, $\mathbf{a}_0$.

	$\mathbf{Q}_\alpha$	$\mathbf{M}(\alpha)$	$\mathbf{R}_T$	$\mathbf{R}_L$	**D**
G	$\mathbf{Q}_1 = \mathbf{G}$	+	+	+	+
$\mathbf{a}_0$	$\mathbf{Q}_2 = \mathbf{a}_0$	+	+	−	−
C, **G**	$\mathbf{Q}_3 = \mathbf{CG}$	+	+	+	+
C, $\mathbf{a}_0$	$\mathbf{Q}_4 = \mathbf{C}\mathbf{a}_0$	+	+	−	−
	$\mathbf{Q}_5 = \mathbf{C}^2\mathbf{a}_0$	+	+	−	−
	$\mathbf{Q}_6 = \mathbf{a}_0 \times \mathbf{C}\mathbf{a}_0$	+	−	−	+
	$\mathbf{Q}_7 = \mathbf{a}_0 \times \mathbf{C}^2\mathbf{a}_0$	+	−	−	+.
G, $\mathbf{a}_0$	$\mathbf{Q}_8 = \mathbf{a}_0 \times \mathbf{G}$	+	−	+	−
C, **G**, $\mathbf{a}_0$	$\mathbf{Q}_9 = \mathbf{a}_0 \times \mathbf{CG}$	+	−	+	−

ample, $I_6\mathbf{Q}_6$ is admissible under the group T_2, although I_6 and $\mathbf{Q}_6$ both change sign under $\mathbf{R}_T$. It is therefore necessary to examine various products of the form $I_\beta\mathbf{Q}_\alpha$ and $K_\beta\mathbf{Q}_\alpha$. Many, but not all, of these products prove to be reducible, in the sense that they can be expressed in terms of scalars and vectors, listed in Tables 1 and 3, which are invariant under the appropriate transformation group. The products which have not been so reduced are listed in Table 4, for each of the symmetry groups T_α. Thus, for each transformation group, the canonical form for **Q** is of the form

$$\mathbf{Q} = \sum_\alpha \psi_\alpha \mathbf{Q}^*_\alpha, \tag{15}$$

where ψ_α is a function of the invariants listed in Table 2, and $\mathbf{Q}^*_\alpha$ are the vectors listed in Table 4. For the group T_2, the results in Tables 2 and 4 agree with those given by Chadwick and Seet (Ref. 2).

The methods employed in eliminating redundant items from the lists of invariants and form-invariant vectors are those which were used to construct minimal integrity bases for isotropic invariants of vectors and tensors, and which were described in Ref. 6. They are based on, first, the identity

$$\epsilon_{ijk}\epsilon_{pqr} = \begin{bmatrix} \delta_{ip} & \delta_{iq} & \delta_{ir} \\ \delta_{jp} & \delta_{jq} & \delta_{jr} \\ \delta_{kp} & \delta_{kq} & \delta_{kr} \end{bmatrix}, \tag{16}$$

where δ_{ij} is the Kronecker delta and ϵ_{ijk} are components of the third-order alternating tensor. This identity facilitates reductions of products of scalar triple

Table 4. Form-invariant vector functions of **C**, **G**, $\mathbf{a}_0$ under transformation groups T_1 to T_5.

T_1	T_2	T_3	T_4	T_5
$\mathbf{Q}_1$	$\mathbf{Q}_1$	$\mathbf{Q}_1$	$\mathbf{Q}_1$	$\mathbf{Q}_1$
$\mathbf{Q}_2$	$\mathbf{Q}_2$			
				$I_6\mathbf{Q}_2$
		$K_3\mathbf{Q}_2$	$K_3\mathbf{Q}_2$	$K_3\mathbf{Q}_2$
		$K_4\mathbf{Q}_2$	$K_4\mathbf{Q}_2$	$K_4\mathbf{Q}_2$
		$K_6\mathbf{Q}_2$		
				$K_8\mathbf{Q}_2$
$\mathbf{Q}_3$	$\mathbf{Q}_3$	$\mathbf{Q}_3$	$\mathbf{Q}_3$	$\mathbf{Q}_3$
$\mathbf{Q}_4$	$\mathbf{Q}_4$			
		$K_3\mathbf{Q}_4$	$K_3\mathbf{Q}_4$	$K_3\mathbf{Q}_4$
$\mathbf{Q}_5$	$\mathbf{Q}_5$			
		$K_3\mathbf{Q}_5$	$K_3\mathbf{Q}_5$	$K_3\mathbf{Q}_5$
$\mathbf{Q}_6$				$\mathbf{Q}_6$
		$K_3\mathbf{Q}_6$		
$\mathbf{Q}_7$				$\mathbf{Q}_7$
		$K_3\mathbf{Q}_7$		
$\mathbf{Q}_8$		$\mathbf{Q}_8$		
				$K_3\mathbf{Q}_8$
				$K_4\mathbf{Q}_8$
				$K_5\mathbf{Q}_8$
$\mathbf{Q}_9$		$\mathbf{Q}_9$		
				$K_3\mathbf{Q}_9$
				$K_4\mathbf{Q}_9$
				$K_5\mathbf{Q}_9$

products (such as I_6K_6) and also of products of a scalar triple product with a vector product (such as $I_6\mathbf{Q}_6$). Second, we employ two identities, namely, the generalized Cayley–Hamilton equation for 3 × 3 matrices (Ref. 9),

$$\begin{aligned}&\mathbf{mnp} + \mathbf{mpn} + \mathbf{npm} + \mathbf{nmp} + \mathbf{pmn} + \mathbf{pnm} - \mathbf{m}(\operatorname{tr}\mathbf{np} - \operatorname{tr}\mathbf{n}\operatorname{tr}\mathbf{p})\\ &- \mathbf{n}(\operatorname{tr}\mathbf{pm} - \operatorname{tr}\mathbf{p}\operatorname{tr}\mathbf{m}) - \mathbf{p}(\operatorname{tr}\mathbf{mn} - \operatorname{tr}\mathbf{m}\operatorname{tr}\mathbf{n}) - (\mathbf{np} + \mathbf{pn})\operatorname{tr}\mathbf{m}\\ &- (\mathbf{pm} + \mathbf{mp})\operatorname{tr}\mathbf{n} - (\mathbf{mn} + \mathbf{nm})\operatorname{tr}\mathbf{p} - \mathbf{I}(\operatorname{tr}\mathbf{m}\operatorname{tr}\mathbf{n}\operatorname{tr}\mathbf{p}\\ &- \operatorname{tr}\mathbf{m}\operatorname{tr}\mathbf{np} - \operatorname{tr}\mathbf{n}\operatorname{tr}\mathbf{mp} - \operatorname{tr}\mathbf{p}\operatorname{tr}\mathbf{mn} + \operatorname{tr}\mathbf{mnp} + \operatorname{tr}\mathbf{pnm}) = \mathbf{0}. \qquad (17)\end{aligned}$$

where tr denotes the trace operation and **m**, **n**, **p** are general 3 × 3 matrices, and the Rivlin–Spencer identity (Ref. 10),

$$\mathbf{uvm} + \mathbf{muv} + \mathbf{um}^{\mathrm{T}}\mathbf{v} = \mathbf{uv}\operatorname{tr}\mathbf{m} + \tfrac{1}{2}\mathbf{m}\operatorname{tr}\mathbf{uv} + \mathbf{I}(\operatorname{tr}\mathbf{uvm} - \tfrac{1}{2}\operatorname{tr}\mathbf{uv}\operatorname{tr}\mathbf{m}), \qquad (18)$$

where **u**, **v** are antisymmetric 3 × 3 matrices and **m** is a general 3 × 3 matrix, together with various other identities derived from (17) and (18) (see Ref. 6). As examples, we find by application of these identities that

$$K_6^2 = (K_1 - K_3^2)I_5 + 2K_3K_4I_4 - K_1I_4^2 - K_4^2,$$
$$I_6\mathbf{Q}_6 = (I_1I_4I_5 - I_2I_4^2 + I_3I_4 - I_5^2)\mathbf{Q}_2 + (I_4I_5 - I_1I_5 + I_2I_4 - I_3)\mathbf{Q}_4 + (I_5 - I_4^2)\mathbf{Q}_5.$$

Many of the reductions involve substantially longer expressions than these. Other examples and details of the reductions are given by O'Neill (Ref. 11).

The results summarized in Tables 2 and 4 are complete, and every effort has been made to eliminate redundant elements from them; we consider it probable that they do not contain redundant quantities, but this is not rigorously proven. For the case in which A and $\mathbf{Q}$ are polynomial functions of their arguments, it is possible in principle to establish irreducibility (assuming that it obtains) by the application of methods from the theory of group representations, as described by Smith (Refs. 12, 13), but this approach has not yet been followed up. The theorems of Wineman and Pipkin (Ref. 14) assure that results based on the premise that A and $\mathbf{Q}$, are polynomial remain complete if A and $\mathbf{Q}$ are general single-valued functions, but may then include redundant items; it is therefore possible that some of the quantities listed in Tables 2 and 4 are redundant if A and $\mathbf{Q}$ are general functions.

4. Transverse Isotropy—Simple Shearing Deformations

To illustrate the physical properties associated with each of the symmetry groups T_α we consider a number of simple shearing deformations, with uniform temperature gradients. It is supposed that the shear plane contains the axis of transverse isotropy, and that the direction of shear is either in the fiber direction or normal to it, and that the temperature gradient is also either in the direction of shear or normal thereto. Thus, we choose $\mathbf{a}_0 = (0, 0, 1)$, and consider the following deformations and temperature fields:

Problem 1,	$x_1 = X_1,$	$x_2 = X_2 + \alpha X_1,$	$x_3 = X_3,$	$\mathbf{G} = (0, 0, G_3);$
Problem 2,	$x_1 = X_1,$	$x_2 = X_2 + \alpha X_1,$	$x_3 = X_3,$	$\mathbf{G} = (0, G_2, 0);$
Problem 3,	$x_1 = X_1,$	$x_2 = X_2 + \alpha X_1,$	$x_3 = X_3,$	$\mathbf{G} = (G_1, 0, 0);$
Problem 4,	$x_1 = X_1,$	$x_2 = X_2,$	$x_3 = X_3 + \beta X_1,$	$\mathbf{G} = (0, 0, G_3);$
Problem 5,	$x_1 = X_1,$	$x_2 = X_2,$	$x_3 = X_3 + \beta X_1,$	$\mathbf{G} = (0, G_2, 0);$
Problem 6,	$x_1 = X_1,$	$x_2 = X_2,$	$x_3 = X_3 + \beta X_1,$	$\mathbf{G} = (G_1, 0, 0).$

Thus, in each case the fibers lie in the shear plane X_1 = const. In Problems 1–3 the shear direction is transverse to the fiber direction, and in Problems 4–6 the shear direction is the fiber direction. In Problems 1 and 4 the temperature gradient is in the fiber direction; in Problems 2 and 5 it is in the shear plane and perpendicular to the fibers; and in Problems 3 and 6 it is perpendicular to the shear plane.

The stress associated with such deformations for the various symmetry groups was discussed in Ref. 5; here we consider the form of the heat flux vector $\mathbf{Q}$. For Problem 1, for example, we have

$$I_1 = I_2 = 3 + \alpha^2, \qquad I_3 = I_4 = I_5 = 1, \qquad I_6 = 0,$$
$$K_1 = K_2 = G_3^2, \quad K_3 = K_4 = K_5 = G_3, \quad K_6 = K_7 = K_8 = 0,$$

and

$$\mathbf{Q}_1 = \mathbf{Q}_3 = (0, 0, G_3), \qquad \mathbf{Q}_2 = \mathbf{Q}_4 = \mathbf{Q}_5 = (0, 0, 1),$$
$$\mathbf{Q}_6 = \mathbf{Q}_7 = \mathbf{Q}_8 = \mathbf{Q}_9 = \mathbf{0}.$$

Hence, for the groups T_1 and T_3, for example, we obtain from (15) and Tables 2 and 4

$$\text{Group } T_1, \qquad \mathbf{Q} = \{0, 0, h_{11}(\alpha^2, G_3)\},$$
$$\text{Group } T_3, \qquad \mathbf{Q} = \{0, 0, G_3 h_{13}(\alpha^2, G_3^2)\}$$

where h_{11} and h_{13} are functions of their respective arguments. Thus, in each case the heat flux is in the X_3 direction (that is, the direction of the temperature gradient), but under the symmetry group T_3 we find that $\mathbf{Q}$ is an odd function of $\mathbf{G}$, whereas this is not necessarily the case under T_1. Similar calculations have been made for each of Problems 1–6 and for each symmetry group T_α. The results are summarized in Table 5, which gives the form of $\mathbf{Q}$ for each of the cases. Of course, the dissipation inequality (9) imposes restrictions on the response functions listed in Table 5. The response functions must also be such that $\mathbf{Q} = \mathbf{0}$ when $\mathbf{G} = \mathbf{0}$.

5. Constitutive Equations for Orthotropic Materials

In this section we consider, in a similar manner to that employed in Section 3, orthotropic materials whose symmetry group is generated by the transformations comprising reflections in three orthogonal planes and rotations through π about the normals to these planes. These materials were considered only briefly in Ref. 5.

We define the preferred directions in the reference configuration by the orthogonal unit vectors $\mathbf{a}_0$, $\mathbf{b}_0$, and $\mathbf{c}_0 = \mathbf{a}_0 \times \mathbf{b}_0$. The symmetry group under consideration is generated by the following tensors:

$$\mathbf{R}_a, \mathbf{R}_b, \mathbf{R}_c = \text{reflections in planes normal to } \mathbf{a}_0, \mathbf{b}_0, \mathbf{c}_0, \text{ respectively},$$
$$\mathbf{D}_a, \mathbf{D}_b, \mathbf{D}_c = \text{rotations through } \pi \text{ about } \mathbf{a}_0, \mathbf{b}_0, \mathbf{c}_0, \text{ respectively}.$$

Table 5. Heat flux vector for simple shearing problems: transverse isotropy group.

Problem		T_1	T_2	T_3	T_4	T_5
1	Q_1	0	0	0	0	0
	Q_2	0	0	0	0	0
	Q_3	$h_{11}(\alpha^2, G_3)$	$h_{12}(\alpha^2, G_3)$	$G_3h_{13}(\alpha^2, G_3^2)$	$G_3h_{14}(\alpha^2, G_3^2)$	$G_3h_{15}(\alpha^2, G_3^2)$
2	Q_1	$G_2f_{21}(\alpha, G_2^2)$	$\alpha G_2f_{22}(\alpha^2, G_2^2)$	$G_2f_{23}(\alpha, G_2^2)$	$\alpha G_2f_{24}(\alpha^2, G_2^2)$	$\alpha G_2f_{25}(\alpha^2, G_2^2)$
	Q_2	$G_2g_{21}(\alpha, G_2^2)$	$G_2g_{22}(\alpha^2, G_2^2)$	$G_2g_{23}(\alpha, G_2^2)$	$G_2g_{24}(\alpha^2, G_2^2)$	$G_2g_{25}(\alpha^2, G_2^2)$
	Q_3	$h_{21}(\alpha, G_2^2)$	$h_{22}(\alpha^2, G_2^2)$	0	0	$\alpha h_{25}(\alpha^2, G_2^2)$
3	Q_1	$G_1f_{31}(\alpha, G_1^2)$	$G_1f_{32}(\alpha^2, G_1^2)$	$G_1f_{33}(\alpha, G_1^2)$	$G_1f_{34}(\alpha^2, G_1^2)$	$G_1f_{35}(\alpha^2, G_1^2)$
	Q_2	$G_1g_{31}(\alpha, G_1^2)$	$\alpha G_1g_{32}(\alpha^2, G_1^2)$	$G_1g_{33}(\alpha, G_1^2)$	$\alpha G_1g_{34}(\alpha^2, G_1^2)$	$\alpha G_1g_{35}(\alpha^2, G_1^2)$
	Q_3	$h_{31}(\alpha, G_1^2)$	$h_{32}(\alpha^2, G_1^2)$	0	0	$\alpha h_{35}(\alpha^2, G_1^2)$
4	Q_1	$\beta f_{41}(\beta^2, G_3)$	$\beta f_{42}(\beta^2, G_3)$	$\beta G_3f_{43}(\beta^2, G_3^2)$	$\beta G_3f_{44}(\beta^2, G_3^2)$	$\beta G_3f_{45}(\beta^2, G_3^2)$
	Q_2	$\beta g_{41}(\beta^2, G_3)$	0	$\beta G_3g_{43}(\beta^2, G_3^2)$	0	$\beta g_{45}(\beta^2, G_3^2)$
	Q_3	$h_{41}(\beta^2, G_3)$	$h_{42}(\beta^2, G_3)$	$G_3h_{43}(\beta^2, G_3^2)$	$G_3h_{44}(\beta^2, G_3^2)$	$G_3h_{45}(\beta^2, G_3^2)$
5	Q_1	$\beta f_{51}(\beta^2, G_2^2, \beta G_2)$ $+ G_2\bar{f}_{51}(\beta^2, G_2^2, \beta G_2)$	$\beta f_{52}(\beta^2, G_2^2)$	$G_2f_{53}(\beta^2, G_2^2)$	0	0
	Q_2	$\beta g_{51}(\beta^2, G_2^2, \beta G_2)$ $+ G_2\bar{g}_{51}(\beta_2, G_2^2, \beta G_2)$	$G_2g_{52}(\beta^2, G_2^2)$	$G_2g_{53}(\beta^2, G_2^2)$	$G_2g_{54}(\beta^2, G_2^2)$	$\beta g_{55}(\beta^2, G_2^2, \beta G_2)$ $+ G_2\bar{g}_{55}(\beta^2, G_2^2, \beta G_2)$
	Q_3	$h_{51}(\beta^2, G_2^2, \beta G_2)$	$h_{52}(\beta^2, G_2^2)$	$\beta G_2h_{53}(\beta^2, G_2^2)$	0	0
6	Q_1	$\beta f_{61}(\beta^2, G_1^2, \beta G_1)$ $+ G_1\bar{f}_{61}(\beta^2, G_1^2, \beta G_1)$	$\beta f_{62}(\beta^2, G_1^2, \beta G_1)$ $+ G_1\bar{f}_{62}(\beta^2, G_1^2, \beta G_1)$	$G_1f_{63}(\beta^2, G_1^2)$	$G_1f_{64}(\beta^2, G_1^2)$	$G_1f_{65}(\beta^2, G_1^2)$
	Q_2	$\beta g_{61}(\beta^2, G_1^2, \beta G_1)$ $+ G_1\bar{g}_{61}(\beta^2, G_1^2, \beta G_1)$	0	$G_1g_{63}(\beta^2, G_1^2)$	0	$\beta g_{65}(\beta^2, G_1^2)$
	Q_3	$h_{61}(\beta^2, G_1^2, \beta G_1)$	$h_{62}(\beta^2, G_1^2, \beta G_1)$	$\beta G_1(\beta^2, G_1^2)$	$\beta G_1h_{64}(\beta^2, G_1^2)$	$\beta G_1h_{65}(\beta^2, G_1^2)$

Thus, if for example we choose

$$\mathbf{a}_0 = (1, 0, 0) \qquad \mathbf{b}_0 = (0, 1, 0), \qquad \mathbf{c}_0 = (0, 0, 1), \tag{19}$$

tthen the matrices of these tensor components become

$$\mathbf{R}_a = \begin{bmatrix} -1 & 0 & 0 \\ 0 & 1 & 0 \\ 0 & 0 & 1 \end{bmatrix}, \quad \mathbf{R}_b = \begin{bmatrix} 1 & 0 & 0 \\ 0 & -1 & 0 \\ 0 & 0 & 1 \end{bmatrix}, \quad \mathbf{R}_c = \begin{bmatrix} 1 & 0 & 0 \\ 0 & 1 & 0 \\ 0 & 0 & -1 \end{bmatrix},$$

$$\mathbf{D}_a = \begin{bmatrix} 1 & 0 & 0 \\ 0 & -1 & 0 \\ 0 & 0 & -1 \end{bmatrix}, \mathbf{D}_b = \begin{bmatrix} -1 & 0 & 0 \\ 0 & 1 & 0 \\ 0 & 0 & -1 \end{bmatrix}, \mathbf{D}_c = \begin{bmatrix} -1 & 0 & 0 \\ 0 & -1 & 0 \\ 0 & 0 & 1 \end{bmatrix},$$

but as in Section 3, the choice of a particular coordinate system is not necessary and $\mathbf{a}_0$, $\mathbf{b}_0$, $\mathbf{c}_0$ may depend on position in the reference configuration. We also employ the central inversion tensor $-\mathbf{I} = \mathbf{R}_a\mathbf{R}_b\mathbf{R}_c$, and note the relations

$$\mathbf{D}_a = \mathbf{R}_b\mathbf{R}_c, \qquad \mathbf{D}_b = \mathbf{R}_c\mathbf{R}_a, \qquad \mathbf{D}_c = \mathbf{R}_a\mathbf{R}_b.$$

The symmetry group for orthotropic symmetry is denoted S, and contains the elements $\mathbf{R}_a$, $\mathbf{R}_b$, $\mathbf{R}_c$, $\mathbf{D}_a$, $\mathbf{D}_b$, $\mathbf{D}_c$, $-\mathbf{I}$, and the identity transformation $\mathbf{I}$. As realizations of this symmetry, we may consider fiber-reinforced materials reinforced by two families of fibers (for example, laminates comprising a large number of unidirectionally reinforced laminae each oriented in one of two directions). If the sense of the fibers does not affect the physical properties, then a cross-ply laminate (that is, one with orthogonal fiber directions) has the symmetry S, with $\mathbf{a}_0$ and $\mathbf{b}_0$ representing the orthogonal fiber directions. A balanced angle-ply laminate also has the symmetry S, with $\mathbf{a}_0$ and $\mathbf{b}_0$ representing the bisectors of the nonorthogonal fiber directions.

For orthotropic symmetry, A and $\mathbf{Q}$ have the forms

$$A = A(\mathbf{C}, \mathbf{a}_0, \mathbf{b}_0, \theta), \qquad \mathbf{Q} = \mathbf{Q}(\mathbf{C}, \mathbf{G}, \mathbf{a}_0, \mathbf{b}_0, \theta).$$

As before, we suppress the scalar argument θ and note (Ref. 8) that A must be an isotropic invariant of $\mathbf{C}$, $\mathbf{a}_0$, $\mathbf{b}_0$ and that, for an arbitrary vector $\mathbf{u}$, $\mathbf{Q} \cdot \mathbf{u}$ is an isotropic invariant of $\mathbf{C}$, $\mathbf{G}$, $\mathbf{a}_0$, $\mathbf{b}_0$, $\mathbf{u}$ and linear in $\mathbf{u}$. Hence, by the procedure used in Section 3, we find from tables of isotropic invariants (Ref. 6) that

$$A = A(I_\beta), \qquad \mathbf{Q} = \sum_\alpha \psi_\alpha(I_\beta, K_\gamma)\mathbf{Q}_\alpha,$$

where I_β and K_γ are the scalars listed in Table 6 and $\mathbf{Q}_\alpha$ are the vectors listed in Table 7. These tables take into account that $\mathbf{a}_0$ and $\mathbf{b}_0$ are orthogonal unit vectors.

Just as in the case of transverse isotropy, the quantities listed in Tables 6

Table 6. Isotropic scalar invariants of $\mathbf{C}$, $\mathbf{G}$, $\mathbf{a}_0$, $\mathbf{b}_0$

	Invariant	$\mathbf{R}_a$	$\mathbf{R}_b$	$\mathbf{R}_c$	$\mathbf{D}_a$	$\mathbf{D}_b$	$\mathbf{D}_c$	$-\mathbf{I}$
$\mathbf{C}$	$I_1 = \operatorname{tr}\mathbf{C}$	+	+	+	+	+	+	+
	$I_2 = \frac{1}{2}\{(\operatorname{tr}\mathbf{C})^2 - \operatorname{tr}\mathbf{C}^2\}$	+	+	+	+	+	+	+
	$I_3 = \det\mathbf{C}$	+	+	+	+	+	+	+
$\mathbf{G}$	$K_1 = \mathbf{G}\cdot\mathbf{G}$	+	+	+	+	+	+	+
$\mathbf{C}, \mathbf{G}$	$K_2 = \mathbf{GCG}$	+	+	+	+	+	+	+
$\mathbf{C}, \mathbf{a}_0$	$I_4 = \mathbf{a}_0\mathbf{C}\mathbf{a}_0$	+	+	+	+	+	+	+
	$I_5 = \mathbf{a}_0\mathbf{C}^2\mathbf{a}_0$	+	+	+	+	+	+	+
	$I_6 = [\mathbf{a}_0, \mathbf{C}\mathbf{a}_0, \mathbf{C}^2\mathbf{a}_0]$	−	−	+	−	−	+	+
$\mathbf{C}, \mathbf{b}_0$	$I_7 = \mathbf{b}_0\mathbf{C}\mathbf{b}_0$	+	+	+	+	+	+	+
	$I_8 = \mathbf{b}_0\mathbf{C}^2\mathbf{b}_0$	+	+	+	+	+	+	+
	$I_9 = [\mathbf{b}_0, \mathbf{C}\mathbf{b}_0, \mathbf{C}^2\mathbf{b}_0]$	−	+	−	−	+	−	+
$\mathbf{G}, \mathbf{a}_0$	$K_3 = \mathbf{a}_0\cdot\mathbf{G}$	+	+	−	−	−	+	−
$\mathbf{C}, \mathbf{G}, \mathbf{a}_0$	$K_4 = \mathbf{a}_0\mathbf{CG}$	+	+	−	−	−	+	−
	$K_5 = \mathbf{a}_0\mathbf{C}^2\mathbf{G}$	+	+	−	−	−	+	−
	$K_6 = [\mathbf{a}_0, \mathbf{G}, \mathbf{C}\mathbf{a}_0]$	−	−	−	+	+	+	−
	$K_7 = [\mathbf{a}_0, \mathbf{G}, \mathbf{C}^2\mathbf{a}_0]$	−	−	−	+	+	+	−
	$K_8 = [\mathbf{a}_0, \mathbf{G}, \mathbf{CG}]$	−	−	+	−	−	+	+
$\mathbf{G}, \mathbf{b}_0$	$K_9 = \mathbf{b}_0\cdot\mathbf{G}$	+	−	+	−	+	−	−
$\mathbf{C}, \mathbf{G}, \mathbf{b}_0$	$K_{10} = \mathbf{b}_0\mathbf{CG}$	+	−	+	−	+	−	−
	$K_{11} = \mathbf{b}_0\mathbf{C}^2\mathbf{G}$	+	−	+	−	+	−	−
	$K_{12} = [\mathbf{b}_0, \mathbf{G}, \mathbf{C}\mathbf{b}_0]$	−	−	−	+	+	+	−
	$K_{13} = [\mathbf{b}_0, \mathbf{G}, \mathbf{C}^2\mathbf{b}_0]$	−	−	−	+	+	+	−
	$K_{14} = [\mathbf{b}_0, \mathbf{G}, \mathbf{CG}]$	−	+	−	−	+	−	+
$\mathbf{a}_0, \mathbf{b}_0, \mathbf{G}$	$K_{15} = [\mathbf{a}_0, \mathbf{b}_0, \mathbf{G}]$	−	+	+	+	−	−	−
$\mathbf{C}, \mathbf{a}_0, \mathbf{b}_0$	$I_{10} = \mathbf{a}_0\mathbf{C}\mathbf{b}_0$	+	−	−	+	−	−	+
	$I_{11} = \mathbf{a}_0\mathbf{C}^2\mathbf{b}_0$	+	−	−	+	−	−	+
	$I_{12} = [\mathbf{a}_0, \mathbf{b}_0, \mathbf{C}\mathbf{a}_0]$	−	+	−	−	+	−	+
	$I_{13} = [\mathbf{b}_0, \mathbf{a}_0, \mathbf{C}\mathbf{b}_0]$	−	−	+	−	−	+	+
$\mathbf{C}, \mathbf{G}, \mathbf{a}_0, \mathbf{b}_0$	$K_{16} = [\mathbf{a}_0, \mathbf{G}, \mathbf{C}\mathbf{b}_0]$	−	+	+	+	−	−	−
	$K_{17} = [\mathbf{b}_0, \mathbf{G}, \mathbf{C}\mathbf{a}_0]$	−	+	+	+	−	−	−

Table 7. Vector functions of $\mathbf{C}$, $\mathbf{G}$, $\mathbf{a}_0$, $\mathbf{b}_0$.

	Vector	R_a	R_b	R_c	D_a	D_b	D_c	$-\mathbf{I}$
$\mathbf{G}$	$\mathbf{Q}_1 = \mathbf{G}$	+	+	+	+	+	+	+
$\mathbf{a}_0$	$\mathbf{Q}_2 = \mathbf{a}_0$	+	+	−	−	−	+	−
$\mathbf{C}, \mathbf{G}$	$\mathbf{Q}_3 = \mathbf{CG}$	+	+	+	+	+	+	+
$\mathbf{C}, \mathbf{a}_0$	$\mathbf{Q}_4 = \mathbf{Ca}_0$	+	+	−	−	−	+	−
	$\mathbf{Q}_5 = \mathbf{C}^2\mathbf{a}_0$	+	+	−	−	−	+	−
	$\mathbf{Q}_6 = \mathbf{a}_0 \times \mathbf{Ca}_0$	−	−	−	+	+	+	−
	$\mathbf{Q}_7 = \mathbf{a}_0 \times \mathbf{C}^2\mathbf{a}_0$	−	−	−	+	+	+	−
$\mathbf{G}, \mathbf{a}_0$	$\mathbf{Q}_8 = \mathbf{a}_0 \times \mathbf{G}$	−	−	+	−	−	+	+
$\mathbf{C}, \mathbf{G}, \mathbf{a}_0$	$\mathbf{Q}_9 = \mathbf{a}_0 \times \mathbf{CG}$	−	−	+	−	−	+	+
$\mathbf{b}_0$	$\mathbf{Q}_{10} = \mathbf{b}_0$	+	−	+	−	+	−	−
$\mathbf{C}, \mathbf{b}_0$	$\mathbf{Q}_{11} = \mathbf{Cb}_0$	+	−	+	−	+	−	−
	$\mathbf{Q}_{12} = \mathbf{C}^2\mathbf{b}_0$	+	−	+	−	+	−	−
	$\mathbf{Q}_{13} = \mathbf{b}_0 \times \mathbf{Cb}_0$	−	−	−	+	+	+	−
	$\mathbf{Q}_{14} = \mathbf{b}_0 \times \mathbf{C}^2\mathbf{b}_0$	−	−	−	+	+	+	−
$\mathbf{G}, \mathbf{b}_0$	$\mathbf{Q}_{15} = \mathbf{b}_0 \times \mathbf{G}$	−	+	−	−	+	−	+
$\mathbf{C}, \mathbf{G}, \mathbf{b}_0$	$\mathbf{Q}_{16} = \mathbf{b}_0 \times \mathbf{CG}$	−	+	−	−	+	−	+
$\mathbf{a}_0, \mathbf{b}_0$	$\mathbf{Q}_{17} = \mathbf{a}_0 \times \mathbf{b}_0$	−	+	+	+	−	−	−
$\mathbf{C}, \mathbf{a}_0, \mathbf{b}_0$	$\mathbf{Q}_{18} = \mathbf{a}_0 \times \mathbf{Cb}_0$	−	+	+	+	−	−	−

and 7 change sign under some of the relevant transformations; these sign changes are also recorded in Tables 6 and 7. Thus, to obtain lists of invariant quantities we have to consider powers and products of the quantities listed in Tables 6 and 7. Many of these prove to be reducible by the methods outlined in Section 3. Lists of scalar invariants and form-invariant vectors under S for which reductions have not been obtained are shown in Tables 8 and 9, respectively. Thus, for orthotropic symmetry, A is a function of the invariants I_1, to I_5, I_7, and I_8, and θ, and $\mathbf{Q}$ has the form

$$\mathbf{Q} = \sum_{\alpha} \psi_\alpha \mathbf{Q}^*_\alpha,$$

where ψ_α is a function of the invariants listed in Table 8 and $\mathbf{Q}^*_\alpha$ are the vectors

Table 8. Invariants of **C** and **G** under transformation group S.

$I_1, I_2, I_3, I_4, I_5, I_7, I_8, K_1, K_2$
$K_3^2, K_3K_4, K_3K_5, K_4^2, K_9^2, K_9K_{10}, K_9K_{11}, K_{10}^2$

Table 9. Form-invariant vector functions of **C**, **G**, $\mathbf{a}_0$, $\mathbf{b}_0$ under transformation group S.

$\mathbf{Q}_1, \mathbf{Q}_3, K_3\mathbf{Q}_2, K_4\mathbf{Q}_2, K_9I_{10}\mathbf{Q}_2, K_3\mathbf{Q}_4,$
$K_3\mathbf{Q}_5, K_9\mathbf{Q}_{10}, K_{10}\mathbf{Q}_{10}, K_3I_{10}\mathbf{Q}_{10}, K_9\mathbf{Q}_{11}, K_9\mathbf{Q}_{12}$

listed in Table 9. Because of the identity (which may be verified directly)

$$(I_1 - I_7 - I_4)\mathbf{Q}_1 + \{K_3(I_1 - I_7) - K_9I_{10}\}\mathbf{Q}_2$$
$$+ \{K_{10} - K_9(I_1 - I_4) - K_3I_{10}\}\mathbf{Q}_{10} - \mathbf{Q}_3 + K_3\mathbf{Q}_4 + K_9\mathbf{Q}_{11} \equiv 0,$$

one of the vectors listed in Table 9 may be omitted. Simplicity suggests deleting $K_9I_{10}\mathbf{Q}_2$ or $K_3I_{10}\mathbf{Q}_{10}$, but this destroys the symmetry of the list with respect to $\mathbf{a}_0$ and $\mathbf{b}_0$. This symmetry may be preserved by deleting $\mathbf{Q}_3$. The form for A has been derived previously, and the stress associated with it is given in Refs. 5, 15, and 16.

The remarks made in Section 3 about completeness and irreducibility also apply in this case.

6. Orthotropy—Simple Shearing Deformations

To illustrate the effects of orthotropic symmetry, we consider the heat flux associated with two simple shearing deformations, and uniform temperature gradients either parallel or perpendicular to the shear direction and the shear plane. As in (19), we choose

$$\mathbf{a}_0 = (1, 0, 0), \qquad \mathbf{b}_0 = (0, 1, 0),$$

and consider the following six problems:

Problem 1, $x_1 = X_1 + \alpha X_2,\quad x_2 = X_2,\quad x_3 = X_3,\quad \mathbf{G} = (0, 0, G_3);$
Problem 2, $x_1 = X_1 + \alpha X_2,\quad x_2 = X_2,\quad x_3 = X_3,\quad \mathbf{G} = (0, G_2, 0);$
Problem 3, $x_1 = X_1 + \alpha X_2,\quad x_2 = X_2,\quad x_3 = X_3,\quad \mathbf{G} = (G_1, 0, 0);$

Table 10. Heat flux vector **Q** for simple shearing problems: orthotropy group S.

	Components of Q		
Problem	Q_1	Q_2	Q_3
1	0	0	$G_3 h_1(\alpha^2, G_3^2)$
2	$\alpha G_2 f_2(\alpha^2, G_2^2)$	$G_2 g_2(\alpha^2, G_2^2)$	0
3	$G_1 f_3(\alpha^2, G_1^2)$	$G_1 \alpha g_3(\alpha^2, G_1^2)$	0
4	$G_3 \beta f_4(\beta^2, G_3^2)$	0	$G_3 h_4(\beta^2, G_3^2)$
5	0	$G_2 g_5(\beta^2, G_2^2)$	0
6	$G_1 f_6(\beta^2, G_1^2)$	0	$G_1 \beta h_6(\beta^2, G_1^2)$

Problem 4, $x_1 = X_1 + \beta X_3,\quad x_2 = X_2,\quad x_3 = X_3,\quad \mathbf{G} = (0, 0, G_3);$
Problem 5, $x_1 = X_1 + \beta X_3,\quad x_2 = X_2,\quad x_3 = X_3,\quad \mathbf{G} = (0, G_2, 0);$
Problem 6, $x_1 = X_1 + \beta X_3,\quad x_2 = X_2,\quad x_3 = X_3,\quad \mathbf{G} = (G_1, 0, 0);$

In Problems 1–3, the shear plane is normal to $\mathbf{b}_0$ and $\mathbf{a}_0$ is the shear direction. In Problems 4 and 5, the planes of $\mathbf{a}_0$ and $\mathbf{b}_0$ are the shear planes, and $\mathbf{a}_0$ is the shear direction. In Problems 1 and 5, the temperature gradient is in the shear plane and perpendicular to the shear direction; in Problems 2 and 4 it is perpendicular to both the shear plane and the shear direction; and in Problems 3 and 6 it is in the shear direction.

It is a straightforward calculation to determine from Tables 8 and 9 the form of the heat flux vector **Q** for each of Problems 1–6. The results are summarized in Table 10.

By interchanging the roles of $\mathbf{a}_0$ and $\mathbf{b}_0$, a further six related problems may be defined and solved.

Acknowledgment

The work of J. M. O'Neill was supported by a Research Studentship awarded by the Science and Engineering Research Council.

References

1. CARLSON, D. E., *Linear Thermoelasticity*, Handbuch der Physik, Edited by C. Truesdell, Springer, Berlin, Germany, Vol. VI a/2, pp. 297–389, 1972.
2. CHADWICK, P., and SEET, L. T. C., *Second-Order Thermoelasticity Theory for Isotropic and Transversely Isotropic Materials*, Trends in Elasticity and Thermoelas-

ticity, Edited by R. E. Czarnota-Bojarski, M. Sokolowski, and H. Zorski, Wolters–Noordhoff, Groningen, The Netherlands, pp. 29–57, 1971.

3. PIPKIN, A. C., and RIVLIN, R. S., *The Formulation of Constitutive Equations in Continuum Physics, I*, Archive for Rational Mechanics and Analysis, Vol. 4, pp. 129–144, 1959.
4. GREEN, A. E., and ADKINS, J. E., *Large Elastic Deformations*, 1st Edition, Clarendon Press, Oxford, England, 1960.
5. SPENCER, A. J. M., *Thermal Stress in Anisotropic Elastic Solids*, Finite Thermoelasticity, Edited by G. Grioli, Accademia Nazionale dei Lincei, Rome, Italy, pp. 219–239, 1986.
6. SPENCER, A. J. M., *Theory of Invariants*, Continuum Physics, Edited by A. C. Eringen, Academic Press, New York, New York, Vol. 1, pp. 239–353, 1971.
7. SMITH, G. F., *On Transversely Isotropic Functions of Vectors, Symmetric Second-Order Tensors and Skew-Symmetric Tensors*, Quarterly of Applied Mathematics, Vol. 39, pp. 509–516, 1982.
8. SPENCER, A. J. M., *The Formulation of Constitutive Equations for Anisotropic Solids*, Mechanical Behavior of Anisotropic Solids, Edited by J.-P. Boehler, Nijhoff, The Hague, The Netherlands, pp. 2–26, 1982.
9. RIVLIN, R. S., *Further Remarks on the Stress-Deformation Relations for Isotropic Materials*, Journal of Rational Mechanics and Analysis, Vol. 4, pp. 681–701, 1955.
10. SPENCER, A. J. M., and RIVLIN, R. S., *Isotropic Integrity Bases for Vectors and Second-Order Tensors, Part I*, Archive for Rational Mechanics and Analysis, Vol. 9, pp. 45–63, 1962.
11. O'NEILL, J. M., *Thermoelastic Stress Analysis of Anisotropic Materials*, PhD Thesis, University of Nottingham, 1987.
12. SMITH, G. F., *On the Minimality of Integrity Bases for Symmetric 3 × 3 Matrices*, Archive for Rational Mechanics and Analysis, Vol. 5, pp. 382–389, 1960.
13. SMITH, G. F., *On Isotropic Integrity Bases*, Archive for Rational Mechanics and Analysis, Vol. 18, pp. 282–292, 1965.
14. WINEMAN, A. S., and PIPKIN, A. C., *Material Symmetry Restrictions on Constitutive Equations*, Archive for Rational Mechanics and Analysis, Vol. 17, pp. 184–214, 1964.
15. SPENCER, A. J. M., *Constitutive Theory for Strongly Anisotropic Solids*, Continuum Theory of the Mechanics of Fibre-Reinforced Composites, CISM Courses and Lectures, Edited by A. J. M. Spencer, Springer, Berlin, Germany, Vol. 282, pp. 1–32, 1984.
16. SPENCER, A. J. M., *Modelling of Finite Deformations of Anisotropic Materials*, Large Deformations of Solids: Physical Basis and Mathematical Modelling, Edited by J. Gittus, J. Zarka, and S. Nemat-Nasser, Elsevier, London, England, pp. 41–52, 1986.

11

Energy Minimization for Membranes

A. C. PIPKIN[†]

Abstract. In membrane theory the structure is treated as having no bending stiffness. However, membrane problems sometimes yield solutions that involve compressive stress. This indicates an instability whose detailed treatment would require consideration of the bending stiffness of the material. To avoid this complication, ordinary membrane theory is replaced by tension field theory in which it is taken as a basic assumption that the membrane cannot sustain a negative principal stress. Membrane problems may be posed as energy minimization problems. Anomalous cases are those in which no particular deformation minimizes the energy even though the energy is bounded below. Instead there is a sequence of more and more finely wrinkled sheets of the membrane that give successively lower values. The limiting state of deformation can be described as one in which there are infinitely many folds in the sheet, spaced infinitesimally close together. This paper reviews some examples that illustrate this sort of situation and shows how the variational problem can be modified to a form in which a minimizing deformation exists.

Key Words. Membrane theory, tension field theory, wrinkles, energy minimization, strain energy, quasiconvexification, relaxed energy density, pure network theory, reinforced networks, finite deformations, energy functional, minimizing sequence, kinematically admissible deformations, theory of reinforced networks, relaxed energy functional.

[†]Deceased.

A. C. Pipkin • Professor, Division of Applied Mathematics, Brown University, Providence, Rhode Island 02912.

Nonlinear Effects in Fluids and Solids, edited by M. M. Carroll and M. Hayes, Plenum Press, New York, 1996.

1. Introduction

The distinction between membrane theory and the theory of plates and shells is that in membrane theory, the structure is treated as having no bending stiffness. Membrane theory is applicable to the cable networks that support hanging roofs, the sidewall of a tire under normal inflation pressure, and various kinds of paneling, if the bending stiffness of the panel makes no important contribution to its load-carrying capacity.

Membrane problems sometimes yield solutions that involve compressive stresses. This indicates an instability whose detailed treatment would require consideration of the bending stiffness of the material. To avoid this complication, ordinary membrane theory is replaced by tension field theory (Refs. 1–13), in which it is taken as a basic assumption that the membrane cannot sustain a negative principal stress. This theory envisages a continuous distribution of wrinkles, spaced infinitesimally close together, with no stress exerted across the wrinkle lines. The relation between the tension parallel to the wrinkles and the stretching along that direction is postulated in some way that is understandable in terms of the originally postulated properties of the material.

The subject can be approached in a systematic way by posing membrane problems as energy minimization problems. In this framework, the anomalous cases are those in which no particular deformation minimizes the energy, even though the energy is bounded below. Instead, there is a sequence of more and more finely wrinkled states of the membrane that give successively lower values of the total energy, approaching the energy's greatest lower bound. The limiting state of deformation can be described as one in which there are infinitely many folds in the sheet, spaced infinitesimally close together. The original mathematical formulation does not contemplate such states, so the problem as originally posed has no solution.

The present paper reviews some simple examples that illustrate this sort of situation, and shows how the variational problem can be modified to a form in which a minimizing deformation exists, even if it represents a continuously wrinkled state. This usually involves replacing the strain energy function by its quasiconvexification, the relaxed energy density. In this approach it is not necessary to postulate a stress–strain relation for wrinkled regions. Instead, the appropriate form arises as a natural consequence of the form of the relaxed energy density, and the basic postulates of tension field theory appear as theorems about energy-minimizing states.

The details of the resulting theory depend on the properties originally assumed for the unwrinkled membrane. The examples given here are based on Rivlin's (Ref. 14) pure network theory, its generalization to reinforced networks by Adkins (Ref. 15), and the theory of isotropic elastic membranes. Finite deformations are considered in all cases.

2. Inextensible Networks

For a first example, let us consider the theory of networks formed from inextensible fibers (Rivlin, Ref. 14). The body considered is a handkerchief, for example, or a network of cables. It initially occupies some region in the x,y plane, with the fibers of the network along lines x = const and y = const. The coordinates x and y are used as Lagrangian coordinates, so that each fiber is permanently identified by its x or y label. In a deformation, the particles of both fibers at the place $\mathbf{x} = (x, y)$ go to $\mathbf{R}(\mathbf{x})$ in three-dimensional space. The derivatives

$$\mathbf{A} = \mathbf{R}_x, \qquad \mathbf{B} = \mathbf{R}_y$$

are tangenital to the deformed fibers. The magnitude of one of these vectors is the ratio of deformed to undeformed length for the fiber considered. The assumption of fiber inextensibility means that these are unit vectors

$$|\mathbf{A}| = 1, \qquad |\mathbf{B}| = 1 \qquad \text{(inextensibility)}.$$

These constraint conditions mean that no segment of any fiber can grow either longer or shorter in a deformation.

In the simplest theory, better applicable to fishnets than to handkerchiefs, we neglect the resistance of the network to distortions either by shearing or by bending, so that there is no strain energy of deformation. Then the energy functional is just the potential of the boundary tractions $\mathbf{T}$, which we consider to be applied as dead loads,

$$E[\mathbf{R}] = -\int_{C_t} \mathbf{R} \cdot \mathbf{T} ds.$$

We specify $\mathbf{R}$ on some part of the boundary C_p,

$$\mathbf{R} = \mathbf{R}_o(x), \quad \text{on } C_p.$$

The two parts C_p and C_t are complementary parts of the boundary. We say that $\mathbf{R}$ is kinematically admissible if it satisfies the boundary condition on C_p and the inextensibility conditions. We wish to minimize E over all kinematically admissible deformations.

With $\mathbf{R}$ fixed on C_p, E has a lower bound and thus a greatest lower bound, E_L say. Let $\mathbf{R}_n$, $n = 1, 2, \ldots$, be a minimizing sequence, a sequence of deformations for which the sequence of energies E_n approaches E_L. For any of these deformations, the final separation of two particles cannot be any larger than the minimum length of fiber connecting them, so for two particles in some convex subset of the sheet,

$$|\mathbf{R}_n(\mathbf{x}_1) - \mathbf{R}_n(\mathbf{x}_2)| \leqq |x_1 - x_2| + |y_1 - y_2|.$$

This Lifshitz condition, along with the boundary condition, guarantees that the minimizing sequence has a convergent subsequence, which converges to $\mathbf{r}(\mathbf{x})$, say. Then apparently $\mathbf{r}(\mathbf{x})$ should be the energy-minimizing deformation. However, its derivatives,

$$\mathbf{a} = \mathbf{r}_x, \qquad \mathbf{b} = \mathbf{r}_y,$$

are not necessarily unit vectors. Instead, the Lifshitz condition on $\mathbf{r}$ implies only that

$$|\mathbf{a}| \leqq 1, \qquad |\mathbf{b}| \leqq 1 \quad \text{(one-sided constraints).}$$

If $\mathbf{a}$ or $\mathbf{b}$ is less than unity in magnitude, $\mathbf{r}(\mathbf{x})$ is not a kinematically admissible deformation within the inextensible theory.

Let us consider an example. The sheet originally occupies a rectangular region and is fixed along the edges $x = 0$ and $y = 0$ (C_p). There is no traction on the top edge $y = H$. On the edge $x = L$ there is a distribution of tractions slanting downward and to the right, as shown in Fig. 1a.

The sheet is in equilibrium in its original undeformed state if each horizontal fiber carries a load equal to the horizontal part of the boundary tension on its end and all of the vertical fibers carry no load, except the boundary fiber $x = L$. This fiber must stand like a post and support the vertical part of the boundary load. A single fiber that carries a finite load all by itself is necessarily allowable if problems are ever to have solutions at all (Ref. 16), but in this case the force

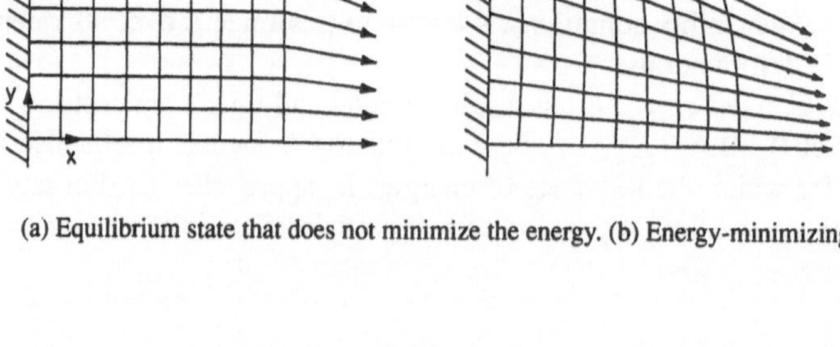

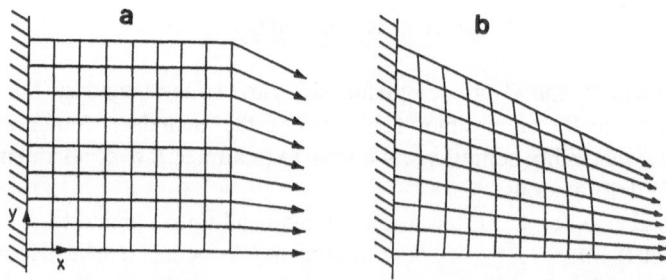

Fig. 1. (a) Equilibrium state that does not minimize the energy. (b) Energy-minimizing state.

carried by the fiber is compressive, and this is too far-fetched. So we have a solution that obviously is not right.

From the energy point of view, the greatest lower bound on the energy is attained when every boundary point goes as far as possible in the direction of the load at that point. With the inextensibility condition, this means that each of the horizontal fibers y = const must swing around into the direction of the load on its end. Then the resulting deformation has the form shown in Fig. 1b,

$$\mathbf{r} = y\mathbf{j} + x\mathbf{a}(y),$$

where $\mathbf{a}(y)$ is a unit vector in the direction of the load on the end of the fiber y = const. For this deformation, $\mathbf{r}_x = \mathbf{a}$ is a unit vector, but $\mathbf{r}_y = \mathbf{b}$ is less than unity in magnitude, so this deformation is not kinematically admissible in the inextensible theory.

However, it is easy to construct a sequence of kinematically admissible deformations $\mathbf{R}_n$ that approaches $\mathbf{r}$ in the limit. Divide the sheet into n horizontal strips of width H/n. First regard the sheet as cut along the dividing lines. Shear each strip so that its top boundary is parallel to the load on its end. Then fold up the bottom edge until it coincides with the top edge of the next strip down. This gives a deformation with $2n - 1$ folds. Figure 2 shows the deformation $\mathbf{R}_4$. For each deformation $\mathbf{R}_n$, the derivatives $\mathbf{A}_n$ and $\mathbf{B}_n$ are discontinuous across the fold lines. As n grows larger, $\mathbf{R}_n$ approaches $\mathbf{r}$ and it happens that $\mathbf{A}_n$ approaches $\mathbf{a} = \mathbf{r}_x$, but the derivatives $\mathbf{B}_n$ approach two distinct limits, as it were, neither equal to $\mathbf{b} = \mathbf{r}_y$.

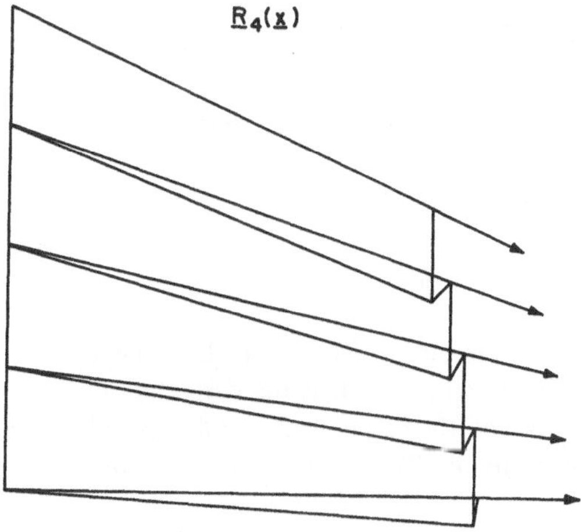

Fig. 2. Kinematically admissible deformation with folds.

In this example there is a convergent energy-minimizing sequence, but its limit is not kinematically admissible, so the problem as originally posed has no solution. We want to repair the statement of the problem so that the new problem actually does have a solution, much like $\mathbf{r}(\mathbf{x})$ if possible.

One way of doing this is to include a strain energy of bending in the energy functional. If the bending stiffness is very small but not zero, in many problems the bending energy will be negligible, but any deformation with a sharp fold will require infinite energy. We would expect the minimizer for such a theory to be close to $\mathbf{r}$, but with a large finite number of small wrinkles. A theory that takes the bending stiffness of fibers into account has been formulated (Refs. 17, 18); it is a special form of finite-deformation plate theory. As in any other plate theory, the equilibrium equations are two orders higher than they are in membrane theory.

But if we do not care to hear about the actual distribution of wrinkles, especially at the cost of solving much more difficult equations, we can take a different approach. Let us declare that $\mathbf{r}(\mathbf{x})$ is the right answer, and formulate the theory in such a way that it does have $\mathbf{r}$ as a solution in the present problem. We want a theory that works in terms of limits of minimizing sequences directly. I call a convergent sequence that is kinematically admissible in the original theory a crinkle, and its limit $\mathbf{r}$ the carrier of the crinkle (Ref. 19). To rephrase the theory, it is necessary to find the constraint conditions that carriers must satisfy, and the energy that is to be ascribed to a carrier. In the present case there is no problem. The energy functional is the same as before,

$$E[\mathbf{r}] = -\int_{C_t} \mathbf{T} \cdot \mathbf{r} ds.$$

A deformation is admissible if $\mathbf{r} = \mathbf{r}_0(\mathbf{x})$ on C_p and, with

$$d\mathbf{r} = \mathbf{a}dx + \mathbf{b}dy,$$

the one-sided constraint conditions

$$|\mathbf{a}| \leqq 1, \qquad |\mathbf{b}| \leqq 1$$

are satisfied. We wish to minimize E over all admissible $\mathbf{r}$.

The step that is not quite trivial is the derivation of the stress–strain relations. Let $\mathbf{t}_a$ and $\mathbf{t}_b$ be the forces per unit length exerted across lines $x =$ const and $y =$ const, respectively. These stress vectors will satisfy the equilibrium equation (Ref. 16)

$$\mathbf{t}_{a,x} + \mathbf{t}_{b,y} = \mathbf{0}.$$

By deriving this equation as the Euler equation for the variational principle, we will find expressions for $\mathbf{t}_a$ and $\mathbf{t}_b$. This turns out to be difficult. The energy functional is a linear functional, defined on a convex and bounded set of admissible deformations. An abstract argument involving dual cones in function space is needed (Ref. 20), but the result is quite simple. Just as in the inextensible theory,

$$\mathbf{t}_a = T_a \mathbf{a}, \qquad \mathbf{t}_b = T_b \mathbf{b},$$

where T_a and T_b are Lagrange multipliers that represent fiber tensions. In the inextensible theory, these tensions can be either positive or negative. For the modified or relaxed theory, in which the one-sided constraints are used, they are necessarily nonnegative,

$$T_a \geqq 0, \quad T_b \geqq 0.$$

Moreover, the tension is positive only if the corresponding fiber is fully extended,

$$T_a = 0, \quad \text{if } |\mathbf{a}| < 1,$$
$$T_b = 0, \quad \text{if } |\mathbf{b}| < 1.$$

Equivalently,

$$|\mathbf{a}| = 1, \quad \text{if } T_a > 0,$$
$$|\mathbf{b}| = 1, \quad \text{if } T_b > 0.$$

These are consequences of the variational formulation, not assumptions. However, the resulting theory has exactly the same form as a theory (Ref. 21) that was postulated earlier on its own merits.

The specification of stress seems so vague that it is surprising that any sort of uniqueness theorem is available (Ref. 21). A load-carrying part of the network can be defined in any particular problem, and the theorem is roughly that the deformation of the load-carrying part is unique. In particular, if a problem is solved by using the two-sided constraints of the original theory and the fiber tensions are positive everywhere, then the deformation is unique within the relaxed theory, and is thus the only deformation with nonnegative tensions in the original theory. If there is no solution in the relaxed theory that is kinematically admissible in the inextensible theory, then in the latter theory there is no solution with fiber tensions nonnegative everywhere.

3. Networks with Shearing Stiffness

Now let us consider materials that are really more like handkerchiefs than like fishnets. We suppose that there is an elastic resistance to changes in the angle between initially orthogonal fibers. For an inextensible material, $\mathbf{A} \cdot \mathbf{B}$ is the sine of the angle of shear. We suppose that the elastic stored energy per unit initial area is a function $W(\mathbf{A} \cdot \mathbf{B})$ of this sine, or thus a function of the amount of shear. Then the energy that is to be minimized has the form

$$E = \iint W(\mathbf{A} \cdot \mathbf{B}) dxdy - \int_{C_t} \mathbf{R} \cdot \mathbf{T} ds,$$

where the area integral extends over the whole initial region. The deformation $\mathbf{R}$ is admissible if it satisifies the displacement boundary conditions and the inextensibility conditions. The resulting theory is a special case of Adkin's theory of reinforced networks (Ref. 15).

Because we have included no energy of bending, again the energy does not necessarily have any kinematically admissible minimizer. Just as in the case of pure networks, we want to rephrase the theory so that it refers directly to carriers of crinkles. If $\mathbf{r}(\mathbf{x})$ is the limit of a sequence of inextensible but perhaps highly folded deformations, its derivatives $\mathbf{r}_x = \mathbf{a}$ and $\mathbf{r}_y = \mathbf{b}$ satisfy the one-sided constraints

$$|\mathbf{a}| \leqq 1, \quad |\mathbf{b}| \leqq 1,$$

just as before.

The new problem is to decide what energy density W to assign when $\mathbf{a}$ and $\mathbf{b}$ are not unit vectors. To do this, we consider a sequence of deformations $\mathbf{R}_n$ that are folded on a very fine scale when n is large. For simple folding, suppose that the sheet is divided into narrow strips labeled (+) and (−) alternatively, with folds at the boundaries between strips. Suppose that the derivatives of $\mathbf{R}_n$ approach $\mathbf{A}_+$, $\mathbf{B}_+$ in one family of strips and $\mathbf{A}_-$, $\mathbf{B}_-$ in the other, and let θ be the limiting fractional density of (+) strips, with

$$0 \leqq \theta(\mathbf{x}) \leqq 1.$$

Then the derivatives of the limit function $\mathbf{r}(\mathbf{x})$ are (Ref. 22)

$$\mathbf{a} = \theta \mathbf{A}_+ + (1 - \theta)\mathbf{A}_-,$$

$$\mathbf{b} = \theta \mathbf{B}_+ + (1 - \theta)\mathbf{B}_-,$$

and the limiting energy density is

$$\theta W(\mathbf{A}_+ \cdot \mathbf{B}_+) + (1 - \theta)W(\mathbf{A}_- \cdot \mathbf{B}_-).$$

We want to express this limiting density in terms of **a** and **b**.

Because $\mathbf{R}_n$ is continuous, its derivative tangential to a fold is the same on either side of the fold, and it follows that the tangential components of **A** and **B** are continuous across the fold (Ref. 22). Because **A** and **B** are unit vectors, the magnitudes of their normal components are then continuous as well, so their inner product is continuous, and in the limit,

$$\mathbf{A}_+ \cdot \mathbf{B}_+ = \mathbf{A}_- \cdot \mathbf{B}_- = \mathbf{A} \cdot \mathbf{B} \qquad \text{(say)}.$$

(There is an exception if the fold is along a fiber.) Thus, the limiting energy density is just $W(\mathbf{A} \cdot \mathbf{B})$. The remaining problem is to find the limiting value of $\mathbf{A} \cdot \mathbf{B}$, given the derivatives **a** and **b** of the limiting deformation **r**.

After some heavy algebra it is found (Ref. 22) that with given values of **a** and **b**, there are two possibilities about the limiting value of $\mathbf{A} \cdot \mathbf{B}$, corresponding to two different ways of viewing the limiting deformation $\mathbf{r}(\mathbf{x})$ as the limit of a crinkle with simple folding. The two possibilities correspond to two diffrent directions for the fold lines.

This puzzling result becomes more understandable when we consider a more complex kind of wrinkling. In complex folding, the sheet is folded along some family of lines and then folded again along some nonparallel family of lines. It turns out that the limiting values that $\mathbf{A} \cdot \mathbf{B}$ can have in this kind of wrinkling fill up the whole range between the two values found for simple folding,

$$\mathbf{a} \cdot \mathbf{b} - [(1 - \mathbf{a} \cdot \mathbf{a})(1 - \mathbf{b} \cdot \mathbf{b})]^{1/2} \leqq \mathbf{A} \cdot \mathbf{B} \leqq \mathbf{a} \cdot \mathbf{b} + [(1 - \mathbf{a} \cdot \mathbf{a})(1 - \mathbf{b} \cdot \mathbf{b})]^{1/2}.$$

Now suppose that $W(\mathbf{A} \cdot \mathbf{B})$ is an even, convex function of its scalar argument, with $W(0) = 0$. For a given state of deformation, defined by **a** and **b**, we choose the wrinkled state that minimizes $W(\mathbf{A} \cdot \mathbf{B})$. If the two bounds on $\mathbf{A} \cdot \mathbf{B}$ have opposite signs, we can choose $\mathbf{A} \cdot \mathbf{B} = 0$, so that $W = 0$. In such cases both principal stretches are less than unity, and the specified values of **a** and **b** can be produced by complex wrinkling of an otherwise undeformed material element (Ref. 22).

When the two bounds on $\mathbf{A} \cdot \mathbf{B}$ have the same sign, we get the least value of $W(\mathbf{A} \cdot \mathbf{B})$ by letting $\mathbf{A} \cdot \mathbf{B}$ be equal to the bound of smaller magnitude. Including the previous result as well, we take

$$\mathbf{A} \cdot \mathbf{B} = k_+ \operatorname{sign}(\mathbf{a} \cdot \mathbf{b}),$$

where

$$k = |\mathbf{a} \cdot \mathbf{b}| - [(1 - \mathbf{a} \cdot \mathbf{a})(1 - \mathbf{b} \cdot \mathbf{b})]^{1/2},$$
$$k_+ = k, \quad \text{if } k \geqq 0,$$
$$k_+ = 0, \quad \text{if } k \leqq 0.$$

The relaxed energy density is then $W(k_+ \text{sign} \mathbf{a} \cdot \mathbf{b})$. At places where both families of fibers are fully extended, this is the same as the original energy density. However, the relaxed density has a feature that the original density did not have: it is convex jointly in **a** and **b**, and thus a convex function of the deformation gradient. The relaxed energy functional is

$$E[\mathbf{r}] = \iint W(k_+ \text{sign} \mathbf{a} \cdot \mathbf{b}) dx dy - \int_{C_1} \mathbf{T} \cdot \mathbf{r} ds,$$

and this is to be minimized over deformations that satisfy the one-sided constraint conditions and the displacement boundary conditions.

In the specialization of this theory to infinitesimal plane deformations (Ref. 23), the constraint conditions in terms of infinitesimal strain components are

$$\epsilon_{xx} \leqq 0, \qquad \epsilon_{yy} \leqq 0.$$

Paradoxically, ϵ_{xx} and ϵ_{yy} can have values that are arbitrarily large in magnitude if they are negative. Moreover, the relaxed energy density is zero in a certain sector of strain space that extends to infinity. For these reasons it is possible to pose problems for which the relaxed energy functional is not bounded below (Ref. 23), even if it is bounded below within the exact finite-deformation theory. In such problems the actual solution is not even approximately a small deformation, and within the infinitesimal theory there is no solution.

4. Rubber Membranes

For an unconstrained, isotropic elastic membrane, the strain energy is some function $W^*(\mathbf{a}, \mathbf{b})$ of the two deformation gradient vectors $\mathbf{a} = \mathbf{r}_x$ and $\mathbf{b} = \mathbf{r}_y$. These vectors are not now subject to any constraint. Any deformation can be viewed on an infinitesimal scale as the stretching of an initially square element

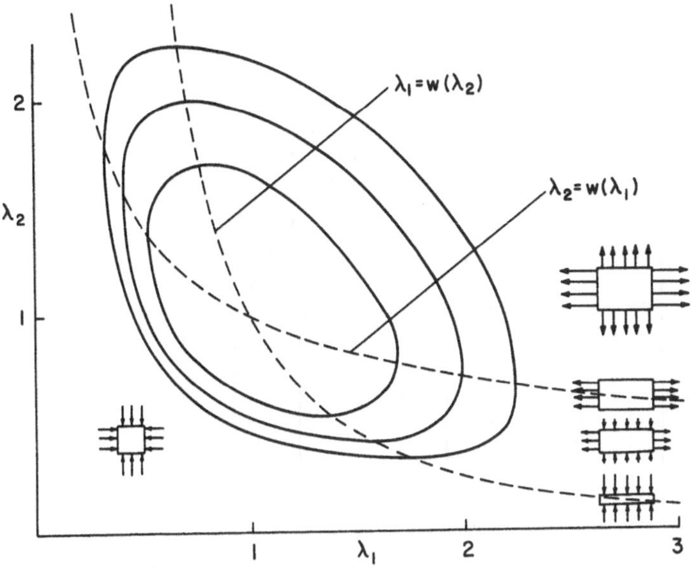

Fig. 3. Contours of constant strain energy and simple tension curves.

into a rectangular shape. Let $W^*(\lambda_1, \lambda_2)$ be the strain energy of an element that was initially a unit square, after it has been stretched to dimensions λ_1 and λ_2. The forces required to do this are $f_a = \partial W^*/\partial\lambda_a$.

Figure 3 shows contours of constant W* for a neo-Hookean material. The dashed lines are the two curves along which $f_1 = 0$ and $f_2 = 0$, respectively. These are the simple tension curves. When f_1 is increased with $f_2 = 0$, λ_1 increases but λ_2 decreases, taking a value $w(\lambda_1)$ that represents the natural width in simple tension. Similarly, when $f_1 = 0$ but $f_2 > 0$ the simple tension curve is $\lambda_1 = w(\lambda_2)$.

In the region where $\lambda_2 < w(\lambda_1)$ the stress f_2 is compressive. However, the stretch λ_2 could be achieved at less expense in energy by first deforming the square to dimensions λ_1 and $w(\lambda_1)$ (for which the energy is minimized over λ_2, given λ_1) and then producing the further reduction of the width to λ_2 by wrinkling. Similarly, when $\lambda_1 < w(\lambda_2)$, the least-energy way of achieving this state of deformation is by letting λ_1 take on its natural value $w(\lambda_2)$ and then reducing λ_1 further by wrinkling instead of compression. The relaxed energy density is the value of W^* at thc natural width. For example,

$$W(\lambda_1, \lambda_2) = W^*[\lambda_1, w(\lambda_1)], \qquad \text{if } \lambda_1 \geqq 1 \text{ and } \lambda_2 \leqq w(\lambda_1).$$

When both λ_1 and λ_2 are less than unity, the given state of deformation can

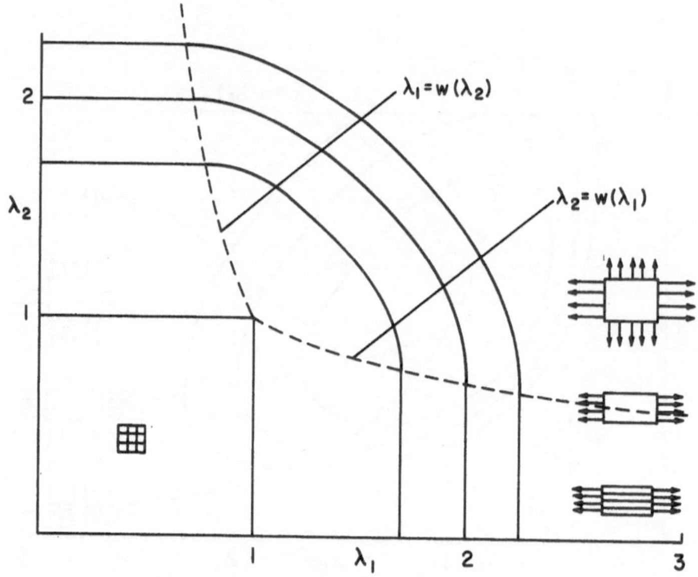

Fig. 4. Contours of the relaxed energy density.

be produced as the limit of a crinkle that involves folding in both directions, with no previous stretching. Within the theory, this requires no work, so the relaxed energy density is zero,

$$W(\lambda_1, \lambda_2) = 0, \qquad \text{if } \lambda_1 \leqq 1 \text{ and } \lambda_2 \leqq 1.$$

Contours of the relaxed energy density are shown in Fig. 4.

Expressing W in terms of the deformation gradients **a** and **b** involves some complicated algebra (Ref. 24). With W^* neo-Hookean as assumed, the relaxed density W is then convex as a function of **a** and **b**. With other forms of W^*, the general requirement is that the relaxed energy density W must be the quasiconvexification of W^*, as defined by Dacorogna (Ref. 25), and this is more than satisfied if W is convex. The energy functional, obtained by replacing W^* by W, is lower semicontinuous when W is quasiconvex (Morrey, Ref. 26). This greatly simplifies the theoretical situation.

5. Stress

The Euler equation for the variational principle always has the form

$$\mathbf{t}_{a,x} + \mathbf{t}_{b,y} = \mathbf{0},$$

and by deriving this equilibrium equation from the variational principal we find expressions for the stress vectors $\mathbf{t}_a$ and $\mathbf{t}_b$. The derivation is most straightforward in the case of elastic membranes, because **a** and **b** are not subject to any constraint conditions. In this case, small changes $d\mathbf{a}$ and $d\mathbf{b}$ in the deformation gradients cause a change of strain energy

$$dW = \mathbf{t}_a \cdot d\mathbf{a} + \mathbf{t}_b \cdot d\mathbf{b},$$

and the stress vectors are derivatives of the strain energy function as usual. What is unusual is the special form of the relaxed density W. Because of this form, the stress–strain relations imply that (Ref 24)

(a) $\mathbf{t}_a = \mathbf{t}_b = \mathbf{0}$ when both principal stretches are no greater than unity;
(b) when $\lambda_2 \leqq w(\lambda_1)$ and $\lambda_1 > 1$, or when $\lambda_1 \leqq w(\lambda_2)$ and $\lambda_2 > 1$, the two stress vectors are parallel and the stress state is a positive simple tension parallel to their common direction;
(c) when $\lambda_1 > w(\lambda_2)$ and $\lambda_2 > w(\lambda_1)$, both principal stresses are positive.

Thus, the basic assumptions of tension field theory come as consequences of the form of the relaxed energy density; tension field theory is just ordinary membrane theory with a special form of strain energy density.

In pure network theory we also found that no compressive stresses can arise from a solution within the relaxed theory, or at least there is a solution with nonnegative fiber tensions. As reactions to one-sided constraints, the fiber tensions are nonnegative Lagrange multipliers.

For reinforced networks, with a strain energy of shearing, the derivation of the Euler equation is complicated for two reasons (Ref. 22). Aside from the one-sided constraint conditions, the relaxed energy density $W(k_+ \text{sign}\mathbf{a} \cdot \mathbf{b})$ is singular with respect to **a** and **b** when $|\mathbf{a}| = 1$ or $|\mathbf{b}| = 1$. The resulting stress–strain relations have the form

$$\mathbf{t}_a = T_a\mathbf{a} + S\mathbf{b}, \qquad \mathbf{t}_b = T_b\mathbf{b} + S\mathbf{a}.$$

where the shearing stress S is the derivative $S = W'$. The fiber tension T_a or T_b is a nonnegative reaction when the corresponding fiber is fully extended. If $|\mathbf{a}| = 1$ but $|\mathbf{b}| < 1$, the tension T_a is nonnegative but $T_b = 0$, and the shearing stress is also zero. The fully extended fibers are the wrinkle lines in this case.

When both $|\mathbf{a}| = 1$ and $|\mathbf{b}| < 1$, all stress components are expressible as derivatives of the relaxed energy density, with no reaction tensions. The stress–strain relations take the forms

$$\mathbf{t}_a = S(\mathbf{a}/m + \mathbf{b}), \qquad \mathbf{t}_b = S(\mathbf{a} + m\mathbf{b}),$$

where

$$m = \text{sign}(\mathbf{a} \cdot \mathbf{b})[(1 - \mathbf{a} \cdot \mathbf{a})/(1 - \mathbf{b} \cdot \mathbf{b})]^{1/2}$$

The coefficient m represents the slope of the wrinkle line in the initial domain, $dy/dx = m$. The vector $\mathbf{a}+m\mathbf{b}$ is tangential to the wrinkle line in the deformed state. As we see, both $\mathbf{t}_a$ and $\mathbf{t}_b$ are parallel to this direction. The stress state is one of uniaxial tension in this direction, and the corresponding principal stress component is tensile rather than compressive (Ref. 22). Thus, again the basic assumptions of tension field theory are recovered as consequences of the form of the relaxed energy density.

Acknowledgment

The work described in this paper was carried out under grant DMS-8702866 from the National Science Foundation. We gratefully acknowledge this support.

References

1. WAGNER, H., *Ebene Blechwandtrager mit sehr dunnem Stegblech*, Flugtechnik und Motorluftschiffahrt, Vol. 20, pp. 200–207, 227–233, 256–262, 279–284, 306–314, 1929.
2. REISSNER, E., *On Tension Field Theory*, Proceedings of the 5th International Congress on Applied Mechanics, pp. 88–92, 1938.
3. KONDO, K., *The Geometry of the Perfect Tension Field*, Journal of the Society for Applied Mechanics of Japan, Vol. 3, pp. 36–39, 85–89, 1950.
4. KONDO, K., IAI, T., MORIGUTI, S., and MURASAKI, T., *Tension-Field Theory*, Memoirs of the Unifying Study of the Basic Problems in Engineering Sciences by Means of Geometry, Gakujutsu Bunken Fukyu-Kai, Tokyo, Japan, Vol. 1C, pp. 61–85, 1955.
5. STEIN, M., and HEDGEPETH, J. M., *Analysis of Partly Wrinkled Membranes*, NASA TN D-813, 1961.
6. CHEREPANOV, G. P., *On the Buckling Under Tension of a Membrane Containing Holes*, Applied Mathematics and Mechanics, Vol 27, pp. 405–420, 1963.
7. MANSFIELD, E. H., *Tension Field Theory*, Proceedings of the 12th International Congress on Applied Mechanics, pp. 305–320, 1968.
8. MANSFIELD, E. H., *Load Transfer via a Wrinkled Membrane*, Proceedings of the Royal Society of London, Vol. A316, pp. 269–289, 1970.
9. WU, C. H., *The Wrinkled Axisymmetric Air Bags Made of Inextensible Membrane*, Journal of Applied Mechanics, Vol. 41, pp. 963–967, 1974.
10. WU, C. H., *Plane Linear Wrinkle Elasticity without Body Force*, Report, Department of Materials Engineering, University of Illinois at Chicago Circle, 1974.

11. WU, C. H., and TING, T. C. T., *Wrinkling of Annular Membanes Subject to Twisting and Stretching*, Report, Department of Materials Engineering, University of Illinois at Chicago Circle, 1974.
12. WU, C. H., *Nonlinear Wrinkling of Nonlinear Membranes of Revolution*, Journal of Applied Mechanics, Vol. 45, pp. 533–538, 1978.
13. ZAK, M., *Statics of Wrinkling Films*, Journal of Elasticity, Vol. 12, pp. 51–63, 1982.
14. RIVLIN, R. S., *Plane Strain of a Net Formed by Inextensible Cords*, Archive for Rational Mechanics and Analysis, Vol. 4, pp. 951–974, 1955.
15. ADKINS, J. E., *Finite Plane Deformation of Thin Elastic Sheets Reinforced with Inextensible Cords*, Philosophical Transactions of the Royal Society of London, Vol. A249, pp. 125–150, 1956.
16. PIPKIN, A. C., *Some Developments in the Theory of Inextensible Networks*, Quarterly of Applied Mathematics, Vol. 38, pp. 343–355, 1980.
17. WANG, W. B., and PIPKIN, A. C., *Inextensible Networks with Bending Stiffness*, Quarterly Journal of Mechanics and Applied Mathematics, Vol. 39, pp. 343–359, 1986.
18. WANG, W. B., and PIPKIN, A. C., *Plane Deformations of Nets with Bending Stiffness*, Acta Mechanica, Vol. 65, pp. 263–279, 1986.
19. PIPKIN, A. C., *Some Examples of Crinkles*, Homogenization and Effective Moduli, Springer Verlag, New York, 1986.
20. PIPKIN, A. C., *Energy Minimization for Nets with Slack*, Quarterly of Applied Mathematics, Vol. 44, pp. 249–253, 1986.
21. PIPKIN, A. C., *Inextensible Networks with Slack*, Quarterly of Applied Mathematics, Vol. 40, pp. 63–71, 1982.
22. PIPKIN, A. C., *Continuously Distributed Wrinkles in Fabrics*, Archive for Rational Mechanics and Analysis, Vol. 95, pp. 93–115, 1986.
23. PIPKIN, A. C., and ROGERS, T. G., *Infinitesimal Plane Wrinkling of Inextensible Networks*, Journal of Elasticity, Vol. 17, pp. 35–52, 1987.
24. PIPKIN, A. C., *The Relaxed Energy Density for Isotropic Elastic Membranes*, IMA Journal of Applied Mathematics, Vol. 36, pp. 85–99, 1986.
25. DACOROGNA, B., *Quasiconvexity and Relaxation of Nonconvex Problems in the Calculus of Variations*, Journal of Functional Analysis, Vol. 46, pp. 102–118, 1982.
26. MORREY, C. B., *Quasiconvexity and the Lower Semicontinuity of Multiple Integrals*, Pacific Journal of Mathematics, Vol. 2, pp. 25–53, 1952.

12

On Boundary Layer-Like Structures in Finite Thermoelasticity

K. R. RAJAGOPAL

Abstract. In this paper inhomogeneous deformations are discussed within the context of finite thermoelasticity with a view toward describing the formation of boundary layer-like structures. Here, attention is confined to static problems and associated with the notion of a boundary layer is a narrow layer adjacent to a boundary wherein the strains are significantly higher than in the far field. The region of concentration of strains is also a region where the deformation is pronouncedly inhomogeneous. It can be shown that in dynamic problems, a boundary layer can also be associated with a region wherein the vorticity is confined, similar to the situation in fluids.

Key Words. Rubberlike solids, thermoelastic response, inhomogeneous deformations, boundary layer, material nonlinearities, finite thermoelasticity, temperature stiffening.

1. Introduction

The seminal papers of R. S. Rivlin (Refs. 1–6) that pertain to finite elasticity have not only left an indelible imprint on that subject but have also helped to shape the mechanics of continua. These fundamental papers of Rivlin, on con-

---

K. R. Rajagopal • Professor, Department of Mechanical Engineering, University of Pittsburgh, Pittsburgh, Pennsylvania 15261.

Nonlinear Effects in Fluids and Solids, edited by M. M. Carroll and M. Hayes, Plenum Press, New York, 1996.

tinua that undergo large deformations, study in great detail the isothermal behavior of rubber. Though the thermomechanics of rubberlike solids have been studied with a view to characterizing the response of such materials (Refs. 7–12), few nonisothermal boundary value problems have been solved, and their results compared with experiments.

Petroski and Carlson (Ref. 13) investigated universal deformations within the context of finite thermoelasticity and later studied specific universal deformations in detail (see Ref. 14). Finite torsion of thermoelastic cylinders and the deformation of spherical sectors, within the context of thermoelasticity, were also analyzed by Petroski (Refs. 15, 16), among other deformations. Martin and Carlson (Ref. 17) studied second-order thermoelastic response of solids. Most of the above studies were restricted to homogeneous deformations wherein the deformation gradient $\mathbf{F}$ and the Almansi-Hamel strain tensor $\mathbf{e}$ have constant entries in a Cartesian coordinate system.

However, an interesting phenomenon manifests itself when we consider the thermoelastic response of bodies undergoing inhomogeneous deformations; in some problems, pronounced boundary layer-like structures develop in the sense that the strains are concentrated adjacent to boundaries, the strains being essentially uniform and small exterior to the boundary layer. Such boundary layers occur even within the context of a purely mechanical theory (Refs. 18–21). The presence of boundary layers within the context of finite thermoelasticity has been explored in recent papers by Rajagopal (Ref. 22) and Huang and Rajagopal (Ref. 23). In these studies, the presence of such boundary layer-like structures within the context of specific nonlinear constitutive theories is investigated, more with a view to glean some information about the structures than with specific applications in mind. These studies also seem to indicate another possible way of delineating the boundary layer by associating a region of inhomogeneous deformation with it, the deformation being essentially homogeneous external to the boundary layer.

The boundary layer solutions that are found here arise due to material nonlinearities and there seems to be no notion of a singular perturbation underlying their manifestation, as is the case in the classical linearly viscous fluid. The terms "strain layer" or "transition layer" might also be appropriate for defining such deformation; however, as these concentrations of strains occur adjacent to the boundary, I shall refer to them as boundary layers.

Here, I shall consider static problems. An interesting possibility exists with regard to localization or the formation of boundary layers within the context of unsteady problems, namely, the concentration of vorticity. Especially in inhomogeneous materials and materials with defects, the intensity of vorticity might initiate and propagate damage.

Now, I shall discuss a very simple deformation which highlights the formation of boundary layers in finite thermoelasticity.

2. Governing Equations

Let us consider the deformation of a cylindrical annulus, such that

$$r = R, \qquad \theta = \Theta, \qquad z = Z + f(R), \tag{1}$$

where (R, Θ, Z) and (r, θ, z) represent a point in the reference and current configuration, in a cylindrical polar coordinate system. Furthermore, suppose that the temperature T is given by

$$T = T(R). \tag{2}$$

We shall suppose that the cylindrical annulus is comprised of a generalized neo-Hookean material, and thus its Cauchy stress $\mathbf{T}$ is given by (Ref. 24)

$$\mathbf{T} = -p\mathbf{I} + \mu\mathbf{B}, \tag{3}$$

where $\mathbf{B}$ is the Cauchy–Green stretch tensor given by

$$\mathbf{B} = \mathbf{F}\mathbf{F}^T, \tag{4}$$

and the shear modulus μ is given by

$$\mu = \bar{\mu}(T)\left\{\left[1 + \frac{b}{n}(I_1 - 3)\right]^{n-1}\right\}, \tag{5}$$

where b and n are positive constants. When $\bar{\mu}(T) = kNT$, where k is the Boltzmann constant and N the number of cross-links, and $n = 1$, the above model reduces to the neo-Hookean model.

We use the above model to illustrate our ideas. The possibility of boundary layers is not restricted to such models. In materials that can soften or stiffen, both due to deformation and temperature, a very rich class of "boundary layer" solutions are possible.

The Helmholtz free energy corresponding to the generalized neo-Hookean model is

$$W = \frac{\bar{\mu}(T)}{2b}\left\{\left[1 + \frac{b}{n}(I_1 - 3)\right]^{n} - 1\right\}, \tag{6}$$

which for $n = 1$ and $\bar{\mu}(T) = kNT$, reduces to

$$W = kNT(I_1 - 3), \tag{7}$$

namely, the Helmholtz free energy for a neo-Hookean solid.

However, such an assumption for the Helmholtz free energy leads to a zero specific heat, and in fact the Helmholtz free energy for a neo-Hookean material is of the form

$$W = kNT(I_1 - 3) + \hat{f}(T), \tag{8}$$

which allows for nonzero specific heat. The model (7) is just an approximation and stems from very rudimentary applications of statistical methods for the development of constitutive relations in polymeric and rubberlike materials; the elementary theory of rubber elasticity ignores the contributions to the specific internal energy due to deformation, all changes in the specific internal energy being attributable to changes in the configurational entropy of the body (see Ref. 8).

It is worthwhile to note that unlike metals, rubber temperature stiffens and the constitutive relation (7) is consistent with the temperature stiffening of the material.

A trivial calculation leads to the following equations of equilibrium:

$$-\frac{\partial p}{\partial R} + \frac{\partial p}{\partial Z} f' = 0, \tag{9}$$

$$\frac{\partial p}{\partial \Theta} = 0, \tag{10}$$

$$-\frac{\partial p}{\partial Z} + \left[knT\{1 + \frac{b}{n}(f')^2\}^{n-1} \right]' + \frac{kNT\{1 + \frac{b}{n}(f')^2\}^{n-1} f'}{R} = 0. \tag{11}$$

As we shall use the concentration of strains adjacent to boundaries as a measure to illustrate the phenomenon of boundary layers, we document the expression for the norm of the Almansi–Hamel strain tensor **e** corresponding to the deformation (1),

$$\| e \| = \frac{1}{2} [(f')^4 + 2(f')^2]^{1/2}. \tag{12}$$

For the purpose of illustrating the main issues, it is sufficient to consider the case corresponding to $n = 1$, i.e., the neo-Hookean case. Of course, when $n \neq 1$, we have the possibility of competition between shear softening due to deformation and stiffening due to increase in temperature (see Ref. 22), or shear stiffening due to both the deformation and temperature leading to boundary layers that are much more pronounced than in the neo-Hookean case.

When $n = 1$, it follows from (9)–(11) that

$$(kNTf')' + \frac{kNTf'}{R} = \lambda = \text{const.} \tag{13}$$

We next derive the equation governing the temperature distribution in the annulus. On assuming that the heat flux $\mathbf{q}$ is given by Fourier's law, i.e.,

$$\mathbf{q} = -K \operatorname{grad} T, \tag{14}$$

where K is a constant, it follows from the energy equation,

$$\rho \frac{d\epsilon}{dt} - \mathbf{T} \cdot \mathbf{L} + \operatorname{div} \mathbf{q} + \rho r = 0, \tag{15}$$

in the absence of radiant heating, the assumption that the thermal conductivity K is a constant, and (1) and (2), that

$$\frac{1}{R}(RT')' = 0. \tag{16}$$

It is possible that the thermal conductivity K depends on both the deformation and temperature. The possibility that the thermal conductivity is a constant while the shear modulus is temperature dependent is similar to the assumption for many liquids that the viscosity is much more sensitive to changes in temperature than the thermal conductivity.

Let us suppose that the outer and inner boundaries of the annulus are maintained at

$$T(R_i) = T_i \quad \text{and} \quad T(R_o) = T_o. \tag{17}$$

Let us introduce nondimensional variables

$$\bar{R} = \frac{R}{R_i} \quad \text{and} \quad \bar{T} = \frac{T}{T_i}. \tag{18}$$

It then follows from (16)–(18) that the nondimensional temperature field $\overline{T}$ is given by

$$\overline{T}(\overline{R}) = \frac{\left[\dfrac{T_i - 1}{T_o}\right]}{\left(\ln \dfrac{R_i}{R_o}\right)} \ln \overline{R} + 1. \tag{19}$$

Henceforth, for the sake of simplicity, we shall drop the overbar on the nondimensional quantities.

On substituting (19) into (13), and carrying out a straightforward computation, we obtain

$$f' = \frac{\left(\hat{\lambda}\overline{R} + \dfrac{C}{\overline{\overline{R}}}\right) \ln \dfrac{R_i}{R_o}}{\left[\left(\dfrac{T_i}{T_o} - 1\right)\ln \overline{R} + \ln\left(\dfrac{R_i}{R_o}\right)\right]}. \tag{20}$$

The displacement $f(R)$ in the z-direction is obtained by integrating the above expression and would lead to another constant of integration, say D. The constants C and D are determined by using appropriate boundary conditions. We shall consider the boundary condition

$$f(1) = \alpha, \qquad f\left(\frac{R_o}{R_i}\right) = 0. \tag{21}$$

The condition (21) implies that the inner boundary is held fixed while the boundary at $R = R_o$ is displaced in the z-direction.

The constant $\hat{\lambda}$ is related to the pressure gradient that develops along the z-direction, and is determined by the constant λ that appears in (13). For the sake of simplicity, let us seek a solution in which $\hat{\lambda} = 0$, which corresponds to a solution for which there is no pressure gradient in the z-direction. Of course, this does not preclude the possibility of other solutions.

The norm of the Almansi–Hamel strain tensor **e** as a function of R is shown in Figs. 1 and 2 for different values of temperatures at which the boundaries of the annulus are held. In both figures, strains are confined in a narrow layer adjacent to the boundary. This structure can be made much more pronouned when the material also shear stiffens with the deformation. Calculations also indicate that if the relationship of the shear modulus μ to the temperature is more pronounced, say an exponential or a polynomial of order $m > 1$, then far sharper boundary layers can develop (see Ref. 22).

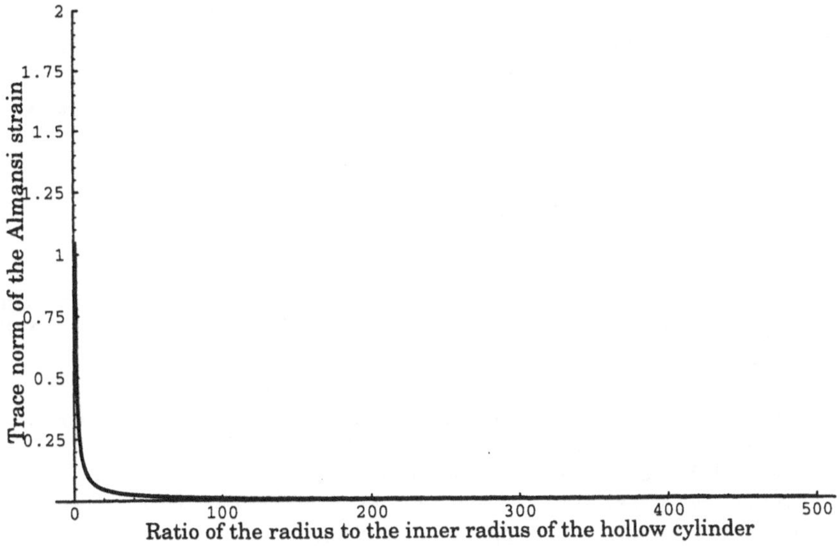

Fig. 1. Norm of the Almansi–Hamel strain tensor **e** as a function of $R(T_o/T_i = 1, \alpha = 4.31)$.

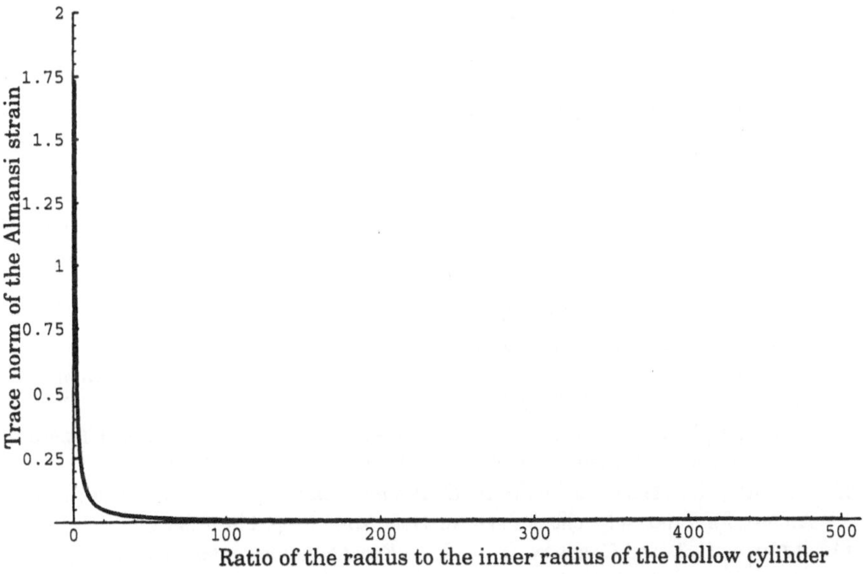

Fig. 2. Norm of the Almansi–Hamel strain tensor e as a function of $R(T_o/T_i = 2, \alpha = 4.31)$.

In this work, we have not addressed the issues of the uniqueness or stability of solutions for the problem that has been considered. We should bear in mind that we have used a semi-inverse method to study the problem, and as the general equilibrium equations are coupled nonlinear partial differential equations, multiple solutions might be possible.

References

1. RIVLIN, R. S., *Large Elastic Deformations of Isotropic Materials, I, Fundamental Concepts*, Philosophical Transactions of the Royal Society of London, Vol. A240, pp. 459–490, 1948.
2. RIVLIN, R. S., *Large Elastic Deformations of Isotropic Materials, II, Some Uniqueness Theorems for Pure Homogeneous Deformations*, Philosophical Transactions of the Royal Society of London, Vol. A240, pp. 491–508, 1948.
3. RIVLIN, R. S., *Large Elastic Deformations of Isotropic Materials, III, Some Simple Problems in Cylindrical Polar Coordinates*, Philosophical Transactions of the Royal Society of London, Vol. A240, pp. 509–525, 1948.
4. RIVLIN, R. S., *Large Elastic Deformations of Isotropic Materials, IV, Further Developments of the General Theory*, Philosophical Transactions of the Royal Society of London, Vol. A241, pp. 379–397, 1948.
5. RIVLIN, R. S., *Large Elastic Deformations of Isotropic Materials, V, The Problem of Flexure*, Proceedings of the Royal Society of London, Vol. A195, pp. 463–473, 1949.
6. RIVLIN, R. S., *Large Elastic Deformations of Isotropic Materials, VI, Further Results in the Theory of Torsion, Shear and Flexure*, Philosophical Transactions of the Royal Society of London, Vol. A242, pp. 173–195, 1949.
7. FLORY, P. J., *Principles of Polymer Chemistry*, Cornell University Press, Ithaca, New York, 1953.
8. TRELOAR, L. R. G., *The Physics of Rubber Elasticity (Third Edition)*, Clarendon Press, Oxford, England, 1975.
9. CHADWICK, P., and SEET, L. T. C., *Trends in Elasticity and Thermoelasticity*, Edited by R. E. Czarnota-Bojarski, M. Sokolowski, and H. Zorski, Wolters–Noordhoff, Groningen, The Netherlands, pp. 29–57, 1971.
10. ALLEN, G., BIANCHI, U., and PRICE, C., *Thermodynamics of Elasticity of Natural Rubber*, Transactions of the Faraday Society, Vol. 59, pp. 2493–2502, 1963.
11. BESSELING, J. F., and VOETMAN, H. H., *Thermo-elastic Effects in Rubber*, Archiwum Mechaniki Stosowanej, Vol. 20, pp. 1789–2202, 1968.
12. CHADWICK, P., *Thermo-mechanics of Rubber-like Materials*, Philosophical Transactions of the Royal Society of London, Vol. A276, pp. 371–403, 1974.
13. PETROSKI, H. J., and CARLSON, D. E., *Controllable States of Elastic Heat Conductors*, Archive for Rational Mechanics and Analysis, Vol. 31, pp. 127–150, 1968.
14. PETROSKI, H. J., and CARLSON, D. E., *Some Exact Solutions to the Equations of Non-Linear Thermoelasticity*, Journal of Applied Mechanics, Vol. 37, pp. 1151–1154, 1970.

15. Petroski, H. J., *On the Finite Torsion and Radial Heating of Thermoelastic Cylinders*, International Journal of Solids and Structures, Vol. 11, pp. 741–749, 1975.
16. Petroski, H. J., *On the Finite Deformation and Heating of Thermoelastic Spherical Sectors*, International Journal of Non-Linear Mechanics, Vol. 10, pp. 327–332, 1975.
17. Martin, S. E., and Carlson, D. E., *The Behavior of Elastic Heat Conductors with Second-Order Response Functions*, Journal of Applied Mathematics and Physics, Vol. 28, pp. 311–328, 1977.
18. Rajagopal, K. R., and Tao, L., *On an Inhomogeneous Deformation of a Generalized Neo-Hookean Material*, Journal of Elasticity, Vol. 28, pp. 165–184, 1992.
19. Zhang, J. P., and Rajagopal, K. R., *Some Inhomogeneous Motions and Deformations Within the Context of a Non-Linear Elastic Solid*, International Journal of Engineering Science, Vol. 30, pp. 919–938, 1992.
20. Tao, L., Rajagopal, K. R., and Wineman, A. S., *Circular Shearing and Torsion of Generalized Neo-Hookean Materials*, IMA Journal of Applied Mathematics, Vol. 48, pp. 23–37, 1992.
21. Haughton, D. M., *Boundary Layer Solutions for Incompressible Elastic Cylinders*, International Journal of Engineering Science, Vol. 30, pp. 1027–1040, 1992.
22. Rajagopal, K. R., *Boundary Layers in Finite Thermoelasticity*, Journal of Elasticity, Vol. 36, pp. 271–301, 1995.
23. Huang, Y., and Rajagopal, K. R., *Finite Circumferential Shearing of Non-Linear Solids within the Context of Thermoelasticity*, IMA Journal of Applied Mathematics, Vol. 50, pp. 22–36, 1994.
24. Knowles, J. K., *The Finite Anti-Plane Shear Field Near the Tip of a Crack for a Class of Incompressible Elastic Solids*, International Journal of Fracture, Vol. 13, pp. 611–639, 1977.

13

On the Thickness Limitation for Euler Buckling

K. N. Sawyers

Abstract. Classical Euler buckling is studied within the context of finite elasticity theory. The condition for stable bifurcation deformations to exist in a plate of general incompressible elastic material is derived; and an asymptotic expansion in terms of the plate's aspect ratio (thickness/height) is obtained. The comparable exact and asymptotic analyses for a plate of neo-Hookean material yield specific numerical values for the limiting aspect ratio which differ by about 6%. Further results show that the neo-Hookean case captures the essential features of Euler buckling for a wide class of elastic materials.

Key Words. Euler column buckling, incompressible elastic material, post-buckling deformations, neutral equilibrium.

1. Introduction

This symposium, honoring a half century of Ronald Rivlin's contributions to Mechanics, is an appropriate setting to describe a research problem that occupied his attention, and the author's, at various times throughout a decade of that period. The problem finds basis in an early triumph of the subject. It addresses the question of how thick a column can be and still exhibit the phenomenon of classical Euler buckling.

K. N. Sawyers • Professor, Department of Mechanical Engineering and Mechanics, Lehigh University, Bethlehem, Pennsylvania 18015.

Nonlinear Effects in Fluids and Solids, edited by M. M. Carroll and M. Hayes, Plenum Press, New York, 1996.

Many workers have investigated the effects of eccentric loading, imperfections, and anelasticity to deduce modifications of the Euler theory for practical thick columns, and much corroborative experimental work has been reported. Our object, however, is to explore the inherent limitation of thickness alone, within the idealized setting of perfect loading and finite elastic behavior. To this end we consider generalized plane deformations in a plate of isotropic, incompressible elastic material whose major side faces are force free and whose top and bottom faces are subjected to equal and opposite dead-load thrust forces. The hallmark of classical Euler buckling is the occurrence of an abrupt bifurcation from the initial perfect state to a neighboring stable equilibrium state; and it is the failure of this feature that gives the limiting thickness.

The body of work described here appears in papers published during the decade 1974–83 (Refs. 1–6). With the benefit of hindsight, it seems that the subject gains coherence if its development is viewed in reverse chronological order. Accordingly, we begin in Section 2 with the energy formulation for the general material, and continue the analysis in Sections 3 and 4 to deduce the equations that govern critical states and postbuckling equilibrium deformations for a plate of arbitrary thickness. This procedure follows closely that proposed by Koiter (Ref. 7). The result obtained in the general setting contains various derivatives of the material strain-energy function with respect to the principal strain invariants of up to the fourth order, and, for this reason, its exact evaluation is not completed. Rather, in Section 5, we give an asymptotic analysis for the limiting case of a thin plate, which result could be used to obtain an approximation to the ultimate thickness. At this stage, however, no basis is available to assess the efficacy of the approximation.

Great simplification accrues from the assumption that the material of the plate is neo-Hookean. This is exploited in Section 6 to carry through a complete and exact evaluation of the relevant energy functional for a plate of arbitrary thickness, and also to develop the asymptotic analysis for a thin plate. A comparison of these results indicates that a two-term approximation to the energy functional gives the ultimate limiting aspect ratio (thickness/height) to within an error of about 6%.

In isolation, the neo-Hookean result might be regarded as merely of theoretical interest. However, with respect to the general bifurcation condition, discussed in Section 7, we find that the neo-Hookean material captures the salient behavior that typifies the entire class of incompressible materials. Whether this observation offers hope for drawing a reasonable inference from the asymptotic analysis of Section 5, similar to the success of the neo-Hookean model, remains an open question. A complicating feature here is the lack of knowledge about what, if any, restrictions arise on the form of the strain-energy function beyond those involving certain combinations of the first and second derivatives, with respect to the strain invariants, required by material stability considerations. It is unlikely that an analyst would be motivated to carry through the rather formidable calculations of Section 4 without some certainty that the energy function made, or could make, sense.

Sufficient detail is given in what follows, particularly in the earlier sections, to allow the reader to gain a good sense of the flavor of the subject while being spared certain unpleasant algebraic necessities. To assist the true devotee, the notation [1(2.12)] is used to refer to equation (2.12) of Reference 1.

2. Potential Energy

We place the origin of a fixed rectangular Cartesian coordinate frame X at the center of the undeformed plate, and orient the axes of X parallel to the edges of the plate in this configuration. We let $\boldsymbol{\xi}$ denote the vector position of a generic particle and denote by ξ_A, $A = 1, 2, 3$, the components of $\boldsymbol{\xi}$ referred to X. The initial bounding surfaces of the plate are the planes $\xi_A = \pm l_A$. Throughout we assume that the major surfaces of the plate, the planes $\xi_2 = \pm l_2$, remain free of all tractions, while the material surfaces $\xi_1 = \pm l_1$, $\xi_3 = \pm l_3$, are constrained to remain parallel to their initial orientations, but points on them can move freely within their respective planes. This latter condition means that tangential tractions on these planes are zero.

We now envisage that a state of pure homogeneous deformation (state I) is attained under the action of normal forces applied to the faces $\xi_1 = \pm l_1$, $\xi_3 = \pm l_3$, their resultants being $\pm R_1$, $\pm R_3$, respectively. The deformation itself is characterized by the extension ratios λ_1, λ_2, λ_3, such that if $\mathbf{X}$ is the vector position in state I of particle $\boldsymbol{\xi}$, then

$$X_A = \lambda_A \xi_A, \qquad A = 1, 2, 3, \qquad \lambda_1\lambda_2\lambda_3 = 1, \tag{1}$$

the latter condition arising from incompressibility.

Subject to the condition that the resultant forces on the faces $\xi_1 = \pm l_1$, $\xi_3 = \pm l_3$ remain constant, a further deformation might be possible wherein the particle $\boldsymbol{\xi}$ moves to occupy the vector position $\mathbf{x}$ whose components satisfy

$$x_1 = X_1 + u_1(\xi_1, \xi_2), \tag{2a}$$

$$x_2 = X_2 + u_2(\xi_1, \xi_2), \tag{2b}$$

$$x_3 = \lambda_3(1 + E)\xi_3, \tag{2c}$$

where E is a constant and, due to incompressibility, $\det(\partial \mathbf{x}/\partial \boldsymbol{\xi}) = 1$. Thus, with (1) and (2),[1]

$$\lambda_1 u_{2,2} + \lambda_2 u_{1,1} + u_{1,1}u_{2,2} - u_{1,2}u_{2,1} + \lambda_3^{-1} E(1 + E)^{-1} = 0. \tag{3}$$

We refer to the resulting state of deformation (2) as state II.

The Finger strain matrix $\|x_{i,m}x_{j,m}\|$ has components C_{ij} and $C_{ij} + c_{ij}$ in states I and II, respectively, where

$$C_{AA} = \lambda_A^2, \qquad A = 1, 2, 3, \tag{4}$$

and

$$c_{11} = (2\lambda_1 + u_{1,1})u_{1,1} + u_{1,2}^2, \qquad c_{22} = (2\lambda_2 + u_{2,2})u_{2,2} + u_{2,1}^2, \tag{5a}$$

$$c_{12} = c_{21} = (\lambda_1 + u_{1,1})u_{2,1} + (\lambda_2 + u_{2,2})u_{1,2}, \tag{5b}$$

$$c_{33} = \lambda_3^2 E(2 + E), \qquad c_{23} = c_{32} = c_{13} = c_{31} = 0. \tag{5c}$$

We let I_μ and $I_\mu + i_\mu$ denote the principal invariants of $\mathbf{C}$ and $\mathbf{C} + \mathbf{c}$, respectively, and employ (3) to find [1(2.12)]

$$I_1 = \lambda_k\lambda_k, \qquad I_2 = \lambda_k^{-1}\lambda_k^{-1}, \tag{6}$$

and

$$i_1 = 2k + \kappa[\mathbf{u}] + (\lambda_3^2 + 2\lambda_2^2)E^2 + \cdots, \tag{7a}$$

$$i_2 = \lambda_3^2(i_1 + j), \tag{7b}$$

$$j = E\{2\Lambda_2 + 4\lambda_1(1 - \lambda^2)u_{1,1} + E\Lambda_1 + 2\kappa[\mathbf{u}]\} + \cdots, \tag{7c}$$

where

$$\lambda = \lambda_2/\lambda_1, \qquad k = \lambda_1(1 - \lambda^2)u_{1,1} + (\lambda_3^2 - \lambda_2^2)E, \tag{8a}$$

$$\Lambda_1 = (\lambda_1^2 - \lambda_3^2)(3\lambda_2^2 + \lambda_3^2)\lambda_3^{-2}, \qquad \Lambda_2 = -(\lambda_1^2 - \lambda_3^2)(\lambda_2^2 - \lambda_3^2)\lambda_3^{-2}, \tag{8b}$$

[1] Differentiation with respect to ξ_α is denoted by $_{,\alpha}$. In the sequel, the usual summation convention applies to repeated lowercase Greek and Latin subscripts which take the values 1, 2 and 1, 2, 3, respectively.

and the operator κ is defined by

$$\kappa[\mathbf{u}] = u_{\alpha,\beta}u_{\alpha,\beta} + 2\lambda(u_{1,2}u_{2,1} - u_{1,1}u_{2,2}). \tag{9}$$

Terms neglected in (7), indicated by . . . , are of order E^3, and these are found not to affect results of the subsequent analysis. Omissions of this type will be made in what follows without specific mention.

If W and w respectively denote the strain energy density at particle $\boldsymbol{\xi}$ in states I and II, then the increase in the strain energy of the plate in passing from state I to state II is given by [1(2.15)]

$$2l_3\iint(w - W)d\xi_1 d\xi_2$$
$$= 2l_3\iint\left(W_\alpha i_\alpha + \frac{1}{2}W_{\alpha\beta}i_\alpha i_\beta + W^{(3)} + W^{(4)} + \cdots\right)d\xi_1 d\xi_2, \tag{10}$$

where $w = w(I_\mu + i_\mu)$, $W = w(I_\mu)$, and we have made a Taylor series expansion about state I. In (10),

$$W_\alpha = \{\partial w/\partial(I_\alpha + i_\alpha)\}\big|_{i_\mu = 0}, \quad \text{etc.},$$

and $W^{(3)} = 1/6\, W_{\alpha\beta\gamma}i_\alpha i_\beta i_\gamma$, etc. The domain of integration for double integrals is the rectangle $(-l_1, l_1) \times (-l_2, l_2)$.

The nonzero components of the Piola–Kirchhoff stress in state I are Π_{11}, Π_{33} given by [1(2.17)]

$$\Pi_{11} = 2\lambda_1(1 - \lambda^2)(W_1 + \lambda_3^2 W_2), \qquad \Pi_{33} = 2\lambda_3^{-1}(\lambda_3^2 - \lambda_2^2)(W_1 + \lambda_1^2 W_2), \tag{11}$$

and we note that

$$R_1 = 4l_2l_3\Pi_{11}, \qquad R_3 = 4l_1l_2\Pi_{33}. \tag{12}$$

Since the surfaces $\xi_1 = \pm l_1$ remain plane, it follows that the displacement u_1 in (2) must satisfy $u_{1,2}(\pm l_1, \xi_2) = 0$ for $-l_2 < \xi_2 < l_2$. To rule out the possibility of a trivial translation of the plate parallel to the 1-axis, we assume that the displacements of the faces $\xi_1 = \pm l_1$ are equal and opposite, i.e.,

$$u_1(l_1, \xi_2) = -u_1(-l_1, \xi_2) = \lambda_1 e l_1, \qquad -l_2 < \xi_2 < l_2, \tag{13}$$

where e is a constant. Thus, the plate undergoes an additional uniform lengthening or shortening in the 1-direction accordingly as e is positive or negative. A similar role is played by the constant E in (2) with respect to the 3-direction.

With (11), (13), and (2c), it follows that the increase in potential energy of the dead loads R_1, R_3 in passing from state I to state II is [1(2.21)]

$$-2(\lambda_1 e l_1 R_1 + \lambda_3 E l_3 R_3) = -8 l_1 l_2 l_3 (\lambda_1 e \Pi_{11} + \lambda_3 E \Pi_{33})$$
$$= -4 l_3 \iint \{\lambda_1 (1 - \lambda^2)(W_1 + \lambda_3^2 W_2) u_{1,1}$$
$$+ (\lambda_3^2 - \lambda_2^2)(W_1 + \lambda_1^2 W_2) E\} d\xi_1 d\xi_2. \quad (14)$$

The sum of the quantities (10) and (14) gives the increase of total potential energy, $G[\mathbf{u}]$, of the system, consisting of the body and dead loads, associated with the deformation from state I to state II. This can be expressed in the form [1(2.22)]

$$G[\mathbf{u}] = G_2[\mathbf{u}] + \mathcal{G}[\mathbf{u}], \quad (15)$$

where, with the further notation

$$k_1 = W_1 + \lambda_3^2 W_2, \qquad k_2 = 2(W_{11} + 2\lambda_3^2 W_{12} + \lambda_3^4 W_{22}), \quad (16a)$$
$$k_{21} = 2\lambda_3^2 (W_{12} + \lambda_3^2 W_{22}), \quad (16b)$$

we find

$$G_2[\mathbf{u}] = 2 l_3 \iint [k_1 \{\kappa[\mathbf{u}] + (\lambda_3^2 + 2\lambda_2^2) E^2\}$$
$$+ \lambda_3^2 W_2 \{4\lambda_1 (1 - \lambda^2) E u_{1,1} + \Lambda_1 E^2\}$$
$$+ k_2 k^2 + 2 k_{21} \Lambda_2 E k + 2\lambda_3^4 W_{22} \Lambda_2^2 E^2] d\xi_1 d\xi_2, \quad (17a)$$

$$\mathcal{G}[\mathbf{u}] = 2 l_3 \iint [2\lambda_3^2 W_2 E \kappa[\mathbf{u}] + k_2 k \{\kappa[\mathbf{u}] + (\lambda_3^2 + 2\lambda_2^2) E^2\}$$
$$+ \frac{1}{4} k_2 (\kappa[\mathbf{u}])^2 + k_{21} E \{4\lambda_1 (1 - \lambda^2) k u_{1,1} + \Lambda_2 \kappa[\mathbf{u}]\}$$
$$+ W^{(3)} + W^{(4)} + \cdots] d\xi_1 d\xi_2. \quad (17b)$$

The stability question turns on whether $G[\mathbf{u}]$ is positive definite for all $\mathbf{u}$ contained in a neighborhood of $\mathbf{u} = \mathbf{0}$ and which satisfy the constraints imposed by (3) and (13). We remark that $G_2[\mathbf{u}]$ is homogeneous of degree 2 in the displacement components u_1, u_2, E, while $\mathcal{G}[\mathbf{u}]$ contains terms of degree three and higher in these components. Thus, $G_2[\mathbf{u}]$ is the second variation of $G[\mathbf{u}]$.

3. Neutral Equilibrium of Critical States

A necessary condition for stability is $G_2[\mathbf{u}] \geq 0$ for all small admissible values of u_1, u_2, E that minimize $G_2[\mathbf{u}]$. With λ_3 held fixed, we are led to seek a value of $\lambda(=\lambda_2/\lambda_1)$ for which a nontrivial displacement $\mathbf{u}$ exists that satisfies $\delta G_2[\mathbf{u}] = 0$ and the relevant conditions of the problem. State I corresponding to such a value of λ is called a critical state.

We note that $\mathbf{u}$ must satisfy the linearized version of (3),

$$\lambda_1^{-1} u_{1,1} + \lambda_2^{-1} u_{2,2} + E = 0. \tag{18}$$

We multiply the left member of (18) by $-4l_3 p(\xi_1, \xi_2)$ and add the result to the integrand of (17a). Forming the variation of the resulting functional $G_2[\mathbf{u}]$ is straightforward, the variations of u_1, u_2, E being arbitrary except that of u_1 must be constant on the faces $\xi_1 = \pm l_1$ [see (13)].

The equilibrium equations are found to be

$$\{k_1 + \lambda_1^2(1 - \lambda^2)^2 k_2\} u_{1,11} + k_1 u_{1,22} - \lambda_1^{-1} p_{,1} = 0, \tag{19a}$$

$$k_1(u_{2,11} + u_{2,22}) - \lambda_2^{-1} p_{,2} = 0, \tag{19b}$$

$$4l_1 l_2 F_1 E + F_2 \iint u_{1,1} d\xi_1 d\xi_2 = \iint p\, d\xi_1 d\xi_2, \tag{19c}$$

where F_1 and F_2 are constants whose specific forms are not required for the present discussion. The natural boundary conditions on the faces $\xi_1 = \pm l_1$ are

$$\int_{-l_2}^{l_2} [\{k_1 + \lambda_1^2(1 - \lambda^2)^2 k_2\} u_{1,1} - \lambda k_1 u_{2,2} + F_3 E - \lambda_1^{-1} p] d\xi_2 = 0, \tag{20a}$$

$$u_{2,1} + \lambda u_{1,2} = 0, \tag{20b}$$

where F_3 is another constant; and those that pertain to the major faces, $\xi_2 = \pm l_2$, are

$$u_{1,2} + \lambda u_{2,1} = 0, \tag{21a}$$

$$k_1(u_{2,2} - \lambda u_{1,1}) - \lambda_2^{-1} p = 0. \tag{21b}$$

By carrying out the integration with respect to ξ_1 in the left member of (19c) and using (13), we find

$$4l_1l_2(F_1E + F_2e) = \iint p\,d\xi_1 d\xi_2. \tag{22}$$

Again with (13), we note that $u_{1,2} \equiv 0$ on $\xi_1 = \pm l_1$, and so (20b) can be replaced by

$$u_{2,1} = 0, \qquad \xi_1 = \pm l_1. \tag{23}$$

We note also that u_1, u_2, E must satisfy (18).

Following Ref. 1, we seek solutions for u_1, u_2, p in separated form as

$$u_2 = \left.\begin{matrix}\cos(\Omega\xi_1)\\ \sin(\Omega\xi_1)\end{matrix}\right\} U(\xi_2), \quad u_1 = \left.\begin{matrix}-\sin(\Omega\xi_1)\\ \cos(\Omega\xi_1)\end{matrix}\right\} \frac{U'}{\lambda\Omega}, \quad p = \left.\begin{matrix}\cos(\Omega\xi_1)\\ \sin(\Omega\xi_1)\end{matrix}\right\} P(\xi_2), \tag{24}$$

and find that (18), (22), and (23) are satisfied provided

$$E = e = 0, \tag{25}$$

$$\Omega = \frac{n\pi}{2l_1}, \qquad \text{with } n = \begin{Bmatrix}2, 4, 6, \ldots\\ 1, 3, 5, \ldots\end{Bmatrix}. \tag{26}$$

The top (bottom) line in (26) corresponds to the top (bottom) choice in (24). In either case, n is the number of half-wavelengths along the 1-direction of the bifurcation mode considered.

The substitution from (24) into (19a, b) yields the governing differential equation for U,

$$U^{(iv)} - \{1 + \lambda^2 + (1 - \lambda)^2 A\}\Omega^2 U'' + \lambda^2\Omega^4 U = 0, \tag{27}$$

and the expression for P,

$$P = (\lambda^3\lambda_3)^{-1/2}\Omega^{-2}k_1\{U'' - \Omega^2[1 + (1 - \lambda)^2 A]U\}', \tag{28}$$

where

$$A = \frac{(1 + \lambda)^2 k_2}{\lambda\lambda_3 k_1} = \frac{2(1 + \lambda)^2(W_{11} + 2\lambda_3^2 W_{12} + \lambda_3^4 W_{22})}{\lambda\lambda_3(W_1 + \lambda_3^2 W_2)}. \tag{29}$$

The use of (24) and (28) in (21) yields boundary conditions for U of the form

$$\left.\begin{aligned} &U'' + \lambda^2\Omega^2 U = 0, && \text{(30a)}\\ &U''' - \{1 + 2\lambda^2 + (1-\lambda)^2 A\}\Omega^2 U' = 0 && \text{(30b)}\end{aligned}\right\}\ \xi_2 = \pm l_2.$$

Also, with (24), (28) we see that (20a) is identically satisfied if U' and U''' are odd functions of ξ_2; but, more generally, we find that (20) can be expressed as

$$\int_{-l_2}^{l_2} \{U'' + \lambda^2\Omega^2 U\}' d\xi_2 = 0, \tag{31}$$

which is identically satisfied in view of (30a).

We shall postpone discussion of the solutions of (27) and (30) until Section 7. Here, we employ $E = 0$ from (25) in (17a) and obtain

$$G_2[\mathbf{u}] = 2l_3 \iint \{k_1\kappa[\mathbf{u}] + \lambda_1^2(1-\lambda^2)^2 k_2(u_{1,1})^2\} d\xi_1 d\xi_2. \tag{32}$$

We cast $\kappa[\mathbf{u}]$ from (9) in the form

$$\begin{aligned}\kappa[\mathbf{u}] = &\{u_1(u_{1,1} - \lambda u_{2,2}) + u_2(u_{2,1} + \lambda u_{1,2})\}_{,1}\\ &+ \{u_1(u_{1,2} + \lambda u_{2,1}) + u_2(u_{2,2} - \lambda u_{1,1})\}_{,2}\\ &- u_1(u_{1,11} + u_{1,22}) - u_2(u_{2,11} + u_{2,22}),\end{aligned} \tag{33}$$

and substitute into (32). From (18), with $E = 0$, we find that

$$0 = \lambda_1^{-1}\{u_1 p_{,1} - (u_1 p)_{,1}\} + \lambda_2^{-1}\{u_2 p_{,2} - (u_2 p)_{,2}\}, \tag{34}$$

which can be added to the integrand of (32). On carrying out the partial integrations, the result is [1(3.22)]

$$\begin{aligned}G_2[\mathbf{u}] = 2l_3\Big\{ &-\iint (u_1\mathcal{F}_1 + u_2\mathcal{F}_2) d\xi_1 d\xi_2\\ &+ \int_{-l_2}^{l_2} [u_1\mathcal{F}_3 + u_2\mathcal{F}_4\Big|_{-l_1}^{l_1} d\xi_2\\ &+ \int_{-l_1}^{l_1} [u_1\mathcal{F}_5 + u_2\mathcal{F}_6\Big|_{-l_2}^{l_2} d\xi_1\Big\},\end{aligned} \tag{35}$$

where $\mathcal{F}_1$, $\mathcal{F}_2$, $\mathcal{F}_4$, $\mathcal{F}_5$, $\mathcal{F}_6$ are the left members of (19a), (19b), (20b), (21a), (21b), respectively, and F_3 is the integrand of (20a), wherein $E = 0$. [Recall $u_1(\pm l_1, \xi_2) = \pm\,\text{const.}$] It follows that $G_2[\mathbf{u}] = 0$ for any displacement field of the form (24), and, accordingly, a critical state is a state of neutral equilibrium.

4. Further Developments for the General Material

We investigate whether $G[\mathbf{u}]$ is positive definite in the neighborhood of $\mathbf{u} = 0$ by letting

$$\mathbf{u} = \epsilon\hat{\mathbf{u}} + \epsilon^2\mathbf{v}, \tag{36}$$

where ϵ is a small parameter, $\hat{\mathbf{u}}$ denotes a solution for u_1, u_2 of (24), (26)–(30), and $\mathbf{v}$ has the form [cf. (2), (13)]

$$v_\alpha = v_\alpha(\xi_1, \xi_2), \tag{37a}$$

$$v_3 = \lambda_3 E\xi_3, \tag{37b}$$

$$v_1(\pm l_1, \xi_2) = \pm\lambda_1 e l_1. \tag{37c}$$

We require that $\hat{\mathbf{u}}$ and $\mathbf{v}$ be orthogonal in the sense that

$$\iint \hat{u}_{\alpha,\beta} v_{\alpha,\beta}\, d\xi_1 d\xi_2 = 0. \tag{38}$$

The use of (36) in (2) yields the incompressibility constraint

$$\lambda_1^{-1} v_{1,1} + \lambda_2^{-2} v_{2,2} + E + \lambda_3(\hat{u}_{1,1}\hat{u}_{2,2} - \hat{u}_{1,2}\hat{u}_{2,1}) - \epsilon\lambda_3\phi[\hat{\mathbf{u}}, \mathbf{v}] = 0, \tag{39}$$

apart from terms of higher degree in ϵ, where the operator ϕ is defined by

$$\phi[\hat{\mathbf{u}}, \mathbf{v}] = \hat{u}_{1,2} v_{2,1} + \hat{u}_{2,1} v_{1,2} - \hat{u}_{1,1} v_{2,2} - \hat{u}_{2,2} v_{1,1}. \tag{40}$$

The result of substitution from (36) into (15) is an expression for $G[\mathbf{u}]$ of the form $\epsilon^2 G_2[\hat{\mathbf{u}}] + \epsilon^3 G_3[\hat{\mathbf{u}}, \mathbf{v}] + \epsilon^4 G_4[\hat{\mathbf{u}}, \mathbf{v}]$, apart from terms of degree 5 and higher in ϵ. It has already been noted that $G_2[\hat{\mathbf{u}}] = 0$. By making use of the fact that $\hat{\mathbf{u}}$ satisfies (19)–(21) and (24), we find that $G_3[\hat{\mathbf{u}}, \mathbf{v}] = 0$ for any field $\mathbf{v}$ that meets (37) and (39). Thus, for any specified field $\hat{\mathbf{u}}$ that satisfies (19)–(21) and (24), we have

$$G[\mathbf{u}] = \epsilon^4 \overline{G}[\mathbf{v}], \qquad \overline{G}[\mathbf{v}] = 2l_3 \iint \bar{g}\, d\xi_1 d\xi_2, \tag{41}$$

where [1(4.13)]

$$\begin{aligned}\bar{g} = {} & k_1\{\kappa[\mathbf{v}] + (\lambda_3^2 + 2\lambda_2^2)E^2\} + \lambda_3^2 W_2 E\{2\kappa[\hat{\mathbf{u}}] + 4\lambda_1(1-\lambda^2)v_{1,1} + \Lambda_1 E\} \\ & + k_2\{\tfrac{1}{4}(\kappa[\hat{\mathbf{u}}])^2 + 2\lambda_1(1-\lambda^2)\hat{u}_{1,1}\phi[\hat{\mathbf{u}},\mathbf{v}] \\ & + 2\lambda_1(1-\lambda^2)\hat{u}_{1,1}\hat{u}_{\alpha,\beta}v_{\alpha,\beta} + \bar{k}\kappa[\hat{\mathbf{u}}] + \bar{k}^2\} \\ & + k_{21}E\{4\lambda_1^2(1-\lambda^2)^2(\hat{u}_{1,1})^2 + \Lambda_2\kappa[\hat{\mathbf{u}}] + 2\Lambda_2\bar{k}\} \\ & + 2\lambda_3^4 W_{22}\Lambda_2^2E^2 + 12\lambda_1^2(1-\lambda^2)^2k_3(\hat{u}_{1,1})^2\{\kappa[\hat{\mathbf{u}}] + 2\bar{k}\} \\ & + 8\lambda_1^2(1-\lambda^2)^2\Lambda_2k_{31}(\hat{u}_{1,1})^2E + 16\lambda_1^4(1-\lambda^2)^4k_4(\hat{u}_{1,1})^4 \\ & + 2\lambda_3\hat{p}\phi[\hat{\mathbf{u}},\mathbf{v}].\end{aligned} \tag{42}$$

The quantities k_1, k_2, k_{21} are given by (16). Additional derivatives of w enter through k_3, k_{31}, k_4 defined by

$$k_3 = (1/6)(W_{111} + 3\lambda_3^2W_{112} + 3\lambda_3^4W_{122} + \lambda_3^6W_{222}), \tag{43a}$$

$$k_{31} = (1/2)\lambda_3^2(W_{112} + 2\lambda_3^2W_{122} + \lambda_3^4W_{222}), \tag{43b}$$

$$k_4 = (1/24)(W_{1111} + 4\lambda_3^2W_{1112} + 6\lambda_3^4W_{1122} + 4\lambda_3^6W_{1222} + \lambda_3^8W_{2222}). \tag{43c}$$

The quantity $\bar{k}$ is the same as k given by (8a), but with $u_{1,1}$ replaced by $v_{1,1}$; and $\hat{p}$ denotes p given by (24).

Now, for a specified first-order field $\hat{u}_1$, $\hat{u}_2$, $\hat{p}$, we seek a field $\mathbf{v}$ which renders $\bar{G}[\mathbf{v}]$ stationary, subject to the constraints implied by orthogonality and incompressibility. We introduce the Lagrange multipliers $4l_3\chi$ (=const), to account for (38), and $-4l_3q(\xi_1, \xi_2)$, to account for (39), wherein the term of order ϵ is neglected. Accordingly, from (41), we consider the quantity [1(5.1)]

$$\bar{G}[v] = 2l_3\iint[\bar{g} - 2q\{\lambda_1^{-1}v_{1,1} + \lambda_2^{-1}v_{2,2} + E + \lambda_3(\hat{u}_{1,1}\hat{u}_{2,2} - \hat{u}_{1,2}\hat{u}_{2,1})\} + 2\chi\hat{u}_{\alpha,\beta}v_{\alpha,\beta}]d\xi_1 d\xi_2. \tag{44}$$

The condition that $\bar{G}[\mathbf{v}]$ be stationary, i.e., $\delta\bar{G}[\mathbf{v}] = 0$, leads to equilibrium and boundary relations that are similar in form to those for u_1, u_2 of the previous section, the main difference being that the equations governing v_1, v_2, E, q, χ are nonhomogeneous.

For the purpose of gaining an idea of the structure of the nonhomogeneous terms, we here show the result of employing (24) to calculate a term in (39). Thus,

$$\hat{u}_{1,1}\hat{u}_{2,2} - \hat{u}_{1,2}\hat{u}_{2,1} = -\frac{1}{2\lambda}\{(UU')' - (-1)^n\alpha\cos(2\Omega\xi_1)\}, \tag{45}$$

where

$$\alpha = UU'' - (U')^2; \tag{46}$$

t follows that v_1, v_2, E must satisfy

$$\lambda v_{1,1} + v_{2,2} + \lambda_2 E = \frac{1}{2\lambda_2}\{(UU')' - (-1)^n \alpha\cos(2\Omega\xi_1)\}. \tag{47}$$

The relevant equilibrium equations [1(6.6)] are found to have the forms

$$k_1\{[1 + (1-\lambda)^2 A]v_{1,11} + v_{1,22}\} - \lambda_1^{-1} q_{,1} = (-1)^n \phi_1 \sin(2\Omega\xi_1), \tag{48a}$$

$$k_1(v_{2,11} + v_{2,22}) - \lambda_2^{-1} q_{,2} = \phi_2 + (-1)^n \phi_3 \cos(2\Omega\xi_1), \tag{48b}$$

where ϕ_1, ϕ_2, ϕ_3 are expressible as quadratic forms in the quantities U, U', U'', U''', with coefficients that depend on the derivatives of w through the quantities k_1, k_2, k_3.

Based on the structure of (47), (48), as well as on the structure of pertinent boundary conditions, we are led to assume a solution for v_1, v_2, q, χ of the form

$$v_2(\xi_1, \xi_2) = V(\xi_2)\cos(2\Omega\xi_1) + \frac{1}{2\lambda_2} UU' - \lambda_2(e + E)\xi_2, \tag{49a}$$

$$v_1 = -\frac{1}{2\lambda\Omega}\left\{V' + \frac{(-1)^n}{2\lambda_2}\alpha\right\}\sin(2\Omega\xi_1) + \lambda_1 e\xi_1, \tag{49b}$$

$$q = Q(\xi_2)\cos(2\Omega\xi_1) + \bar{Q}(\xi_2), \qquad \chi = 0. \tag{49c}$$

This choice for v_1, v_2 satisfies (47) and the constraint on v_1 expressed in (37c). With (24) and (26), we find that the orthogonality condition (38) is satisfied. The counterpart of (23), for this second-order problem, is $v_{2,1}(\pm l_1, \xi_2) = 0$, which expresses the condition that tangential tractions on the faces $\xi_1 = \pm l_1$ vanish. This, too, is satisfied by (49a).

The substitution from (49) in (48) yields expressions for Q and $\bar{Q}$, in terms of V and U, and a nonhomogeneous fourth-order equation for V [1(6.8)],

$$V^{(iv)} - \{1 + \lambda^2 + (1-\lambda)^2 A\}(2\Omega)^2 V'' + \lambda^2(2\Omega)^4 V = (-1)^n \phi_4, \tag{50}$$

where ϕ_4 is some linear combination of ϕ_1, ϕ_2, ϕ_3 and their derivatives wrt ξ_2. Accordingly, since U satisfies (27), ϕ_4 can be cast into the quadratic-form structure shared by the other ϕ's, but with coefficients that depend on derivatives of w through the ratios k_2/k_1 and k_3/k_1.

Two of the four remaining conditions that arise from the variational problem, $\delta\bar{G}[\mathbf{v}] = 0$, express the pointwise vanishing of all tractions on the major surfaces of the plate, and, with (49), these can be cast in the forms [1(6.12)]

$$\left.\begin{aligned} &V'' + \lambda^2(2\Omega)^2 V = (-1)^n \sqrt{\lambda_3}\phi_5, \\ &V''' - (2\Omega)^2\{1 + 2\lambda^2 + (1-\lambda)^2 A\}V' = (-1)^n \sqrt{\lambda_3}\phi_6 \end{aligned}\right\} \xi_2 = \pm l_2. \quad \begin{aligned}(51a)\\(51b)\end{aligned}$$

On making use of the fact that U satisfies (30), we find it possible to express ϕ_5 as C_1UU' and ϕ_6 as $C_2U^2 + C_3(U')^2$, where the C's contain the ratios k_2/k_1, k_3/k_1. The other two conditions contain integrals and express the requirement of dead loading on the faces $\xi_1 = \pm l_1$, $\xi_3 = \pm l_3$. These give rise to a pair of nonhomogeneous linear equations for the determination of the additional purehomogeneous strains e and E in the 1- and 3-directions, respectively. These quantities are found to depend on derivatives of w through k_1, k_2, k_3 as well as through W_2, k_{21}, W_{22}, and k_{31} [1(6.19)].

In Section 7 of Ref. 1 we employ the expressions for $\hat{u}_1$, $\hat{u}_2$, $\hat{p}$, given by (24), and those for v_1, v_2, given by (49), to gain an expression for $\bar{G}[\mathbf{v}]$ in terms of the functions U and V. While this could be accomplished directly, merely by evaluating each term of $\bar{g}$ successively and integrating the result, we follow a more structured approach. In particular, we initially use the fact that v_1, v_2, E, q satisfy (39) with $\epsilon = 0$), and the equilibrium equations (48), to write $\bar{g}$ in a form that, when integrated, removes certain trivial contributions to $\bar{G}[\mathbf{v}]$. This step is similar in spirit to that used in passing from (32) to (35). What emerges here is an expression for $\bar{G}[\mathbf{v}]$ of the form [1(7.9)]

$$\bar{G}[\mathbf{v}] = 2l_3 \iint \{H_{\alpha\beta}v_{\alpha,\beta} + H_{33}E + H + \lambda_3 q(\hat{u}_{1,2}\hat{u}_{2,1} - \hat{u}_{1,1}\hat{u}_{2,2})\}d\xi_1 d\xi_2, \quad (52)$$

where $H_{\alpha\beta}$ and H_{33} are expressions that can be written as quadratic forms in the quantities $\hat{u}_{1,1}$, $\hat{u}_{1,2}$, $\hat{u}_{2,1}$, $\hat{u}_{2,2}$, $\hat{p}$; and H is homogeneous of fourth degree in the quantities $\hat{u}_{\alpha,\beta}$. It is at this stage where we make use of the separated forms for $\hat{u}_1$, $\hat{u}_2$, $\hat{p}$ given by (24) to find

$$H_{AA} = h_{AA} + \bar{h}_{AA}\cos 2\Omega\xi_1, \qquad H_{AB} = \bar{h}_{AB}\sin 2\Omega\xi_1, \qquad A \neq B, \quad (53a)$$

$$H = h_0 + \bar{h}_1\cos 2\Omega\xi_1 + \bar{h}_2\cos 4\Omega\xi_1, \quad (53b)$$

where the h's and $\bar{h}$'s are functions of ξ_2 through their dependence on $U, U', \ldots$. Then, after substituting for v_1, v_2, q from (49) and carrying out the integration wrt ξ_1 in (52), we obtain an expression for $\bar{G}[\mathrm{v}]$ of the form [1(7.10)]

$$\bar{G}[\mathrm{v}] = 4l_1l_3\int_{l_2}^{l_2} \{\Gamma_1 V''' + \Gamma_2 V'' + \Gamma_3 V' + \Gamma_4 V + \Gamma_5 e + \Gamma_6 E + \Gamma_7\} d\xi_2, \tag{54}$$

where $\Gamma_1, \ldots, \Gamma_6$ are homogeneous of degree two in U, U', U'', U''', and Γ_7 is homogeneous of degree four in these quantities.

The remainder of Section 7 of Ref. 1 is devoted to further manipulation of the terms of the integrand of (54). In particular, on exploiting the fact that V satisfies (50) and boundary conditions (51), and that U satisfies (27) and boundary conditions (30), we are able to carry out several steps of integration by parts and arrive at a form for the integrand that depends on V only through the single function V'. The result of this analysis is the expression [1(7.27)]

$$\bar{G}[\mathrm{v}] = 4l_1l_2l_3\lambda_3k_1(G_1 - G_2 + G_3 - G_4), \tag{55}$$

where

$$G_1 = \frac{1}{32\lambda^3\Omega^2 l_2}\int_{-l_2}^{l_2} [(-1)^n g_1 V' + g_2] d\xi_1, \tag{56a}$$

$$G_2 = \frac{1}{2\lambda l_2}[UU'\{(U')^2 + \lambda^2\Omega^2U^2\} \,|_{\xi_2 = l_2}, \tag{56b}$$

$$G_3 = \frac{(-1)^n}{8\lambda^{5/2}\lambda_3^{1/2}l_2}[\{7\lambda^2 + 1 + (\lambda-1)^2A\}\mathrm{UU'V'} - 4\{(\lambda^2 + 1 - \bar{A})(U')^2 - \lambda^2\bar{A}\Omega^2U^2\}V \,|_{\xi_2 = l_2}], \tag{56c}$$

and where G_4 is the contribution to $\bar{G}[\mathrm{v}]$ arising from the strains e and E. This term has the form

$$G_4 = \frac{2}{\lambda(a_1a_2 - a^2)}(a_1b_2^2 + a_2b_1^2 - 2ab_1b_2), \tag{57}$$

where

$$b_1 = -\frac{1}{2\lambda\Omega^2 l_2}\int_{-l_2}^{l_2} \{(U'')^2 + \Omega^2(U')^2B_1\} d\xi_2, \tag{58a}$$

$$b_2 = \frac{1}{2}\left[b_1 + \frac{B_2}{2\lambda l_2}\int_{-l_2}^{l_2} (U')^2 d\xi_2\right], \tag{58b}$$

in which the a's and B's are constants that contain dependence on w through W_α, $W_{\alpha\beta}$, $W_{\alpha\beta\gamma}$. In (56), g_1 and g_2, respectively, are homogeneous of degrees two and four in the quantities U, U', U'', U'''; and

$$\bar{A} = \frac{\lambda - 1}{\lambda + 1} A = \frac{(\lambda^2 - 1)}{\lambda\lambda_3} \frac{k_2}{k_1}. \tag{59}$$

To resolve the stability question, we must evaluate $\bar{G}[\mathrm{v}]$ in (55) using the solution for V of (50) and (51) for a particular choice of a solution U of (27) and (30). This exercise is made much less formidable on choosing the neo-Hookean form for w, and these results are briefly described in Section 6.

5. Asymptotic Analysis for a Thin Plate

For a plate having small aspect ratio, l_2/l_1, the possibility arises for flexural-type deformations to bifurcate from a critical state in response to small values of thrust loading applied to the faces $\xi_1 = \pm l_1$. The corresponding critical value of $\lambda(=\lambda_2/\lambda_1)$ is just slightly larger than unity. In fact, it has been found (Refs. 2, 3) that

$$\lambda - 1 = \frac{2}{3}\eta^2 + \frac{16}{45}\eta^4 + \frac{2}{27}\left(A^{(0)} + \frac{18}{7}\right)\eta^6 + \frac{2}{27}\left(A^{(0)} + \frac{113}{175}\right)\eta^8 + 0(\eta^{10}), \tag{60}$$

where η is defined by

$$\eta = \Omega l_2 = n\pi l_2/(2l_1), \tag{61}$$

and

$$A^{(0)} = A|_{\lambda = 1}. \tag{62}$$

The result expressed by (60) is somewhat remarkable since it shows the relative insensitivity of the critical value of the deformation parameter to details of material properties at least through terms of order $(l_2/l_1)^8$. Perhaps more remarkable is the fact that we need employ only the first two terms in the right-hand member of (60) in order to complete the stability analysis for the general material; and these terms contain no reference to any material properties.[2] We retain the mode number n in (61) for generality, at no undue analytical expense, with the understanding that n should ultimately be assigned the smallest value consistent with any passive constraints that might be operative. If no flexural modes are suppressed, then $n = 1$, and the bifurcation deformation consists of just one half-wavelength parallel to the 1-direction.

Since our goal here is to obtain asymptotic solutions for U and V, in the limit of vanishing η, we first expand relevant quantities in powers of $(\lambda - 1)$. Typical of these is the expansion of the important material function A, which has the form

$$A = A^{(0)} + \left\{\frac{1}{8}A^{(0)}(2 - A^{(0)}) + B^{(0)}\right\}(\lambda - 1)^2[1 - (\lambda - 1)] + 0(\lambda - 1)^4, \quad (63)$$

where

$$A^{(0)} = \frac{4k_2^{(0)}}{\lambda_3 k_1^{(0)}}, \qquad B^{(0)} = \frac{48k_3^{(0)}}{\lambda_3^2 k_1^{(0)}}, \quad (64)$$

the superscript 0 denoting evaluation of the specified quantity at $\lambda = 1$. The required results are compiled in Section 8 of Ref. 1.

To proceed, we employ (63), with (60), in (27) and (30), and introduce the dimensionless thickness coordinate t,

$$t = \xi_2/l_2, \quad (65)$$

and note that the operator $d/d\xi_2$ is given by

$$\frac{d}{d\xi_2} = \frac{\Omega}{\eta}\frac{d}{dt}. \quad (66)$$

A direct analysis, using the equations described immediately above, yields the result that U, expressed in terms of the variable t, has the form [1(9.8)]

[2]An alternative approach would be to make the Ansatz $\lambda - 1 = \eta_1\eta^2 + \eta_2\eta^4 + \cdots$, in the development described later in this section, and to deduce the values for η_1 and η_2 given in (60). By using (60) *ab initio*, two equations in the asymptotic development are found to be identically satisfied.

$$U = 1 - \frac{1}{2}\eta^2 t^2 + \eta^4\left(\frac{1}{3}t^2 - \frac{1}{8}t^4\right) + \eta^6\left(\frac{4}{45}t^2 - \frac{1}{18}t^4 - \frac{1}{144}t^6\right) + 0(\eta^8), \tag{67}$$

which is seen to be independent of both λ_3 and any specific material properties, at least through terms of order η^6. The form (67) reflects the normalization employed, viz., $U(0) = 1$.

We next employ (63) and (60), first to substitute in the left members of (50), (51), and then employ (63) and similar expansions of other material functions, as well as (67), to calculate ϕ_4, ϕ_5, ϕ_6 in the right members of the equations governing V. Again, a direct analysis of these equations, based on the matching of terms of comparable powers of η, yields an expression for V as a function of t of the form [1(10.6)]

$$V = \Omega\eta t\left\{\frac{1}{4} + \eta^2\left(-\frac{7}{6} + \frac{1}{2}t^2\right) + \eta^4\left(\frac{23}{72} - \frac{1}{12}A^{(0)} - \frac{2}{9}t^2 + \frac{1}{6}t^4\right) + \eta^6\left[D - \left(\frac{23}{60} + \frac{1}{54}A^{(0)}\right)t^2 + \left(\frac{2}{9} + \frac{1}{30}A^{(0)}\right)t^4 + \frac{1}{45}t^6\right]\right\} + 0(\eta^9), \tag{68}$$

where D is a constant that is not required for the calculation of $\bar{G}[\mathbf{v}]$.

What remains is the calculation of the quantities $G_1, \ldots, G_4$ that appear in (55) by using (67), (68), (60), (63), and expressions comparable to (63), in (56), (57), the integrals being over the interval $(-1, 1)$, and boundary values being taken at $t = 1$. We find that the results for G_1, G_2, G_3 depend on the extension ratio λ_3 only implicitly, to the extent that the quantity $A^{(0)}$ depends on λ_3. However, G_4 is found to depend on λ_3 explicitly, as well as on combinations of $W^{(0)}_{\alpha}$, $W^{(0)}_{\alpha\beta}$ other than through the material constant $A^{(0)}$. Although we could retain an arbitrary value of λ_3, we make the simplifying assumption $\lambda_3 = 1$ and find that $\bar{G}[\mathbf{v}]$ has the form [1(11.25)]

$$\bar{G}[\mathbf{v}] = \frac{l_1 l_3 K_0 \eta^6}{3 l_2^3}\left\{1 - \eta^2\left(\frac{98}{9} - \delta\right) + 0(\eta^4)\right\}, \tag{69}$$

where

$$K_0 = k_1^{(0)}\Big|_{\lambda_3 = 1}, \qquad \delta = \frac{9}{5}A_0 + \frac{16}{9}\frac{\omega_0}{K_0}\left(1 - 2\frac{\omega_0}{K_0}\right), \tag{70a}$$

$$A_0 = A^{(0)}\Big|_{\lambda_3=1}, \qquad \omega_0 = W_2^{(0)}\Big|_{\lambda_3=1}. \tag{70b}$$

6. Results for a Neo-Hookean Plate

The simplifications that accrue from the assumption that the material of the plate is neo-Hookean have been exploited in Refs. 4 and 5, where complete solutions for the second-order problem are derived for a plate of arbitrary thickness, and where an exact evaluation of $\bar{G}[\mathbf{v}]$ is carried out. Since only one material constant appears, we employ a dimensionless formulation wherein the Piola–Kirchhoff stresses in state I are [cf. (11)]

$$\Pi_{11} = \lambda_1 - \lambda_2^2/\lambda_1, \qquad \Pi_{33} = \lambda_3 - \lambda_2^2/\lambda_3, \tag{71}$$

and the strain energy density is

$$W = \tfrac{1}{2}(\lambda_1^2 + \lambda_2^2 + \lambda_3^2 - 3). \tag{72}$$

On taking $A \equiv 0$ in (27), (30), we obtain solutions for U that can be written in the forms [5(2.14)]

$$U(\xi_2) = \begin{cases} M(\cosh(\lambda\Omega\xi_2) - m\cosh(\Omega\xi_2)) & \text{(flexure)}, \\ M(\sinh(\lambda\Omega\xi_2) - m\sinh(\Omega\xi_2)) & \text{(barreling)}, \end{cases} \tag{73}$$

where

$$m = (\lambda \sinh 2\lambda\eta/\sinh 2\eta)^{1/2}, \tag{74}$$

and the relationship between $\eta(=\Omega l_2)$ and the critical λ value is the solution of

$$\frac{\tanh(\lambda\eta)}{\tanh\eta} = \left\{\frac{4\lambda^3}{(\lambda^2+1)^2}\right\}^{\nu}. \tag{75}$$

The M's in (73) are arbitrary constants, and the exponent in (75) is $+1$ for the flexural case and -1 for barreling. The bifurcation condition (75) can be cast in the more convenient form [4(3.18)]

$$\frac{\sinh[(\lambda+1)\eta]}{\sinh[(\lambda-1)\eta]} = \nu\frac{(\lambda+1)\{\lambda(\lambda+1)^2 + (\lambda-1)^2\}}{(\lambda-1)\{(\lambda+1)^2 - \lambda(\lambda-1)^2\}}. \tag{76}$$

Apart from the trivial case $\lambda = 1$, the denominator in the right-hand member of (76) vanishes for just one other positive value of λ, the positive root of

$$\lambda^3 - 3\lambda^2 - \lambda - 1 = 0. \tag{77}$$

Denoting this by $\bar{\lambda}_0$, we find $\bar{\lambda}_0 = 3.3829. \ldots$ The analysis of (76), for the flexural case, shows that as the aspect ratio η increases on the interval $(0, \infty)$, the critical λ value increases monotonically from unity to $\bar{\lambda}_0$. For the barreling case, as η decreases from ∞ to 0, the critical λ value increases monotonically on the interval $(\bar{\lambda}_0, \infty)$. No other solutions of (76) are possible for positive values of η. Thus, for any specified η, one critical λ value exists on each of the intervals $(1, \bar{\lambda}_0)$ and $(\bar{\lambda}_0, \infty)$. In particular, no bifurcations are possible for any $\lambda < 1$, i.e., if the load R_1 is tensile.

The solution of (50), (51) for the neo-Hookean material contains just four terms and can be written as [4(5.14)]

$$V = (-1)^n \Omega M^2 \sqrt{\lambda_3/\lambda}\, \{\mu_1 \sinh(2\lambda\Omega\xi_2) + \mu_2 \sinh(2\Omega\xi_2) + \mu_3 \sinh[(\lambda + 1)\Omega\xi_2] + \mu_4 \sinh[(\lambda - 1)\Omega\xi_2]\}, \tag{78}$$

where the μ's are constants that depend on λ, η, and ν. We note that V is inherently an odd function of ξ_2 for both flexure and barreling cases. In Section 6 of Ref. 4 we develop an expression for $\bar{G}[\mathbf{v}]$ [the counterparts of (55)–(58)] and proceed with its explicit evaluation through the use of (73) and (78). The result can be expressed as [5(4.2)]

$$\bar{G}[\mathbf{v}] = \frac{l_1 l_3 \lambda_3 \eta^6}{6 l_2^3} G^*, \tag{79a}$$

$$G^* = \frac{3M^4}{2\eta^3}(\gamma_1 - \gamma\gamma_2), \tag{79b}$$

where γ_1, γ_2 depend on λ, η, ν, and

$$\gamma = \frac{7\lambda^2 + 4\lambda\lambda_3^3 + 1}{5\lambda^3 + 3\lambda(\lambda\lambda_3^3 + 1) + \lambda_3^3}. \tag{80}$$

We recall that λ, η, ν are related through the bifurcation condition (76).

The stability question is decided by the values of G^*, which are calculated in Refs. 4 and 5. For this, we choose an appropriate value for the arbitrary constant M by imposing the normalizing condition that the maximum value of U on the interval $[-l_2, l_2]$ be unity. For the flexural case this maximum occurs at $\xi_2 = 0$, provided $1 < \lambda < 2.05$, and then moves toward the outer edge, $\xi_2 = l_2$, as λ increases from 2.05 to $\bar{\lambda}_0$. For barreling bifurcations, the maximum occurs at the outer edge if $\lambda = \bar{\lambda}_0$, and then moves inward as λ increases from $\bar{\lambda}_0$. The choices for M that accomplish the required normalization are given in equation (4.3) of Ref. 5.

For the flexural case, the one that pertains to Euler buckling, we find

$$\lim_{\eta \to 0} G^* = 1, \tag{81}$$

and that G^* decreases monotonically to $-\infty$ as $\eta \to \infty$. Of particular interest is the observation that $G^* = 0$ at $\eta \approx 0.32$, which sets the limit on the largest aspect ratio for which classical Euler buckling obtains. For the lowest-order mode, this is $l_2/l_1 = 2\eta/\pi \approx 0.20$, and the corresponding critical λ value is approximately 1.072.[3] The derivation in Ref. 4 of asymptotic results for a thin plate, comparable to that of Section 5, yields the expression [4(7.19)]

$$G^* = 1 - \tfrac{2}{3}\{16 + (2 + \lambda_3^3)^{-1}\}\eta^2 + 0(\eta^4), \tag{82}$$

or, with $\lambda_3 = 1$,

$$G_0^* = 1 - (98/9)\eta^2 + 0(\eta^4). \tag{83}$$

With the neglect of $0(\eta^4)$ terms, we see that $G_0^* = 0$ for $\eta \approx 0.30$, which compares favorably (within about 6%) with the exact value given above. Flexural results for a thin plate of general material can be specialized to the neo-Hookean case by taking $K_0 = \frac{1}{2}$ and $\delta = 0$ in (69). On using (83) in (79a), we find agreement between the two results.

The above analysis indicates that stable buckling of the Euler type (i.e., flexure) does not occur for underlying deformations characterized by critical λ values in the interval $(1.072, \bar{\lambda}_0)$. The result in Ref. 5, however, shows that, for $\lambda > \bar{\lambda}_0$, the quantity G^* of (79b) is positive for all values of η, with $G^* \to \infty$ in the limiting cases $\eta \to \infty$ and $\eta \to 0$. The conclusion is that barreling bifurcations are stable; but to achieve them would require the suppression of all flexural-type modes as the thrust force on the faces $\xi_1 = \pm l_1$ is increased. One (perhaps artificial) method to accomplish this would be to have the midplane, $\xi_2 = 0$, remain in contact with a rigid frictionless platen. Then a barreling-type bifurcation of high (theoretically infinite) order would be reached first. This phenomenon has been interpreted as surface wrinkling (Refs. 8, 9).

---

[3]Although this result pertains to the special case $\lambda_3 = 1$, we find that the location of the zero crossing is quite insensitive to the value of λ_3. Specifically, the corresponding critical λ values are within 0.2% of 1.072 in the extreme limiting cases $\lambda_3 \to 0$ and $\lambda_3 \to \infty$.

7. Bifurcation Conditions for a Thick Plate

In the first paper of the series (Ref. 6) we employ the standard small-on-large theory (Ref. 10) to derive necessary conditions for bifurcations from state I to occur. The relevant equilibrium and boundary conditions [6(3.11), (3.14), (4.1)], although differing in form from their counterparts derived in Section 3, are found to admit the separable solutions (24), with (26), where the governing equation and boundary conditions for U are identical to (27), (30).

The material function A, given by (29), is key in determining the nature of the solutions of (27). Three distinct and mutually exhaustive regimes arise naturally, and these are conveniently described by

$$\text{(i)} \quad A \geq -1, \tag{84a}$$

$$\text{(ii)} \quad -1 > A > -(\lambda + 1)^2/(\lambda - 1)^2, \tag{84b}$$

$$\text{(iii)} \quad -(\lambda + 1)^2/(\lambda - 1)^2 \geq A, \tag{84c}$$

as λ assumes all positive values. On the basis of a separate investigation (Ref. 11) we rule out the need to consider (84c) further since any such material would violate fundamental stability requirements. The class of materials covered by (84a) is of particular interest, since it includes the Mooney–Rivlin and neo-Hooken materials $\{A \equiv 0\}$, and this is analyzed in Ref. 6. The remaining case is considered in Ref. 2.

With the notation

$$R = \tfrac{1}{2}\{(\lambda + 1)^2 + A(\lambda - 1)^2\}^{1/2}, \tag{85a}$$

$$T = \tfrac{1}{2}|\lambda - 1|(A + 1)^{1/2}, \tag{85b}$$

we find, for $A > -1$, that the bifurcation condition relating η and critical λ values can be written as [6(5.11), (5.12)]

$$\frac{\sinh(2R\eta)}{\sinh(2T\eta)} = \nu \frac{R}{T} \frac{4T^2 + \lambda(\lambda + 1)^2}{4R^2 - \lambda(\lambda - 1)^2}, \tag{86}$$

where, again, ν is $+1(-1)$ for flexure (barreling). We note that (86) reduces to (76) on setting $A = 0$. We prove in Ref. 6 that (86) admits no critical λ value on the interval (0, 1) for any value of η, which rules out the possibility of bifurcations under tensile loading applied to the faces $\xi_1 = \pm l_1$ for any material of case (i). If the denominator in the right-hand member of (86) is positive (negative), only flexural (barreling) bifurcations are possible. The vanishing of this denominator, for $\lambda > 1$, thus defines a curve in the λA-plane, viz.,

$$A = \lambda - \frac{(\lambda + 1)^2}{(\lambda - 1)^2} \equiv \bar{A}(\lambda), \qquad \lambda > 1, \tag{87}$$

which separates the flexure and barreling regimes. (See Fig. 3 of Ref. 6.) For a given material, and for a specified value of λ_3, the material function A is expressible as a function of λ, which can be compared with $\bar{A}(\lambda)$ in (87). If $A > \bar{A}(\lambda)$, then only flexural bifurcations are possible. Otherwise, if A crosses the separatrix, at $(\bar{\lambda}, \bar{A})$, say, we prove in Ref. 6 that $\eta \rightarrow \infty$ as $\lambda \rightarrow \bar{\lambda}$, and vice versa. We note that $\bar{\lambda} = 3, \lambda_0, 4, 5$, etc. if $A(\lambda) = -1, 0, 11/9, 11/4$, etc., respectively. For any specified constant value of $A(\geq -1)$ we find that (86) yields a solution for critical λ values for flexural bifurcations, which increase monotonically on $(1, \bar{\lambda})$ as η increases on $(0, \infty)$, and, for barreling bifurcations, which increase monotonically on $(\bar{\lambda}, \infty)$ as η decreases from ∞ to 0. We show in Ref. 6 how these results can be used to determine critical λ values corresponding to any given η for a plate of general material, i.e., where A is not constant.

With respect to the bifurcation condition (86), and for the full range of possible λ values, we find that the neo-Hookean material captures certain essential features that apply to the broad class of materials characterized by the requirement $A \geq -1$. This observation serves to justify the effort expended in carrying out the exact calculation described in Section 6.

For (84b) we see that T in (85b) is imaginary, which fact can be accommodated, formally, by replacing T with $i|T|$ in (86). The resulting bifurcation condition is [2(8.10)]

$$\frac{\sinh(2R\eta)}{\sin(2|T|\eta)} = \nu \frac{R}{|T|} \frac{\lambda(\lambda + 1)^2 + (A + 1)(\lambda - 1)^2}{(A - \lambda)(\lambda - 1)^2 + (\lambda + 1)^2}, \tag{88}$$

whose analysis reveals two significant differences from that of case (i). One of these is the possibility that, for certain values of η, a barreling bifurcation can occur before a flexural bifurcation as the thrust loading applied to the faces $\xi_1 = \pm l_1$ is increased from zero. The other difference is the possibility of bifurcations under conditions of tensile loading, which shall not be considered further in the present discussion.

We note that the denominator of the right-hand member of (88) vanishes if the material function A has the value $\bar{A}(\bar{\lambda})$, for some $\bar{\lambda}$, where $\bar{A}(\lambda)$ is defined in (87). At the same time, the denominator of the left-hand member vanishes for all values of η given by

$$\eta = m\bar{\eta}, \qquad m = 1, 2, 3, \ldots, \qquad \bar{\eta} = \pi/(2|\bar{T}|), \tag{89}$$

where $\bar{T}$ denotes the value of T, evaluated at $\bar{\lambda}$. From (85), (87), we obtain

$$2|\bar{T}| = \{\bar{\lambda}(\bar{\lambda} + 1)(3 - \bar{\lambda})\}^{1/2}, \qquad \bar{\lambda} < 3. \tag{90}$$

A detailed analysis (Ref. 2) shows that the flexural and barreling curves cross each other at each of the points $(\bar{\lambda}, m\bar{\eta})$, where they intersect, if indeed such intersections can occur at all.[4] In any event, it is shown in Ref. 2 that the smallest possible $\bar{\eta}$ value is approximately $2\pi/5$, which gives an aspect ratio $(l_2/l_1) = 4/5$ for the lowest-order flexural mode. It is unlikely that such a large aspect ratio would be relevant for stable Euler buckling.

With respect to flexure, we find that (88) admits a solution that emanates from the point $(\lambda, \eta) = (1, 0)$ in the $\lambda\eta$-plane, and rises steeply from this point for increasing λ values. This feature is reflected, of course, in the asymptotic result (60), which is valid for both of (84a, b). Again, the neo-Hookean model is found to be quite typical of the entire class of incompressible materials, at least as far as the bifurcation condition for thin members is concerned.

References

1. Sawyers, K. N., and Rivlin, R. S., *Further Results on the Stability of a Thick Elastic Plate under Thrust*, Journal de Mecanique Theorique et Appliquée, Vol. 2, pp. 663–698, 1983.
2. Sawyers, K. N., *Material Stability and Bifurcation*, Finite Elasticity, Edited by R. S. Rivlin, ASME, New York, New York, AMD Vol. 27, pp. 103–123, 1977.
3. Sawyers, K. N., and Rivlin, R. S., *The Flexural Bifurcation Condition for a Thin Plate under Thrust*, Mechanics Research Communications, Vol. 3, pp. 203–207, 1976.
4. Sawyers, K. N., and Rivlin, R. S., *Stability of a Thick Elastic Plate under Thrust*, Journal of Elasticity, Vol. 12, pp. 101–125, 1982.
5. Sawyers, K. N., *Stability of a Thick Neo-Hookean Plate*, Proceedings of the IUTAM Symposium on Finite Elasticity, Edited by D. E. Carlson and R. T. Shield, Martinus Nijhoff Publishers, The Hague, The Netherlands, pp. 319–330, 1981.
6. Sawyers, K. N., and Rivlin, R. S., *Bifurcation Conditions for a Thick Elastic Plate under Thrust*, International Journal of Solids and Structures, Vol. 10, pp. 483–501, 1974.
7. Koiter, W. T., *On the Stability of Elastic Equilibrium*, Thesis, Delft, The Netherlands, 1945.

[4]To decide this, one could plot $\bar{A}(\lambda)$ versus λ from (87) and then graph the relevant material function $A(\lambda)$ on the same plot. Unless these two curves intersect, no $\bar{\lambda}$ values appear.

8. Biot, M. A., *Exact Theory of Buckling of a Thick Slab*, Applied Scientific Research, Vol. A12, pp. 183–198, 1963.
9. Usmani, S. A., and Beatty, M. F., *On the Surface Instability of a Highly Elastic Half-Space*, Journal of Elasticity, Vol. 4, pp. 249–263, 1974.
10. Green, A. E., Rivlin, R. S., and Shield, R. T., *General Theory of Small Elastic Deformations Superposed on Finite Elastic Deformations*, Proceedings of the Royal Society of London, Vol. A211, pp. 128–154, 1952.
11. Sawyers, K. N., and Rivlin, R. S., *Instability of an Elastic Material*, International Journal of Solids and Structures, Vol. 9, pp. 607–613, 1973.

14

Irreducible Constitutive Expressions

G. F. SMITH AND G. BAO

Abstract. The general expression for a polynomial function of a tensor which is invariant under a symmetry group is given by a polynomial in the elements $I_1, \ldots, I_n$ of an integrity basis. This expression will contain in general a number of redundant terms such as $I_1 I_2 - I_3^2 = 0$ which are referred to as syzgies. A procedure is discussed which yields an expression in which all redundant terms are removed. The resulting expression is said to be irreducible.

Key Words. Canonical forms, crystallographic group, integrity basis, syzygy, polynomial invariant, irreducible expression, irreducible representations, group, inequivalent irreducible representations, generating function, tetragonal–scalenohedral crystal class, characters, polynomial function, vector-valued functions, tensor-valued functions.

1. Introduction

We consider the problem of determining the canonical forms of a vector-valued function **P**(**E**) and a symmetric second-order tensor-valued function **T**(**E**) of a single symmetric second-order tensor **E** which are invariant under a given crystallograhpic group $\{\mathbf{A}\} = \{\mathbf{A}_1, \ldots, \mathbf{A}_N\}$. The restrictions imposed by the

---

G. F. Smith • Director, Center for the Application of Mathematics, Lehigh University, Bethlehem, Pennsylvania 18015. **G. Bao** • Professor, Department of Mechanics, Peking University, Beijing, People's Republic of China 100871; *currently* Associate Professor, Department of Mechanical Engineering, Johns Hopkins University, Baltimore, Maryland 21218.

Nonlinear Effects in Fluids and Solids, edited by M. M. Carroll and M. Hayes, Plenum Press, New York, 1996.

invariance requirement are given by

$$\mathbf{A}_K\mathbf{P}(\mathbf{E}) = \mathbf{P}(\mathbf{A}_K\mathbf{E}\mathbf{A}_K^T), \qquad K = 1, \ldots, N, \tag{1a}$$

$$\mathbf{A}_K\mathbf{T}(\mathbf{E})\mathbf{A}_K^T = \mathbf{T}(\mathbf{A}_K\mathbf{E}\mathbf{A}_K^T), \qquad K = 1, \ldots, N. \tag{1b}$$

We may follow the procedure outlined by Pipkin and Rivlin (Ref. 1) to obtain expressions such that any polynomial functions $\mathbf{P}(\mathbf{E})$ and $\mathbf{T}(\mathbf{E})$ which satisfy the restrictions (1) are given by

$$\mathbf{P}(\mathbf{E}) = a_1\mathbf{J}_1(\mathbf{E}) + \cdots + a_r\mathbf{J}_r(\mathbf{E}), \tag{2a}$$

$$\mathbf{T}(\mathbf{E}) = b_1\mathbf{N}_1(\mathbf{E}) + \cdots + b_s\mathbf{N}_s(\mathbf{E}), \tag{2b}$$

where the $\mathbf{J}_i(\mathbf{E})$ and $\mathbf{N}_i(\mathbf{E})$ satisfy (1a) and (1b), respectively, and where the a_i , $i = 1, \ldots, r$, and b_i , $i = 1, \ldots, s$, are the scalar-valued functions of $\mathbf{E}$ which are invariant under the group $\{\mathbf{A}\}$. Thus, the a_i and b_i satisfy restrictions of the form

$$a_i(\mathbf{E}) = a_i(\mathbf{A}_K\mathbf{E}\mathbf{A}_K^T), \qquad b_i(\mathbf{E}) = b_i(\mathbf{A}_K\mathbf{E}\mathbf{A}_K^T), \tag{3}$$

for all $\mathbf{A}_K$ belonging to the group $\{\mathbf{A}\}$. The general form of scalar-valued polynomial functions of $\mathbf{E}$ has been obtained by Smith and Rivlin (Ref. 2) for the cases where $\{\mathbf{A}\}$ is any of the 32 crystallographic groups. From Ref. 2, we see that

$$a_i(\mathbf{E}) = k^i_{i_1i_2\cdots i_n} I_1^{i_1}I_2^{i_2}\cdots I_n^{i_n}, \qquad b_i(\mathbf{E}) = h^i_{i_1i_2\cdots i_n} I_1^{i_1}I_2^{i_2}\cdots I_n^{i_n}, \tag{4}$$

where the $I_1, I_2, \ldots, I_n$ are elements of an integrity basis for scalar-valued functions of $\mathbf{E}$ which are invariant under $\{\mathbf{A}\}$. We recall that the elements $I_1, I_2, \ldots, I_n$ of an integrity basis are functions of $\mathbf{E}$, each of which is invariant under $\{\mathbf{A}\}$, such that any polynomial function $W(\mathbf{E})$ which is invariant under $\{\mathbf{A}\}$ is expressible as a polynomial in the $I_1, I_2, \ldots, I_n$. The determination of an integrity basis constitutes the first main problem of invariant theory. In general, the elements of an integrity basis are not functionally independent. For example, it may occur that $I_1I_2 - I_3^2 = 0$. This relation is referred to as a syzygy. A syzygy is a relation $K(I_1, I_2, \ldots, I_n) = 0$ which is not an identity in $I_1, \ldots, I_n$ but which becomes an identity when the I_j are written as functions of $\mathbf{E}$. The second main problem of invariant theory requires the determination of a set of syzygies $K_i(I_1, I_2, \ldots, I_n) = 0$, $i = 1, \ldots, p$, such that every syzygy $K(I_1, I_2, \ldots, I_n) = 0$ relating the elements of an integrity basis is expressible in the form

$$K(I_1, I_2, \ldots, I_n) = d_1K_1(I_1, I_2, \ldots, I_n) + \cdots + d_pK_p(I_1, I_2, \ldots, I_n), \tag{5}$$

where the d_i are polynomials in $I_1, \ldots, I_n$. Consideration of the first main problem enables us to express the general polynomial invariant $W(\mathbf{E})$ as

$$W(\mathbf{E}) = c_{i_1 i_2 \cdots i_n} I_1^{i_1} I_2^{i_2} \cdots I_n^{i_n}, \tag{6}$$

where the $c_{i_1 i_2 \cdots i_n}$ are constants. We may employ the relations $K_i(I_1, I_2, \ldots, I_n) = 0$ to remove redundant terms from the expression (6). The objective is to produce a general expression for $W(\mathbf{E})$ which does not contain any redundant terms. An expression for $W(\mathbf{E})$ which contains no redundant terms is said to be irreducible. A simple case occurs when the group $\{\mathbf{A}\}$ is the orthogonal group which defines the symmetry of an isotropic material. An integrity basis for functions of $\mathbf{E}$ which are invariant under the orthogonal group is comprised of the invariants

$$I_1 = \operatorname{tr} \mathbf{E}, \qquad I_2 = \operatorname{tr} \mathbf{E}^2, \qquad I_3 = \operatorname{tr} \mathbf{E}^3. \tag{7}$$

This would give the solution of the first main problem. There are no syzygies relating the I_1, I_2, I_3, given by (7) so that the second main problem does not arise. The expression

$$W(\mathbf{E}) = c_{i_1 i_2 i_3} I_1^{i_1} I_2^{i_2} I_3^{i_3}, \qquad i_1, i_2, i_3 = 1, 2, \ldots, \tag{8}$$

is then irreducible, i.e., it contains no redundant terms. Smith (Ref. 3) has given irreducible expressions for scalar-valued functions $W(\mathbf{E})$ which are invariant under a group $\{\mathbf{A}\}$ for the cases where $\{\mathbf{A}\}$ is any of the 32 crystallographic groups.

The solution of the first main problem as applied to vector-valued functions $\mathbf{P}(\mathbf{E})$ and symmetric second-order tensor-valued functions $\mathbf{T}(\mathbf{E})$ yields expressions of the forms (2). In general, there will be redundant terms appearing in the expressions (2). The second main problem of invariant theory requires the determination of relations $\mathbf{L}_i(I_1, \ldots, I_n; \mathbf{J}_1, \ldots, \mathbf{J}_r) = \mathbf{0}$ and $\mathbf{M}_i(I_1, \ldots, I_n; \mathbf{N}_1, \ldots, \mathbf{N}_s) = \mathbf{0}$ such that all redundant terms appearing in (2) are consequences of these relations. We may then employ the relations $\mathbf{L}_i(\ldots) = \mathbf{0}$, $\mathbf{M}_i(\ldots) = \mathbf{0}$ to eliminate the redundant terms in (2) and thus replace (2) by equivalent expressions which are irreducible. We give below a procedure similar to that employed in Ref. 3 for generating irreducible expressions for $\mathbf{P}(\mathbf{E})$ and $\mathbf{T}(\mathbf{E})$.

2. Decomposition Procedure

We may replace the problem of determining the canonical form of the function $\mathbf{T}(\mathbf{E})$ which is invariant under $\{\mathbf{A}\}$ by a number of simpler problems. Similar considerations apply for $\mathbf{P}(\mathbf{E})$. Application of the transformation $\mathbf{A}_K = \|A_{ij}^{(K)}\|$ belonging to the group $\{\mathbf{A}\}$ to the symmetric second-order tensor $\mathbf{T} = \|T_{ij}\|$

yields the result

$$T_{ij}^{(K)} = A_{ip}^{(K)} A_{jq}^{(K)} T_{pq}. \tag{9}$$

We employ the notion

$$\mathbf{t} = \|T_{11}, T_{22}, T_{33}, T_{23}, T_{31}, T_{12}\|^T = \|t_1, t_2, \ldots, t_6\|^T. \tag{10}$$

We may then rewrite (9) as

$$\mathbf{t}^{(K)} = \mathbf{S}(\mathbf{A}_K)\mathbf{t}, \tag{11}$$

where the entries in the matrix $\mathbf{S}(\mathbf{A}_K)$ may be obtained from (9) provided we observe that $T_{12} = T_{21}$, $T_{13} = T_{31}$, and $T_{23} = T_{32}$. The set of matrices $\mathbf{S}(\mathbf{A}_K)$, $K = 1, \ldots, N$, defines the transformation properties of the column vector $\mathbf{t}$ under the group $\{\mathbf{A}\} = \{\mathbf{A}_1, \mathbf{A}_2, \ldots, \mathbf{A}_N\}$ and is said to form a matrix representation of the group $\{\mathbf{A}\}$. We may rewrite (11) as

$$\mathbf{Q}\mathbf{t}^{(K)} = \mathbf{Q}\mathbf{S}(\mathbf{A}_K)\mathbf{Q}^{-1}\mathbf{Q}\mathbf{t}, \tag{12}$$

where $\mathbf{Q}$ is nonsingular. Appropriate choice of the matrix $\mathbf{Q}$ enables us to decompose the matrix representation $\mathbf{S}(\mathbf{A}_K)$, $K = 1, \ldots, N$, into the direct sum of irreducible representations of the group $\{\mathbf{A}\}$. There are only finitely many inequivalent irreducible representations associated with a finite group $\{\mathbf{A}\} = \{\mathbf{A}_1, \mathbf{A}_2, \ldots, \mathbf{A}_N\}$. We denote these by $\mathbf{\Gamma}_1(\mathbf{A}_K), \mathbf{\Gamma}_2(\mathbf{A}_K), \ldots, \mathbf{\Gamma}_r(\mathbf{A}_K)$, $K = 1, \ldots, N$. Thus, we have

$$\begin{aligned}\mathbf{Q}\mathbf{S}(\mathbf{A}_K)\mathbf{Q}^{-1} &= n_1\mathbf{\Gamma}_1(\mathbf{A}_K) \dot{+} n_2\mathbf{\Gamma}_2(\mathbf{A}_K) \\ &\dot{+} \cdots \dot{+} n_r\mathbf{\Gamma}_r(\mathbf{A}_K), \qquad K = 1, \ldots, N,\end{aligned} \tag{13}$$

where, for example,

$$2\mathbf{\Gamma}_1(\mathbf{A}_K) \dot{+} \mathbf{\Gamma}_2(\mathbf{A}_K) = \begin{Vmatrix} \mathbf{\Gamma}_1(\mathbf{A}_K) & \cdot & \cdot \\ \cdot & \mathbf{\Gamma}_1(\mathbf{A}_K) & \cdot \\ \cdot & \cdot & \mathbf{\Gamma}_2(\mathbf{A}_K) \end{Vmatrix}. \tag{14}$$

The column vector $\mathbf{Q}\mathbf{t}$ may be written as

$$\mathbf{Q}\mathbf{t} = \begin{Vmatrix} \mathbf{t}_1 \\ \cdot \\ \cdot \\ \cdot \\ \mathbf{t}_r \end{Vmatrix}, \qquad \mathbf{t}_j = \begin{Vmatrix} \mathbf{t}_{j1} \\ \cdot \\ \cdot \\ \cdot \\ \mathbf{t}_{jn_j} \end{Vmatrix}. \tag{15}$$

With the notation (15), the equation (12) may be written as

$$\begin{aligned} \mathbf{t}_{1i}^{(K)} &= \mathbf{\Gamma}_1(\mathbf{A}_K)\mathbf{t}_{1i}, \qquad i = 1, \ldots, n_1, \ldots, \\ \mathbf{t}_{ri}^{(K)} &= \mathbf{\Gamma}_r(\mathbf{A}_K)\mathbf{t}_{ri}, \qquad i = 1, \ldots, n_r. \end{aligned} \tag{16}$$

The column vectors $\mathbf{t}_{ji}$, $i = 1, \ldots, n_j$, are referred to as quantities of type Γ_j. The transformation properties under the group $\{\mathbf{A}\}$ of a quantity of type Γ_j are defined by the set of matrices $\mathbf{\Gamma}_j(\mathbf{A}_K)$ comprising an irreducible representation of $\{\mathbf{A}\}$.

Thus, the problem of determining the general form of $\mathbf{T}(\mathbf{E})$ consistent with the restrictions (1b) may be reduced to the problem of determining the general forms of expressions $\mathbf{t}_j(\mathbf{E})$ which are subject to the restrictions that

$$\mathbf{\Gamma}_j(\mathbf{A}_K)\mathbf{t}_j(\mathbf{E}) = \mathbf{t}_j(\mathbf{A}_K\mathbf{E}\mathbf{A}_K^T), \qquad K = 1, \ldots, N. \tag{17}$$

For example, for a given group $\{\mathbf{A}\}$, suppose that we have determined a 6×6 matrix $\mathbf{Q}$ such that $\mathbf{Qt} = \mathbf{t}_1 \dot{+} \mathbf{t}_2 \dot{+} \mathbf{t}_3$ where $\mathbf{t}_1$, $\mathbf{t}_2$, $\mathbf{t}_3$, are quantities of types Γ_1, Γ_2, Γ_3, respectively. We then determine the general forms of the $\mathbf{t}_1(\mathbf{E}), \ldots, \mathbf{t}_3(\mathbf{E})$ consistent with (17). This would give the general form of $\mathbf{Qt}(\mathbf{E})$ and hence the general forms of $\mathbf{t}(\mathbf{E})$ and $\mathbf{T}(\mathbf{E})$.

3. Generating Functions

We consider the problem of generating the general form of a quantity $\mathbf{T}(\mathbf{E})$ of type Γ_ν which is invariant under the group $\{\mathbf{A}\}$. Thus, $\mathbf{T}(\mathbf{E})$ must satisfy

$$\Gamma_\nu(\mathbf{A}_K)\mathbf{T}(\mathbf{E}) = \mathbf{T}(\mathbf{A}_K\mathbf{E}\mathbf{A}_K^T). \tag{18}$$

Suppose that we have determined an expression for $\mathbf{T}(\mathbf{E})$ consistent with (18). We proceed by writing this expression as

$$\mathbf{T}(\mathbf{E}) = \mathbf{T}_1(\mathbf{E}) + \mathbf{T}_2(\mathbf{E}) + \cdots + \mathbf{T}_f(\mathbf{E}) + \cdots, \tag{19}$$

where $\mathbf{T}_i(\mathbf{E})$ denotes the terms in the expression for $\mathbf{T}(\mathbf{E})$ which are of degree i in the components of $\mathbf{E}$. We may compute the number n_i of linearly independent terms of type Γ_ν and of degree i in the components of $\mathbf{E}$. If there are m_i terms appearing in the expression $\mathbf{T}_i(\mathbf{E})$, then $m_i - n_i$ of these terms are redundant. We must eliminate these redundant terms and thus replace $\mathbf{T}_i(\mathbf{E})$ by $\mathbf{T}_i^*(\mathbf{E})$ where all terms appearing in $\mathbf{T}_i^*(\mathbf{E})$ are linearly independent. This must be accomplished for i = 1, 2, . . . , n, where n is arbitrarily large. The resulting expression $\mathbf{T}_1^*(\mathbf{E})$ + $\mathbf{T}_2^*(\mathbf{E}) + \cdots + \mathbf{T}_n^*(\mathbf{E}) + \cdots$ would be the irreducible expression required. We next outline the procedure employed in the determination of the number n_i of linearly independent terms of type Γ_ν and of degree i in $\mathbf{E}$.

Let $\mathbf{e}$ denote a column vector whose entries are the six independent components of the symmetric second-order tensor $\mathbf{E} = \|E_{ij}\|$ where $E_{ij} = E_{ji}$. Thus,

$$\mathbf{e} = \|E_{11}, E_{22}, E_{33}, E_{23}, E_{31}, E_{12}\|^T = \|e_1, e_2, e_3, e_4, e_5, e_6\|^T. \tag{20}$$

Let $\mathbf{S}(\mathbf{A}_K)$ denote the matrix representation which defines the transformation properties of $\mathbf{e}$ under the group $\{\mathbf{A}\} = \{\mathbf{A}_1, \mathbf{A}_2, \ldots, \mathbf{A}_N\}$. Let $\mathbf{S}_f(\mathbf{A}_K)$ denote the matrix representation which defines the transformation properties of the monomials

$$e_1^{r_1} e_2^{r_2} \cdots e_6^{r_6}, \qquad r_1 + r_2 + \cdots + r_6 = f, \tag{21}$$

of degree f in the components of $\mathbf{e}$ under the group $\{\mathbf{A}\}$. The matrix $\mathbf{S}_f(\mathbf{A}_K)$ is referred to as the symmetrized Kronecker f^{th} power of the matrix $\mathbf{S}(\mathbf{A}_K)$. For example, the set of 21×21 matrices $\mathbf{S}_2(\mathbf{A}_K)$, $K = 1, \ldots, N$, defines the transformation properties under $\{\mathbf{A}\}$ of the 21 quantities $e_1^2, e_2^2, \ldots, e_6^2, e_1e_2, e_1e_3, \ldots, e_5e_6$. We suppose that there are r inequivalent irreducible representations associated with the group $\{\mathbf{A}\}$ which are given by $\mathbf{\Gamma}_1(\mathbf{A}_K)$, $K = 1, \ldots, N; \ldots;$ $\Gamma_r(\mathbf{A}_K)$, $K = 1, \ldots, N$. We note that $\mathbf{\Gamma}_1(\mathbf{A}_K)$ is the identity representation for which each of the N matrices $\mathbf{\Gamma}_1(\mathbf{A}_K)$, $K = 1, \ldots, N$, is the 1×1 identity matrix. The character of the irreducible matrix representation $\mathbf{\Gamma}_\nu(\mathbf{A}_K)$ of $\{\mathbf{A}\}$ is given by the set of N numbers

$$(\chi_1^{(\nu)}, \cdots, \chi_N^{(\nu)}), \qquad \chi_K^{(\nu)} = \mathrm{tr}\mathbf{\Gamma}_\nu(\mathbf{A}_K). \tag{22}$$

The matrix representation $\mathbf{S}(\mathbf{A}_K) = \mathbf{S}_1(\mathbf{A}_K), \mathbf{S}_2(\mathbf{A}_K), \ldots$ may be decomposed into the direct sum of irreducible representations $\mathbf{\Gamma}_1(\mathbf{A}_K), \ldots, \mathbf{\Gamma}_r(\mathbf{A}_K)$. Thus, for each $f, f = 1, 2, \ldots$, there is a matrix $\mathbf{Q}_f$ such that

$$\mathbf{Q}_f\mathbf{S}_f(\mathbf{A}_K)\mathbf{Q}_f^{-1} = n_1^{(f)}\mathbf{\Gamma}_1(\mathbf{A}_K) \dot{+} n_2^{(f)}\mathbf{\Gamma}_2(\mathbf{A}_K) \dot{+} \cdots \dot{+} n_r^{(f)}\mathbf{\Gamma}_r(\mathbf{A}_K). \tag{23}$$

The number of times the irreducible representation $\mathbf{\Gamma}_\nu(\mathbf{A}_K)$ appears in the decomposition of the representation $\mathbf{S}_f(\mathbf{A}_K)$ is given by

$$n_\nu^{(f)} = \frac{1}{N}\sum_{K=1}^{N} \bar{\chi}_K^{(\nu)}\, \mathrm{tr}\mathbf{S}_f(\mathbf{A}_K), \tag{24}$$

where $\bar{\chi}_K^{(\nu)}$ denotes the complex conjugate of $\chi_K^{(\nu)}$. We observe then that the number $n_\nu^{(f)}$ gives the linearly independent quantities of type Γ_ν which are of degree f in the components of $\mathbf{e}$ (or $\mathbf{E}$). We further note that the quantity $\mathrm{tr}\mathbf{S}_f(\mathbf{A}_K)$ is given by the coefficient of x^f in the expansion of the quantity $\{\det[I - x\mathbf{S}(\mathbf{A}_K)]\}^{-1}$. Thus, we have

$$\frac{1}{\det[I - x\mathbf{S}(\mathbf{A}_K)]} = 1 + x\,\mathrm{tr}\mathbf{S}_1(\mathbf{A}_K) + x^2\mathrm{tr}\mathbf{S}_2(\mathbf{A}_K) + x^3\mathrm{tr}\mathbf{S}_3(\mathbf{A}_K) + \cdots. \quad (25)$$

With (24) and (25), we see that the number of linearly independent quantities of type Γ_ν which are of degree f in the components of $\mathbf{e}$ is given by the coefficient of x^f in the expansion of the quantity

$$G_\nu(x) = \frac{1}{n}\sum_{k=1}^{N} \frac{\overline{\chi}_K^{(\nu)}}{\det[I - x\mathbf{S}(\mathbf{A}_K)]}. \quad (26)$$

This is referred to as the generating function for the number of linearly independent quantities of type Γ_ν.

The formula (24) is a consequence of the orthogonality properties of the characters of the irreducible representations of a group $\{\mathbf{A}\}$. A discussion of these properties may be found in Wigner (Ref. 4). We next give a simple example to show how (25) arises. Suppose that $\mathbf{S}(\mathbf{A}_K)$ is the 2×2 diagonal matrix given by

$$\mathbf{S}(\mathbf{A}_K) = \begin{Vmatrix} \epsilon_1 & 0 \\ 0 & \epsilon_2 \end{Vmatrix}, \quad (27)$$

where $\mathbf{S}(\mathbf{A}_K)$ defines the transformation properties of the column vector $\|Z_1, Z_2\|^T$ under the transformation $\mathbf{A}_K$. We have

$$\begin{Vmatrix} Z_1^{(K)} \\ Z_2^{(K)} \end{Vmatrix} = \begin{Vmatrix} \epsilon_1 & 0 \\ 0 & \epsilon_2 \end{Vmatrix} \begin{Vmatrix} Z_1 \\ Z_2 \end{Vmatrix}. \quad (28)$$

The transformation properties under $\mathbf{A}_K$ of the three monomials Z_1^2, Z_1Z_2, Z_2^2 of degree 2, the four monomials $Z_1^3, Z_1^2Z_2, Z_1Z_2^2, Z_2^3$ of degree 3, . . . are given by

$$\begin{Vmatrix} Z_1^{(K)}Z_1^{(K)} \\ Z_1^{(K)}Z_2^{(K)} \\ Z_2^{(K)}Z_2^{(K)} \end{Vmatrix} = \begin{Vmatrix} \epsilon_1^2 & 0 & 0 \\ 0 & \epsilon_1\epsilon_2 & 0 \\ 0 & 0 & \epsilon_2^2 \end{Vmatrix} \begin{Vmatrix} Z_1^2 \\ Z_1Z_2 \\ Z_2^2 \end{Vmatrix}, \quad (29a)$$

$$\begin{Vmatrix} Z_1^{(K)}Z_1^{(K)}Z_1^{(K)} \\ Z_1^{(K)}Z_1^{(K)}Z_2^{(K)} \\ Z_1^{(K)}Z_2^{(K)}Z_2^{(K)} \\ Z_2^{(K)}Z_2^{(K)}Z_2^{(K)} \end{Vmatrix} = \begin{Vmatrix} \epsilon_1^3 & 0 & 0 & 0 \\ 0 & \epsilon_1^2\epsilon_2 & 0 & 0 \\ 0 & 0 & \epsilon_1\epsilon_2^2 & 0 \\ 0 & 0 & 0 & \epsilon_2^3 \end{Vmatrix} \begin{Vmatrix} Z_1^3 \\ Z_1^2Z_2 \\ Z_1Z_2^2 \\ Z_2^3 \end{Vmatrix}, \ldots. \quad (29b)$$

The 3×3 and 4×4 matrices appearing in (29) are the symmetrized Kronecker square $\mathbf{S}_2(\mathbf{A}_K)$ and the symmetrized Kronecker cube $\mathbf{S}_3(\mathbf{A}_K)$, respectively, of the matrix $\mathbf{S}(\mathbf{A}_K)$ given by (27). We see from (28) and (29) that the quantities $\mathrm{tr}\mathbf{S}(\mathbf{A}_K) = \mathrm{tr}\mathbf{S}_1(\mathbf{A}_K), \mathrm{tr}\mathbf{S}_2(\mathbf{A}_K), \ldots$ are given by

$$\mathrm{tr}\mathbf{S}_1(\mathbf{A}_K) = \epsilon_1 + \epsilon_2 , \tag{30a}$$

$$\mathrm{tr}\mathbf{S}_2(\mathbf{A}_K) = \epsilon_1^2 + \epsilon_1\epsilon_2 + \epsilon_2^2 , \tag{30b}$$

$$\mathrm{tr}\mathbf{S}_3(\mathbf{A}_K) = \epsilon_1^3 + \epsilon_1^2\epsilon_2 + \epsilon_1\epsilon_2^2 + \epsilon_2^3, \ldots . \tag{30c}$$

We observe that these quantities are the coefficients of x, x^2, x^3, respectively, in the expansion of

$$\begin{aligned}\frac{1}{\det[\mathbf{I} - x\mathbf{S}(\mathbf{A}_K)]} &= \frac{1}{(1 - x\epsilon_1)(1 - x\epsilon_2)}\\ &= (1 + x\epsilon_1 + x^2\epsilon_1^2 + x^3\epsilon_1^3 + \cdots)(1 + x\epsilon_2 + x^2\epsilon_2^2 + x^3\epsilon_2^3 + \cdots)\\ &= 1 + x(\epsilon_1 + \epsilon_2) + x^2(\epsilon_1^2 + \epsilon_1\epsilon_2 + \epsilon_2^2)\\ &\quad + x^3(\epsilon_1^3 + \epsilon_1^2\epsilon_2 + \epsilon_1\epsilon_2^2 + \epsilon_2^3) + \cdots,\end{aligned} \tag{31}$$

where the matrix $\mathbf{S}(\mathbf{A}_K)$ is given by (27). This is the result (25) for the special case where $\mathbf{S}(\mathbf{A}_K)$ is given by (27).

4. Procedure for Generating Irreducible Expressions

We consider the problem of determining irreducible expressions for vector-valued functions $\mathbf{P}(\mathbf{E})$ and symmetric second-order tensor-valued functions $\mathbf{T}(\mathbf{E})$ of the symmetric second-order tensor $\mathbf{E}$ which are invariant under the crystallographic group D_{2d} associated with the tetragonal–scalenohedral crystal class. The set of matrices $\{\mathbf{A}_1, \ldots, \mathbf{A}_8\}$ defining the group D_{2d} are given by

$$\{\mathbf{A}_1, \ldots, \mathbf{A}_8\} = \{\mathbf{I}, \mathbf{D}_1, \mathbf{D}_2, \mathbf{D}_3, \mathbf{T}_3, \mathbf{D}_1\mathbf{T}_3, \mathbf{D}_2\mathbf{T}_3, \mathbf{D}_3\mathbf{T}_3\}, \tag{32}$$

where

$$\mathbf{I} = \begin{Vmatrix} 1 & 0 & 0 \\ 0 & 1 & 0 \\ 0 & 0 & 1 \end{Vmatrix}, \quad \mathbf{D}_1 = \begin{Vmatrix} 1 & 0 & 0 \\ 0 & -1 & 0 \\ 0 & 0 & -1 \end{Vmatrix}, \tag{33a}$$

$$\mathbf{D}_2 = \begin{Vmatrix} -1 & 0 & 0 \\ 0 & 1 & 0 \\ & & \end{Vmatrix}, \quad \mathbf{D}_3 = \begin{Vmatrix} -1 & 0 & 0 \\ 0 & -1 & 0 \\ & & \end{Vmatrix}, \tag{33b}$$

$$\mathbf{T}_3 = \begin{Vmatrix} 0 & 1 & 0 \\ 1 & 0 & 0 \\ 0 & 0 & 1 \end{Vmatrix}, \quad \mathbf{D}_1\mathbf{T}_3 = \begin{Vmatrix} 0 & 1 & 0 \\ -1 & 0 & 0 \\ 0 & 0 & -1 \end{Vmatrix}, \tag{33c}$$

$$\mathbf{D}_2\mathbf{T}_3 = \begin{Vmatrix} 0 & -1 & 0 \\ 1 & 0 & 0 \\ 0 & 0 & -1 \end{Vmatrix}, \quad \mathbf{D}_3\mathbf{T}_3 = \begin{Vmatrix} 0 & -1 & 0 \\ -1 & 0 & 0 \\ 0 & 0 & 1 \end{Vmatrix}. \tag{33d}$$

There are five inequivalent irreducible representations associated with the group D_{2d}. These are given by

$$\boldsymbol{\Gamma}_1(\mathbf{A}_1), \ldots, \boldsymbol{\Gamma}_1(\mathbf{A}_8) = 1, 1, 1, 1, 1, 1, 1, 1, \tag{34a}$$

$$\boldsymbol{\Gamma}_2(\mathbf{A}_1), \ldots, \boldsymbol{\Gamma}_2(\mathbf{A}_8) = 1, -1, -1, 1, -1, 1, 1, -1, \tag{34b}$$

$$\boldsymbol{\Gamma}_3(\mathbf{A}_1), \ldots, \boldsymbol{\Gamma}_3(\mathbf{A}_8) = 1, -1, -1, 1, 1, -1, -1, 1, \tag{34c}$$

$$\boldsymbol{\Gamma}_4(\mathbf{A}_1), \ldots, \boldsymbol{\Gamma}_4(\mathbf{A}_8) = 1, 1, 1, 1, -1, -1, -1, -1, \tag{34d}$$

$$\boldsymbol{\Gamma}_5(\mathbf{A}_1), \ldots, \boldsymbol{\Gamma}_5(\mathbf{A}_8) = \mathbf{E}, \mathbf{F}, -\mathbf{F}, -\mathbf{E}, \mathbf{K}, \mathbf{L}, -\mathbf{L}, -\mathbf{K}, \tag{34e}$$

where

$$\mathbf{E} = \begin{Vmatrix} 1 & 0 \\ 0 & 1 \end{Vmatrix}, \quad \mathbf{F} = \begin{Vmatrix} 1 & 0 \\ 0 & -1 \end{Vmatrix}, \quad \mathbf{K} = \begin{Vmatrix} 0 & 1 \\ 1 & 0 \end{Vmatrix}, \quad \mathbf{L} = \begin{Vmatrix} 0 & 1 \\ -1 & 0 \end{Vmatrix}. \tag{35}$$

The characters $(\chi_1^{(\nu)}, \ldots, \chi_N^{(\nu)})$, where $\chi_K^{(\nu)} = \operatorname{tr}\boldsymbol{\Gamma}_\nu(\mathbf{A}_K)$, of the irreducible representations $\boldsymbol{\Gamma}_1(\mathbf{A}_K), \ldots, \boldsymbol{\Gamma}_5(\mathbf{A}_K)$ are seen from (34) and (35) to be given by

$$(\chi_1^{(1)}, \ldots, \chi_8^{(1)}) = 1, 1, 1, 1, 1, 1, 1, 1, \tag{36a}$$

$$(\chi_1^{(2)}, \ldots, \chi_8^{(2)}) = 1, -1, -1, 1, -1, 1, 1, -1, \tag{36b}$$

$$(\chi_1^{(3)}, \ldots, \chi_8^{(3)}) = 1, -1, -1, 1, 1, -1, -1, 1, \tag{36c}$$

$$(\chi_1^{(4)}, \ldots, \chi_8^{(4)}) = 1, 1, 1, 1, -1, -1, -1, -1, \tag{36d}$$

$$(\chi_1^{(5)}, \ldots, \chi_8^{(5)}) = 2, 0, 0, -2, 0, 0, 0, 0. \tag{36e}$$

We list below the linear combinations of the components P_i and T_{ij} of the vector $\mathbf{P}$ and the symmetric second-order tensor $\mathbf{T}$ whose transformation properties under the group D_{2d} are defined by the irreducible representations $\boldsymbol{\Gamma}_1(\mathbf{A}_K), \ldots, \boldsymbol{\Gamma}_5(\mathbf{A}_K)$:

$$\boldsymbol{\Gamma}_1(\mathbf{A}_K){:}\ T_{33}, T_{11} + T_{22}, \tag{37a}$$

$$\boldsymbol{\Gamma}_2(\mathbf{A}_K):\ , \tag{37b}$$

$$\boldsymbol{\Gamma}_3(\mathbf{A}_K):\ P_3, T_{12}, \tag{37c}$$

$$\boldsymbol{\Gamma}_4(\mathbf{A}_K):\ T_{11} - T_{22}, \tag{37d}$$

$$\boldsymbol{\Gamma}_4(\mathbf{A}_K):\ \left\| \begin{matrix} P_1 \\ P_2 \end{matrix} \right\|, \left\| \begin{matrix} T_{23} \\ T_{31} \end{matrix} \right\|. \tag{37e}$$

We refer to the quantities listed above as quantities of types $\Gamma_1, \ldots, \Gamma_5$ respectively. We observe that the matrix representation $\mathbf{R}(\mathbf{A}_K)$ defining the transformation properties of the column vector $\|P_1, P_2, P_3\|^T$, where the P_i are the components of a vector $\mathbf{P}$, is given by $\mathbf{R}(\mathbf{A}_K) = \mathbf{A}_K$. From (32), (33), and (34) we readily see that

$$\mathbf{A}_K = \boldsymbol{\Gamma}_5(\mathbf{A}_K) \dotplus \boldsymbol{\Gamma}_3(\mathbf{A}_K). \tag{38}$$

Thus, the transformation properties of $\|P_1, P_2\|^T$ and P_3 are defined by the irreducible representations $\boldsymbol{\Gamma}_5(\mathbf{A}_K)$ and $\boldsymbol{\Gamma}_3(\mathbf{A}_K)$, respectively, and we say that $\|P_1, P_2\|^T$ and P_3 are quantities of types Γ_5 and Γ_3, respectively.

Let $\mathbf{S}(\mathbf{A}_K)$ denote the set of 6×6 matrices comprising the matrix representation which defines the transformation properties under D_{2d} of the column vector

$$\mathbf{e} = \|E_{11}, E_{22}, E_{33}, E_{23}, E_{31}, E_{12}\|. \tag{39}$$

The matrices $\mathbf{S}(\mathbf{A}_K)$ are readily determined from inspection of the equations

$$E_{ij}^{(K)} = A_{ip}^{(K)} A_{jq}^{(K)} E_{pq}, \qquad E_{ij} = E_{ji}, \qquad K = 1, \ldots, N, \tag{40}$$

which define the manner in which the components of $\mathbf{E}$ transform under $\mathbf{A}_K$. We obtain the results that

$$\mathbf{S}(\mathbf{A}_1) = \left\| \begin{matrix} 1 & \cdot & \cdot & \cdot & \cdot & \cdot \\ \cdot & 1 & \cdot & \cdot & \cdot & \cdot \\ \cdot & \cdot & 1 & \cdot & \cdot & \cdot \\ \cdot & \cdot & \cdot & 1 & \cdot & \cdot \\ \cdot & \cdot & \cdot & \cdot & 1 & \cdot \\ \cdot & \cdot & \cdot & \cdot & \cdot & 1 \end{matrix} \right\|, \qquad \mathbf{S}(\mathbf{A}_2) = \left\| \begin{matrix} 1 & \cdot & \cdot & \cdot & \cdot & \cdot \\ \cdot & 1 & \cdot & \cdot & \cdot & \cdot \\ \cdot & \cdot & 1 & \cdot & \cdot & \cdot \\ \cdot & \cdot & \cdot & 1 & \cdot & \cdot \\ \cdot & \cdot & \cdot & \cdot & -1 & \cdot \\ \cdot & \cdot & \cdot & \cdot & \cdot & -1 \end{matrix} \right\|, \tag{41a}$$

$$\mathbf{S}(\mathbf{A}_3) = \left\| \begin{matrix} 1 & \cdot & \cdot & \cdot & \cdot & \cdot \\ \cdot & 1 & \cdot & \cdot & \cdot & \cdot \\ \cdot & \cdot & 1 & \cdot & \cdot & \cdot \\ \cdot & \cdot & \cdot & -1 & \cdot & \cdot \\ \cdot & \cdot & \cdot & \cdot & 1 & \cdot \\ \cdot & \cdot & \cdot & \cdot & \cdot & -1 \end{matrix} \right\|, \qquad \mathbf{S}(\mathbf{A}_4) = \left\| \begin{matrix} 1 & \cdot & \cdot & \cdot & \cdot & \cdot \\ \cdot & 1 & \cdot & \cdot & \cdot & \cdot \\ \cdot & \cdot & 1 & \cdot & \cdot & \cdot \\ \cdot & \cdot & \cdot & -1 & \cdot & \cdot \\ \cdot & \cdot & \cdot & \cdot & -1 & \cdot \\ \cdot & \cdot & \cdot & \cdot & \cdot & 1 \end{matrix} \right\|, \tag{41b}$$

$$
\mathbf{S}(\mathbf{A}_5) = \begin{Vmatrix} \cdot & 1 & \cdot & \cdot & \cdot & \cdot \\ 1 & \cdot & \cdot & \cdot & \cdot & \cdot \\ \cdot & \cdot & 1 & \cdot & \cdot & \cdot \\ \cdot & \cdot & \cdot & \cdot & 1 & \cdot \\ \cdot & \cdot & \cdot & 1 & \cdot & \cdot \\ \cdot & \cdot & \cdot & \cdot & \cdot & 1 \end{Vmatrix}, \qquad \mathbf{S}(\mathbf{A}_6) = \begin{Vmatrix} \cdot & 1 & \cdot & \cdot & \cdot & \cdot \\ 1 & \cdot & \cdot & \cdot & \cdot & \cdot \\ \cdot & \cdot & 1 & \cdot & \cdot & \cdot \\ \cdot & \cdot & \cdot & \cdot & 1 & \cdot \\ \cdot & \cdot & \cdot & -1 & \cdot & \cdot \\ \cdot & \cdot & \cdot & \cdot & \cdot & -1 \end{Vmatrix}, \tag{41c}
$$

$$
\mathbf{S}(\mathbf{A}_7) = \begin{Vmatrix} \cdot & 1 & \cdot & \cdot & \cdot & \cdot \\ 1 & \cdot & \cdot & \cdot & \cdot & \cdot \\ \cdot & \cdot & 1 & \cdot & \cdot & \cdot \\ \cdot & \cdot & \cdot & \cdot & -1 & \cdot \\ \cdot & \cdot & \cdot & 1 & \cdot & \cdot \\ \cdot & \cdot & \cdot & \cdot & \cdot & -1 \end{Vmatrix}, \qquad \mathbf{S}(\mathbf{A}_8) = \begin{Vmatrix} \cdot & 1 & \cdot & \cdot & \cdot & \cdot \\ 1 & \cdot & \cdot & \cdot & \cdot & \cdot \\ \cdot & \cdot & 1 & \cdot & \cdot & \cdot \\ \cdot & \cdot & \cdot & \cdot & -1 & \cdot \\ \cdot & \cdot & \cdot & -1 & \cdot & \cdot \\ \cdot & \cdot & \cdot & \cdot & \cdot & 1 \end{Vmatrix}, \tag{41d}
$$

The quantities $\det[I - x\mathbf{S}(\mathbf{A}_K)]$ which appear in generating functions (26) are given by

$$\det[\mathbf{I} - x\mathbf{S}(\mathbf{A}_K)] = (1-x)^6, \qquad K = 1, \tag{42a}$$

$$\det[\mathbf{I} - x\mathbf{S}(\mathbf{A}_K)] = (1-x)^2(1-x^2)^2, \qquad K = 2, 3, 4, 5, 8, \tag{42b}$$

$$\det[\mathbf{I} - x\mathbf{S}(\mathbf{A}_K)] = (1-x^2)(1-x^4), \qquad K = 6, 7, \tag{42c}$$

With (26), (36), and (42), we see that the generating functions $G_\nu(x)$ for the number of linearly independent quantities of types Γ_ν which are invariant under the group D_{2d} are given by

$$G_1(x) = \frac{1 + 2x^3 + x^6}{(1-x)^2(1-x^2)^3(1-x^4)}, \tag{43a}$$

$$G_2(x) = \frac{x^2 + 2x^3 + x^4}{(1-x)^2(1-x^2)^3(1-x^4)}, \tag{43b}$$

$$G_3(x) = \frac{x + x^2 + x^4 + x^5}{(1-x)^2(1-x^2)^3(1-x^4)}, \tag{43c}$$

$$G_4(x) = \frac{x + x^2 + x^4 + x^5}{(1-x)^2(1-x^2)^3(1-x^4)}, \tag{43d}$$

$$G_5(x) = \frac{x + 2x^2 + 2x^3 + 2x^4 + x^5}{(1-x)^2(1-x^2)^3(1-x^4)}. \tag{43e}$$

We now turn to the determination of the irreducible expressions for quantities of type Γ_1, i.e., invariants. We see from Ref. 2 that an integrity basis for functions $W(\mathbf{E})$ which are invariant under the group D_{2d} is given by

$$K_1 = E_{11} + E_{22}, \qquad K_2 = E_{11}E_{22}, \qquad K_3 = E_{33}, \tag{44a}$$
$$K_4 = E_{23}^2 + E_{31}^2, \qquad K_5 = E_{23}^2E_{31}^2, \qquad K_6 = E_{12}^2, \tag{44b}$$
$$L_1 = E_{23}E_{31}E_{12}, \qquad L_2 = (E_{11} - E_{22})(E_{31}^2 - E_{23}^2). \tag{44c}$$

With (44), we see that any polynomial function $W(\mathbf{E})$ which is invariant under D_{2d} is expressible in the form

$$W(\mathbf{E}) = c_{i_1i_2} \cdots {}_{i_8} K_1^{i_1}K_2^{i_2} \cdots K_6^{i_6}L_1^{i_7}L_2^{i_8}, \tag{45}$$

where the $c_{i_1i_2} \cdots {}_{i_8}$ are constants. Alternatively, we may express $W(\mathbf{E})$ as

$$W(\mathbf{E}) = d_{i_1i_2}L_1^{i_1}L_2^{i_2}, \tag{46}$$

where the $d_{i_1i_2}$ are polynomials in $K_1, \ldots, K_6$. We observe that

$$L_1^2 = K_5K_6, \qquad L_2^2 = (K_1^2 - 4K_2)(K_4^2 - 4K_5), \tag{47}$$

and thus from (45) we see that $W(\mathbf{E})$ is expressible in the form

$$W(\mathbf{E}) = S_0 + S_1L_1 + S_2L_2 + S_3L_1L_2, \tag{48}$$

where the $S_0, \ldots, S_3$ are polynomials in $K_1, \ldots, K_6$. The invariants $K_1, \ldots, K_6$ are functionally independent, and hence there are no polynomial relations of the form $K(K_1, \ldots, K_6) = 0$ other than identities such as $K_1^2 = K_1^2$. We wish to determine the number of monomials of degree n in $\mathbf{E}$ which appear in the expression (48). We first consider the expression $S_0(K_1, \ldots, K_6)$. The distinct monomial terms appearing in S_0 are given by the monomial terms appearing in the expression

$$(1 + K_1 + K_1^2 + K_1^3 + \cdots)(1 + K_2 + K_2^2 + K_2^3 + \cdots) \\ \cdots (1 + K_6 + K_6^2 + K_6^3 + \cdots). \tag{49}$$

The number of distinct monomials appearing in S_0 which are of degree n in $\mathbf{E}$ is given by the number of terms of degree n in x in the function obtained from (49) on replacing K_i by x^j, where j denotes the degree of K_i in $\mathbf{E}$. Thus, the number of terms of degree n in $\mathbf{E}$ appearing in S_0 is given by the coefficient of x^n in the expression

$$(1 + x + x^2 + x^3 + \cdots)(1 + x^2 + x^4 + x^6 + \cdots) \\ (1 + x + x^2 + x^3 + \cdots)(1 + x^2 + x^4 + x^6 + \cdots) \\ (1 + x^4 + x^8 + x^{12} + \cdots)(1 + x^2 + x^4 + x^6 + \cdots), \tag{50}$$

where we have noted from (44) that the degree in **E** of K_1 is one, the degree of K_2 is two, ... Alternatively, we may say that the number of terms of degree n in **E** appearing in S_0 is the coefficient of x^n in the formal expansion of

$$(1 - x)^{-1}(1 - x^2)^{-1}(1 - x)^{-1}(1 - x^2)^{-1}(1 - x^4)^{-1}(1 - x^2)^{-1}. \qquad (51)$$

Similarly, the number of monomials of degree n in **E** appearing in S_1L_1, S_2L_2, $S_3L_1L_2$ is given by the coefficient of x^n in the expressions

$$\frac{x^3}{(1 - x)^2(1 - x^2)^3(1 - x^4)}, \qquad (52a)$$

$$\frac{x^3}{(1 - x)^2(1 - x^2)^3(1 - x^4)}, \qquad (52b)$$

$$\frac{x^6}{(1 - x)^2(1 - x^2)^3(1 - x^4)}, \qquad (52c)$$

where we have noted from (44) that L_1 and L_2 are of degree 3 in **E**. Combining (51) and (52), we see that the number of monomial terms of degree n in **E** apeparing in the expression (48) is given by the coefficient of x^n in the expansion of

$$\frac{1 + 2x^3 + x^6}{(1 - x)^2(1 - x^2)^3(1 - x^4)}. \qquad (53)$$

We see from (43) that this coincides with the expression for $G_1(x)$. We recall that the coefficient of x^n in the expansion of $G_1(x)$ gives the number of linearly independent functions of type Γ_1 which are of degree n in **E**. Thus, we arrive at the result that the number of terms of degree n in **E** appearing in the expression (48) is equal to the number of linearly independent terms of degree n in **E** which are type Γ_1. We conclude that the expression (48) is an irreducible expression.

We next consider the problem of determining the general irreducible expression for a function of type Γ_3. We may employ the results due to Kiral and Smith (Ref. 5) to show that a function $V(\mathbf{E})$ which is of type Γ_3 is expressible in the form

$$V(\mathbf{E}) = T_0E_{12} + T_1E_{23}E_{31}, \qquad (54)$$

where T_0, T_1 are polynomial functions of **E** which are invariant under D_{2d}, i.e., functions of type Γ_1. With (48), we may express (54) in the form

$$V(\mathbf{E}) = S_0E_{12} + S_1L_1E_{12} + S_2L_2E_{12} + S_3L_1L_2E_{12} \\ + S_4E_{23}E_{31} + S_5L_1E_{23}E_{31} + S_6L_2E_{23}E_{31} + S_7L_1L_2E_{23}E_{31}, \qquad (55)$$

where the $S_0, \ldots, S_7$ are polynomials in $K_1, \ldots, K_6$ and where $K_1, \ldots, K_6, L_1, L_2$ are defined by (44). We have the following identities:

$$L_1E_{12} = K_6E_{23}E_{31}, \qquad L_1E_{23}E_{31} = K_5E_{12}, \qquad (56a)$$

$$L_1L_2E_{12} = K_6L_2E_{23}E_{31}, \qquad L_1L_2E_{23}E_{31} = K_5L_2E_{12}. \qquad (56b)$$

With (55) and (56), we see that the general expression for functions $V(\mathbf{E})$ of type Γ_3, is given by

$$V(\mathbf{E}) = (S_1^* + S_2^*L_2)E_{12} + (S_3^* + S_4^*L_2)E_{23}E_{31}, \qquad (57)$$

where S_i^*, $i = 1, \ldots, 4$, are polynomial functions of $K_1, \ldots, K_6$. Proceeding as above, we may show that the number of monomial terms of degree n in $\mathbf{E}$ appearing in the expression (57) is given by the coefficient of x^n in the expansion of the function

$$\frac{x + x^2 + x^4 + x^5}{(1 - x)^2(1 - x^2)^3(1 - x^4)}. \qquad (58)$$

Since the expression (58) is identical to the generating function $G_3(x)$ given by (43), we may conclude that the expression (57) is irreducible.

We see from results given in Ref. 5 that any polynomial function $X(\mathbf{E})$ which is of type Γ_4 is expressible in the form

$$X(\mathbf{E}) = T_0(E_{11} - E_{22}) + T_1(E_{23}^2 - E_{31}^2), \qquad (59)$$

where T_0, T_1 are polynomial functions of $\mathbf{E}$ which are invariant under D_{2d}. With (48), we may express (59) as

$$X(\mathbf{E}) - S_0(E_{11} - E_{22}) + S_1L_1(E_{11} - E_{22}) + S_2L_2(E_{11} - E_{22}) \\ + S_3L_1L_2(E_{11} - E_{22}) + S_4(E_{23}^2 - E_{31}^2) + S_5L_1(E_{23}^2 - E_{31}^2) \\ + S_6L_2(E_{23}^2 - E_{31}^2) + S_7L_1L_2(E_{23}^2 - E_{31}^2), \qquad (60)$$

where $S_0, \ldots, S_7$ are polynomials in $K_1, \ldots, K_6$ and where $K_1, \ldots, K_6, L_1, L_2$ are defined by (44). We have the following identities:

$$L_2(E_{11} - E_{22}) = (4K_2 - K_1^2)(E_{23}^2 - E_{31}^2), \qquad (61a)$$

$$L_1L_2(E_{11} - E_{22}) = (4K_2 - K_1^2)L_1(E_{23}^2 - E_{31}^2), \tag{61b}$$

$$L_2(E_{23}^2 - E_{31}^2) = (4K_5 - K_4^2)(E_{11} - E_{22}), \tag{61c}$$

$$L_1L_2(E_{23}^2 - E_{31}^2) = (4K_5 - K_4^2)L_1(E_{11} - E_{22}). \tag{61d}$$

With (60) and (61), we see that the general expression for functions $X(\mathbf{E})$ of type Γ_4, is given by

$$X(\mathbf{E}) = (S_1^* + S_2^*L_1)(E_{11} - E_{22}) + (S_3^* + S_4^*L_1)(E_{23}^2 - E_{31}^2), \tag{62}$$

where S_i^*, $i = 1, \ldots, 4$, are polynomial functions of $K_1, \ldots, K_6$. We readily see that the number of monomial terms of degree n in $\mathbf{E}$ appearing in (62) is given by the coefficient of x^n in the expansion of

$$\frac{x + x^2 + x^4 + x^5}{(1 - x)^2(1 - x^2)^3(1 - x^4)}. \tag{63}$$

Since (63) is identical to the generating function $G_4(x)$ given in (43), we may conclude that (62) is irreducible.

We see from Ref. 5 that a polynomial function $Y(\mathbf{E})$ of type Γ_5 is expressible in the form

$$Y(\mathbf{E}) = T_1\mathbf{N}_1 + T_2\mathbf{N}_2 + T_3\mathbf{N}_3 + T_4\mathbf{N}_4 + T_5\mathbf{N}_5, \tag{64}$$

where the $T_i(\mathbf{E})$ are functions of type Γ_1 and are expressible in the form (48). The quantities N_i, $i = 1, \ldots, 5$, are quantities of type Γ_5 and are defined as

$$\mathbf{N}_1 = \left\| \begin{matrix} E_{23} \\ E_{31} \end{matrix} \right\|, \quad \mathbf{N}_2 = (E_{11} - E_{22})\left\| \begin{matrix} -E_{23} \\ E_{31} \end{matrix} \right\|, \quad \mathbf{N}_3 = E_{12}\left\| \begin{matrix} E_{31} \\ E_{23} \end{matrix} \right\|, \tag{65a}$$

$$\mathbf{N}_4 = (E_{11} - E_{22})E_{12}\left\| \begin{matrix} E_{31} \\ -E_{23} \end{matrix} \right\|, \quad \mathbf{N}_5 = E_{23}E_{31}\left\| \begin{matrix} E_{31} \\ E_{23} \end{matrix} \right\|. \tag{65b}$$

We have the following identities:

$$L_1\mathbf{N}_3 = K_6\mathbf{N}_5, \quad 2L_1\mathbf{N}_4 = K_6(L_2\mathbf{N}_1 - K_4\mathbf{N}_2), \quad L_1\mathbf{N}_5 = K_5\mathbf{N}_3, \tag{66a}$$

$$L_2\mathbf{N}_2 = (K_1^2 - 4K_2)(K_4\mathbf{N}_1 - 2\mathbf{N}_5), \quad L_2\mathbf{N}_3 = K_4\mathbf{N}_4 + 2L_1\mathbf{N}_2, \tag{66b}$$

$$L_2\mathbf{N}_4 = (K_1^2 - 4K_2)(K_4\mathbf{N}_3 - 2L_1\mathbf{N}_1), \tag{66c}$$

$$2L_2\mathbf{N}_5 = K_4(L_2\mathbf{N}_1 - K_4\mathbf{N}_2) + 4K_5\mathbf{N}_2, \tag{66d}$$

$$L_1L_2\mathbf{N}_1 = K_4L_1\mathbf{N}_2 + 2K_5\mathbf{N}_4, \tag{66e}$$

$$L_1L_2\mathbf{N}_2 = (K_1^2 - 4K_2)(K_4L_1\mathbf{N}_1 - 2K_5\mathbf{N}_3), \tag{66f}$$

$$2L_1L_2\mathbf{N}_3 = K_4K_6(L_2\mathbf{N}_1 - K_4\mathbf{N}_2) + 4K_5K_6\mathbf{N}_2, \tag{66g}$$

$$L_1L_2\mathbf{N}_4 = (K_1^2 - 4K_2)K_6(K_4\mathbf{N}_5 - 2K_5\mathbf{N}_1), \tag{66h}$$

$$L_1L_2\mathbf{N}_5 = K_5(K_4\mathbf{N}_4 + 2L_1\mathbf{N}_2), \tag{66i}$$

With (64) and (66), we see that type Γ_5 function $Y(\mathbf{E})$ is expressible in the form

$$Y(\mathbf{E}) = \sum_{i=1}^{5} S_i\mathbf{N}_i + S_6L_1\mathbf{N}_1 + S_7L_1\mathbf{N}_2 + S_8L_2\mathbf{N}_1, \tag{67}$$

where S_i, $i = 1, \ldots, 8$, are polynomial functions of $K_1, \ldots, K_6$.

We may now give general irreducible expressions for vector-valued functions $\mathbf{P}(\mathbf{E})$ and symmetric second-order tensor-valued functions $\mathbf{T}(\mathbf{E})$ which are invariant under D_{2d}. We observe from (37) that $\|P_1, P_2\|^T$ and P_3 are quantities of types Γ_5 and Γ_3, respectively. With (57) and (67), we have

$$\begin{Vmatrix} P_1 \\ P_2 \end{Vmatrix} = \sum_{i=1}^{5} a_i\mathbf{N}_i + a_6L_1\mathbf{N}_1 + a_7L_1\mathbf{N}_2 + a_8L_2\mathbf{N}_1, \tag{68a}$$

$$P_3 = (b_1 + b_2\mathrm{L}_2)E_{12} + (b_3 + b_4L_2)E_{23}E_{31}, \tag{68b}$$

where $a_1, \ldots, a_8, b_1, \ldots, b_4$ are polynomials in $K_1, \ldots, K_6$ and $\mathbf{N}_1, \ldots, \mathbf{N}_5$ are defined in (65). From (37), we see that $T_{11} + T_{22}$, T_{33}, T_{12}, $T_{11} - T_{22}$, $\|T_{23}, T_{31}\|^T$ are of types Γ_1, Γ_1, Γ_3, Γ_4, and Γ_5, respectively. With (48), (57), (62), and (67), we have

$$T_{11} + T_{22} = c_0 + c_1L_1 + c_2L_2 + c_3L_1L_2, \tag{69a}$$

$$T_{33} = d_0 + d_1L_1 + d_2L_2 + d_3L_1L_2, \tag{69b}$$

$$T_{12} = (e_1 + e_2L_2)E_{12} + (e_3 + e_4L_2)E_{23}E_{31}, \tag{69c}$$

$$T_{11} - T_{22} = (f_1 + f_2L_1)(E_{11} - E_{22}) + (f_3 + f_4L_1)(E_{23}^2 - E_{31}^2), \tag{69d}$$

$$\begin{Vmatrix} T_{23} \\ T_{31} \end{Vmatrix} = \sum_{i=1}^{5} g_i\mathbf{N}_i + g_6L_1\mathbf{N}_1 + g_7L_1\mathbf{N}_2 + g_8L_2\mathbf{N}_1, \tag{69e}$$

where $c_0, \ldots, g_8$, are polynomials in $K_1, \ldots, K_6$. $\mathbf{N}_1, \ldots, \mathbf{N}_5$ are defined in (65). The quantities $K_1, \ldots, K_6, L_1, L_2$ appearing in (68) and (69) are defined in (44).

We did not need to determine the general irreducible expression for a function $\mathcal{R}(\mathbf{E})$ of type Γ_2. If we had done so, we would be in a position to give the general irreducible expression for an nth order tensor-valued function $T_{i_1 \ldots i_n}$ (E) which is invariant under D_{2d} where $n = 3, 4, 5, \ldots$. This would require that we be able to decompose $T_{i_1 \ldots i_n}$ into a number of quantities of types $\Gamma_1, \ldots, \Gamma_5$. We have written a computer program (see Xu, Smith, and Smith, Ref. 6) which will automatically carry out the decomposition for any of the crystallographic groups. Results of the type given above have been obtained for almost all of the 32 crystallographic groups by Bao (Ref. 7).

References

1. PIPKIN, A. C. and RIVLIN, R. S., *The Formulation of Constitutive Equations in Continuum Physics, I*, Archive for Rational Mechanics and Analysis, Vol. 4, pp. 129–144, 1959.
2. SMITH, G. F., and RIVLIN, R. S., *The Strain-Energy Function for Anisotropic Elastic Materials*, Transactions of the American Mathematical Society, Vol. 88, pp. 175–193, 1958.
3. SMITH, G. F., *Further Results on the Strain-Energy Function for Anisotropic Elastic Materials*, Archive for Rational Mechanics and Analysis, Vol. 10, pp. 108–118, 1962.
4. WIGNER, E. P., *Group Theory and Its Application to the Quantum Mechanics of Atomic Spectra*, Academic Press, New York, New York, 1959.
5. KIRAL, E., and SMITH, G. F., *On the Constitutive Relations for Anisotropic Materials—Triclinic, Monoclinic, Rhombic, Tetragonal and Hexagonal Crystal Systems*, International Journal of Engineering Science, Vol. 12, pp. 471–490, 1974.
6. XU, Y., SMITH, M. M., and SMITH, G. F., *Computer Aided Generation of Anisotropic Constitutive Expressions*, International Journal of Engineering Science, Vol. 25, pp. 711–722, 1987.
7. BAO, G., PhD Dissertation, Lehigh University, 1987.

15

Determination of the Strain Energy Density Function for Compressible Isotropic Nonlinear Elastic Solids by Torsion–Normal Force Experiments

A. S. WINEMAN AND G. B. MCKENNA

Abstract. Penn and Kearsley derived expressions for the derivatives of the strain energy density function for an incompressible elastic material in terms of the measured torque–twist and axial force–twist relations obtained during experiments involving torsion of a circular cylinder. The present work is concerned with a method for determining the derivatives of the strain energy density function for a compressible isotropic elastic material by means of experiments involving torsion of a circular cylinder. The problem is made more difficult in the compressible case because cylindrical surfaces undergo deformation, which is then used to show why the Penn and Kearsley result cannot be extended to compressible materials. We then discuss the material identification method for determining the derivatives of the strain energy density function from the torque–twist, axial force–twist, and volume change–twist relations, which can be measured using the torsional dilatometer developed by Duran and McKenna. In this method, a polynomial representation for the strain energy density function is assumed and the constants are determined from the measured data.

Key Words. Torsion experiment, nonlinear elasticity, solid circular cylinder, compressible isotropic nonlinear elastic solid, volumetric strain, axial force, twisting moment, Runge–Kutta method, strain energy density function, method of material identification.

A. S. Wineman • Professor, Department of Mechanical Engineering and Applied Mechanics, University of Michigan, Ann Arbor, Michigan 48109. **G. B. McKenna** • Polymers Division, NIST, Gaithersburg, Maryland 20899.

Nonlinear Effects in Fluids and Solids, edited by M. M. Carroll and M. Hayes, Plenum Press, New York, 1996.

1. Introduction

One of the fundamental problems in the mechanics of elastic materials is the determination of the strain energy density function from experimental data. In the case of incompressible isotropic materials, Rivlin and Saunders (Ref. 1) and others developed extensively the analyses necessary to use homogeneous deformations to determine the derivatives of the strain energy density function $W(I_1, I_2)$ with respect to the invariants I_1 and I_2 of the deformation tensor. Early experiments by Treloar (Ref. 2), Kawabata and Kawaii (Ref. 3), and Becker (Ref. 4) used sheets of rubber that were stretched in different directions in order to map out the stress–deformation response in $I_1 - I_2$ space and used these analyses to obtain $\partial W/\partial I_1$ and $\partial W/\partial I_2$. Rivlin's solution of the problem of the torsion of a right circular cylinder of incompressible elastic material (Ref. 5) has made a significant impact on the determination of the strain energy density function. Rivlin's solution is valid for all isotropic incompressible elastic materials, and therefore can be useful determining the strain energy density function of such materials. Of particular interest are the expressions for the twisting moment and axial force in terms of a general strain energy density function.

Subsequently, Penn and Kearsley (Ref. 6) established a very interesting and useful method for using Rivlin's result to determine the strain energy density function for an incompressible isotropic elastic solid. In the torsion of a circular cylinder, the twisting moment and axial force are given by integrals over the cross section of linear combinations of derivatives of the strain energy density function (Ref. 7). If the twisting moment versus angle of twist and the axial force versus angle of twist are known by measurement, these can be interpreted as integral equations for the derivatives of the strain energy density function. Penn and Kearsley solved these equations and obtained explicit expressions for the derivatives of the strain energy density function in terms of the measured data. This result is made possible because (1) the deformation is known *a priori* from the constraint of incompressibility and (2) it is a controllable deformation, i.e., can be shown to be possible in every nonlinear incompressible isotropic elastic solid.

In the case of compressible materials, the determination of the strain energy density function is far more complicated due to volume change. Hence, in the case of torsion, the results of Penn and Kearsley are no longer valid. The deformation is *a priori* unknown, and its form depends on the material properties (strain energy density function). However, there have been recent efforts to experimentally measure the twisting moment, axial force, and volume change in torsion of a cylinder made of a polymeric solid (Wang *et al.*, Ref. 8; Duran and McKenna, Ref. 9; Pixa *et al.*, Ref 10) and it seems that there is an incentive to explore the possibility that such measurements are useful in the determination of the strain energy density function for compressible isotropic materials.

McKenna and Zapas (Ref. 11) attempted to extract $\partial W/\partial I_1$ and $\partial W/\partial I_2$ from a solution of torsion of a compressible cylinder and experimental data for poly(methyl methacrylate), and then compare results with those obtained using the Penn and Kearsley analysis and the assumption of an isochoric deformation. Unfortunately there was an error in their solution to the compressible cylinder problem.

In the present work, we revisit the problem of torsion of a cylinder of compressible isotropic elastic material, correct the prior errors, and explore the potential for using data from a torsion experiment to determine the strain energy density function. We will first establish some properties of the deformation which must hold for all compressible materials. These are used to show that the method of Penn and Kearsley cannot be used to obtain explicit expressions for the derivatives of the strain energy function in terms of the measured twisting moment and axial force response. We then discuss the method of material identification as a means of using these data to determine the derivatives of the strain energy density function as well as the deformation of the cylinder. Material identification, as applied to nonlinear elasticity, was introduced by Iding *et al.* (Ref. 12). The method consists of four elements: (1) measured force and deformation quantities in some experiment; (2) a representation for the strain energy density function in terms of a set of arbitrary parameters; (3) a method for calculating the quantities measured in the experiment for any reasonable choice of the parameters; (4) a method for adjusting the parameters in order to minimize the error between the measured and computed quantities.

Iding and colleagues illustrated the identification procedure with a hypothetical uniaxial extension experiment in which the specimen is a flat sheet of decreasing width and the material is incompressible. The displacements of points along the edges were taken as the measured quantities. Wineman *et al.* (Ref. 13) showed that the inflation of a circular membrane is a particularly useful experiment for the identification procedure when the material is incompressible. The measured quantities were the particle coordinates in the inflated membrane. In this paper, we describe the identification of the strain energy density function for a compressible material through the use of the torsion of a circular cylinder.

The boundary value problem for the deformation of a twisted cylinder of a nonlinear compressible isotropic elastic material is formulated in Section 2. In Section 3, it is shown that the radial deformation must be expressed in terms of specific variables formed from the particle radii in the current and reference configurations, the undeformed radius of the cylinder, and the angle of twist per unit length. A consequence of this result, as shown in Section 4, is that the method of Penn and Kearsley cannot be used to express the derivatives of the strain energy density function in terms of the measured data. Section 5 contains a discussion of the four elements of the identification methods as they pertain to the torsion experiment.

2. Governing Equations

Consider a solid circular cylinder which in its undeformed state has radius R_0 and length L_0. The cylinder undergoes a torsional deformation in which planes normal to the axis in the undeformed state remain plane, stay at a fixed distance from the end of the cylinder, and undergo a rotation proportional to their distance from the end of the cylinder. Cylinder surfaces stay cylindrical, but may undergo radial expansion or contraction.

Let the axis of a cylindrical polar coordinate system coincide with the axis of cylinder, with the plane $Z = 0$ at one end. A material particle has coordinates (R, Θ, Z) in the reference state and (r, θ, z) in the current state. The deformation is described by the mapping

$$r = r(R), \tag{1a}$$

$$\theta = \Theta + \psi Z, \tag{1b}$$

$$z = Z, \tag{1c}$$

where ψ denotes the angle of twist per unit length.

The physical components of the deformation gradient with respect to the cylindrical coordinate system are given by

$$\mathbf{F} = \begin{bmatrix} dr/dR & 0 & 0 \\ 0 & r/R & \psi r \\ 0 & 0 & 1 \end{bmatrix}. \tag{2}$$

The physical components of the left Cauchy–Green strain tensor $\mathbf{B} = \mathbf{F}\mathbf{F}^{\mathrm{T}}$ are

$$\mathbf{B} = \begin{bmatrix} (dr/dR)^2 & 0 & 0 \\ 0 & (r/R)^2 + (\psi r)^2 & \psi r \\ 0 & \psi r & 1 \end{bmatrix}. \tag{3}$$

Also,

$$\mathbf{B}^2 = \begin{bmatrix} (dr/dR)^4 & 0 & 0 \\ 0 & [(r/R)^2 + (\psi r)^2]^2 + (\psi r)^2 & \psi r[1 + (r/R)^2 + (\psi r)^2] \\ 0 & \psi r[1 + (r/R)^2 + (\psi r)^2] & 1 + (\psi r)^2 \end{bmatrix}. \tag{4}$$

The invariants of the strain tensor are given by

$$I_1 = tr\,\mathbf{B} = 1 + \left(\frac{dr}{dR}\right)^2 + \left(\frac{r}{R}\right)^2 + (\psi r)^2, \tag{5a}$$

$$I_2 = \frac{1}{2}[I_1^2 - tr\,\mathbf{B}^2] = \left(\frac{r}{R}\right)^2 + \left(\frac{dr}{dR}\right)^2\left[1 + \left(\frac{r}{R}\right)^2 + (\psi r)^2\right]. \tag{5b}$$

$$I_3 = (\det \mathbf{F})^2 = \left(\frac{r}{R}\right)^2\left(\frac{dr}{dR}\right)^2. \tag{5c}$$

Let $W(I_1, I_2, I_3)$ denote the strain energy density of a compressible isotropic nonlinear elastic solid. The Cauchy stress $\boldsymbol{\sigma}$ is given by the constitutive relation (Ref. 14),

$$\boldsymbol{\sigma} = \frac{2}{\sqrt{I_3}}[I_3 W_3 \mathbf{I} + (W_1 + I_1 W_2)\mathbf{B} - W_2 \mathbf{B}^2], \tag{6}$$

in which $W_i = \partial W/\partial I_1$, $i = 1, 2, 3$. The physical components of stress induced by the deformation (1) are then given by

$$\sigma_{rr} = \frac{2}{\frac{r}{R}\frac{dr}{dR}}\left(\frac{dr}{dR}\right)^2\left[W_1 + \left(1 + \left(\frac{r}{R}\right)^2 + (\psi r)^2\right)W_2 + \left(\frac{r}{R}\right)^2 W_3\right], \tag{7}$$

$$\sigma_{\theta\theta} = \frac{2}{\frac{r}{R}\frac{dr}{dR}}\left[\left(\left(\frac{r}{R}\right)^2 + (\psi r)^2\right)W_1 + \left(\left(\frac{r}{R}\right)^2 + \left(\frac{r}{R}\frac{dr}{dR}\right)^2 + \left(\frac{dr}{dR}\right)^2(\psi r)^2\right)W_2 + \left(\frac{r}{R}\frac{dr}{dR}\right)^2 W_3\right], \tag{8}$$

$$\sigma_{zz} = \frac{2}{\frac{r}{R}\frac{dr}{dR}}\left[W_1 + \left(\left(\frac{r}{R}\right)^2 + \left(\frac{dr}{dR}\right)^2\right)W_2 + \left(\frac{r}{R}\frac{dr}{dR}\right)^2 W_3\right], \tag{9}$$

$$\sigma_{z\theta} = \frac{2}{\frac{r}{R}\frac{dr}{dR}}(\psi r)\left[W_1 + \left(\frac{dr}{dR}\right)^2 W_2\right], \tag{10}$$

and $\sigma_{r\theta} = \sigma_{rz} = 0$.

The only nontrivial equilibrium condition is
The only nontrivial equilibrium condition is

$$\frac{d\sigma_{rr}}{dr} + \frac{\sigma_{rr} - \sigma_{\theta\theta}}{r} = 0. \tag{11}$$

In (11), the derivative is taken with respect to the current radius r, whereas in the expressions for the stresses in (7), the kinematic quantities depend on the reference radius R. Because of this, the equilibrium equation is transformed to

$$\frac{d\sigma_{rr}}{dr} + \frac{dr/dR}{r/R}\frac{\sigma_{rr} - \sigma_{\theta\theta}}{r} = 0. \tag{12}$$

A nonlinear second-order ordinary differential equation for $r(R)$ is obtained by substituting σ_{rr} and $\sigma_{\theta\theta}$ from (7) and (8) into (12). Two boundary conditions on $r(R)$ are established by requiring: (i) a point on the centerline does not move,

$$r(0) = 0, \tag{13}$$

and (ii) the outer cylindrical surface is free of traction,

$$\sigma_{rr}(R_0) = 0. \tag{14}$$

Equations (12)–(14) define a boundary value problem for the radial mapping in (1).

3. Change of Variables

Let the change of variables

$$x = \psi r, \qquad X = \psi R \tag{15}$$

be introduced. This is motivated by a similar change in variables introduced by Penn and Kearsley (Ref. 6) in the case of incompressible materials. Then,

$$r/R = x/X, \qquad dr/dR = dx/dX. \tag{16}$$

Accordingly, $\mathbf{F}$ in (2) becomes

$$\mathbf{F} = \begin{bmatrix} dx/dX & 0 & 0 \\ 0 & x/X & x \\ 0 & 0 & 1 \end{bmatrix}. \tag{17}$$

Note that ψ no longer explicitly appears in $\mathbf{F}$, and hence $\mathbf{B}$ and the invariants. The expressions for the stresses in (7) and (8) can then be denoted as $\sigma_{rr} = \hat{\sigma}_{rr}(x/X, dx/dX, X)$ and $\sigma_{\theta\theta} = \hat{\sigma}_{\theta\theta}(x/X, dx/dX, X)$. Equilibrium equation (12) becomes

$$\frac{d\hat{\sigma}_{rr}(x/X, dx/dX, X)}{dX} + \frac{dx/dX}{x/X}\frac{A(x/X, dx/dX, X)}{X} = 0, \tag{18}$$

in which

$$A(x/X, dx/dX, X) = \hat{\sigma}_{rr}(x/X, dx/dX, X) - \hat{\sigma}_{\theta\theta}(x/X, dx/dX, X). \tag{19}$$

Note that ψ does not explicitly appear in the differential equation defined by (18) and (19). Thus, regardless of the form of the strain energy density W, the solution can be written as

$$x = f(X; c_1, c_2), \tag{20}$$

in which c_1, c_2 denote constants of integration. Boundary condition (13) implies that (20) must satisfy

$$f(0; c_1, c_2) = 0. \tag{21}$$

When (20) is substituted into the expression for σ_{rr}, the latter can be written as

$$\sigma_{rr} = \hat{\sigma}_{rr}(X; c_1, c_2). \tag{22}$$

Using (15) and (22), boundary condition (14) can be expressed as

$$\hat{\sigma}_{rr}(\psi R_0; c_1, c_2) = 0. \tag{23}$$

Let $X_0 = \psi R_0$. It follows from (21) and (23) that the solution (20) of the boundary value problem has the form

$$x = g(X; X_0), \tag{24}$$

or, in view of (15),

$$\psi r = g(\psi R; \psi R_0). \tag{25}$$

Thus, for any compressible isotropic elastic solid, the angle of twist appears only in the combinations ψr, ψR, or ψR_0. Moreover, the general solution depends explicitly on the radius R_0 of the undeformed cylinder through the combination ψR_0.

A number of comments can be made about result (24) or (25):

(i) Boundary condition (13) implies

$$g(0; \psi R_0) = 0. \tag{26}$$

(ii) Suppose that the direction of twist is reversed, so that $\psi \to -\psi$. Then, according to (15), $X \to -X$ and $x \to -x$. It is readily seen from (5), (7), (8),

(18), and (19) that the invariants, the stresses σ_{rr} and $\sigma_{\theta\theta}$, and finally the governing differential equation are unchanged. Thus, if $x = g(X; X_0)$ satisfies (18) and (19), then so does $-x = g(-X; -X_0)$. It follows that

$$g(-X; -X_0) = -g(X; X_0), \tag{27}$$

or

$$g(-\psi R; -\psi R_0) = -g(\psi R; \psi R_0), \tag{28}$$

which, along with (25), implies

$$r|_{-\psi} = \frac{g(-\psi R; -\psi R_0)}{-\psi} = r|_{\psi}. \tag{29}$$

(iii) The change in volume per unit volume of the cylinder in its reference state is

$$\frac{\pi r_0^2 L_0 - \pi R_0^2 L_0}{\pi R_0^2 L_0} = \frac{g(\psi R_0, \psi R_0)^2}{(\psi R_0)^2} - 1, \tag{30}$$

where r_0 denotes the outer radius of the cylinder in its current twisted state. Thus, the volumetric strain depends on the angle of twist ψ and undeformed radius R_0 only through the combination ψR_0. By (28), the volumetric strain is an even function of ψR_0. This result can be subjected to experimental verification.

(iv) Finally, Levinson (Ref. 15) determined the radial deformation function for several models of the Blatz–Ko type (Ref. 16). His results could be presented in the form (25).

4. Axial Force, Twisting Moment

Let M and N denote, respectively, the twisting moment and axial force applied to an end of the cylinder. Also, let $r_0 = r(R_0)$, the deformed outer radius. Then,

$$M = 2\pi \int_0^{r_0} r^2 \sigma_{z\theta} dr, \tag{31}$$

$$N = 2\pi \int_0^{r_0} r\sigma_{zz} dr. \tag{32}$$

Let the integrals over the deformed cross section be transformed into integrals over the cross section of the reference state. Then (31) and (32) can be written as

$$M = 2\pi \int_0^{R_0} \left(\frac{r}{R}\right)^2 \left(\frac{dr}{dR}\right) R^2 \sigma_{z\theta} dR, \tag{33}$$

$$N = 2\pi \int_0^{R_0} \left(\frac{r}{R}\right) \left(\frac{dr}{dR}\right) R\sigma_{zz} dR. \tag{34}$$

On substituting from the stresses in (9) and (10), (33) and (34) become

$$\frac{M}{4\pi} = \int_0^{R_0} \psi r^2 \left[W_1 + \left(\frac{dr}{dR}\right)^2 W_2 \right] RdR, \tag{35}$$

$$\frac{N}{4\pi} = \int_0^{R_0} \left[W_1 + \left(\left(\frac{r}{R}\right)^2 + \left(\frac{dr}{dR}\right)^2 \right) W_2 + \left(\frac{r}{R}\frac{dr}{dR}\right)^2 W_3 \right] RdR. \tag{36}$$

Using (15) we obtain

$$\frac{M\psi^3}{4\pi} = \int_0^{X_0} \left[W_1 + \left(\frac{dx}{dX}\right)^2 W_2 \right] x^2 XdX, \tag{37}$$

$$\frac{N\psi^2}{4\pi} = \int_0^{X_0} \left[W_1 + \left(\left(\frac{x}{X}\right)^2 + \left(\frac{dx}{dX}\right)^2 \right) W_2 + \left(\frac{x}{X}\frac{dx}{dX}\right)^2 W_3 \right] XdX. \tag{38}$$

By (17) and (24), the integrands of (37) and (38) depend only on X and $X_0 = \psi R_0$. Letting these integrands be denoted by $F_M(X; \psi R_0)$ and $F_N(X; \psi R_0)$, respectively, (37) and (38) can be written as

$$\frac{M\psi^3}{4\pi} = \int_0^{\psi R_0} F_M(X; \psi R_0) dX, \tag{39}$$

$$\frac{N\psi^2}{4\pi} = \int_0^{\psi R_0} F_N(X; \psi R_0)dX. \tag{40}$$

Equations (37) and (38) [or (39) and (40)] can be regarded as two integral equations relating the normal force and twisting moment as functions of ψ to the strain energy density and radial deformation. When the material is incompressible, the radial deformation is known, ($x = X$ or $r = R$), and the integrands are independent of X_0 [or ψR_0]. Then, as shown by Penn and Kearsley (Ref. 6), the integral equations can be solved by direct application of Leibniz's rule of calculus. This leads to an explicit expression for each derivative of the strain energy density in terms of derivatives of the left-hand sides of (37) and (38) with respect to ψ.

When the material is compressible, the integrand must be regarded as depending on the parameter $X_0 = \psi R_0$. Application of Leibniz's rule no longer leads to explicit expressions for the integrands. In fact, it results in higher derivatives of W_1, W_2, W_3 in the integrand, and thus greater complexity. Other means must be employed to solve these equations for both the radial deformation and the strain energy density.

5. Material Identification by Torsion Experiments

The elements of the material identification method outlined in the Introduction will now be discussed in detail in the context of the torsion experiment.

5.1. Specification of the Experiment and Measured Quantities. In the torsion experiments of Duran and McKenna (Ref. 9), three quantities were measured as functions of ψ: the twisting moment M, the axial force N, and the change in volume per unit volume in the reference state. The latter is defined by (30) and is denoted by V. The measured quantities are denoted, respectively, by $M^{\text{meas}}(\psi)$, $N^{\text{meas}}(\psi)$, and $V^{\text{meas}}(\psi)$.

5.2. Representation for *W*. There are many ways to express the strain energy density function W in terms of the invariants I_1, I_2, and I_3. Iding *et al.* (Ref. 12) and Wineman *et al.* (Ref. 13) discussed the representation of W, for incompressible materials, through the use of finite element bases in I_1–I_2 space. In their torsion experiments, Duran and McKenna (Ref. 9) obtained data up to a maximum shear strain of about 0.1. For this reason, in this study, W is approximated by a single expression which is to apply over the entire range of data. In particular, W is represented by the polynomial,

$$W = c_1(I_1 - 3) + c_2(I_2 - 3) + c_3(I_1 - 3)^2 + c_4(I_2 - 3)^2 + c_5(I_1 - 3)(I_2 - 3) + c_6(I_3 - 3), \tag{41}$$

in which $c_1, c_2, \ldots, c_6$ are constants to be determined by the identification method. This set of constants is denoted by $\mathbf{c}$. The derivatives of W with respect to the invariants are

$$W_1 = c_1 + 2c_3(I_1 - 3) + c_5(I_2 - 3), \tag{42a}$$

$$W_2 = c_2 + 2c_4(I_2 - 3) + c_5(I_1 - 3), \tag{42b}$$

$$W_3 = c_6. \tag{42c}$$

5.3. Calculation of M, N, V. When a set of values is chosen for the constants $\mathbf{c}$, W in (41) defines a specific strain energy density function. The identification method presented here is not concerned with satisfying restrictions imposed on W by inequalities such as the Baker–Ericksen inequality (Ref. 17). However, these could be incorporated into the method. W_1, W_2, and W_3 from (42), (7), (8), and (12)–(14) then define a boundary value problem for the corresponding radial deformation. Recalling (25), the solution to the boundary value problem is denoted by

$$r = \bar{g}(\psi R; \psi R_0; \mathbf{c})/\psi. \tag{43}$$

The equations of the boundary value problem are nonlinear, and must be solved for many different choices for constants $\mathbf{c}$. Thus, a numerical method of solution is used, which is as follows.

Let the expressions for the stresses on the right-hand sides of (7) and (8) be denoted by $\tilde{\sigma}_{rr}(\lambda_r, \lambda_\theta, R)$ and $\tilde{\sigma}_{\theta\theta}(\lambda_r, \lambda_\theta, R)$, where $\lambda_r = dr/dR$ and $\lambda_\theta = r/R$. Then

$$\frac{d\sigma_{rr}}{dr} = \frac{\partial \tilde{\sigma}_{rr}}{\partial \lambda_r}\frac{d\lambda_r}{dr} + \frac{\partial \tilde{\sigma}_{rr}}{\partial \lambda_\theta}\frac{d\lambda_\theta}{dr} + \frac{\partial \tilde{\sigma}_{rr}}{\partial R}. \tag{44}$$

Also, note that

$$\frac{d\lambda_\theta}{dR} = \frac{\lambda_r - \lambda_\theta}{R}. \tag{45}$$

Then the second-order nonlinear ordinary differential equation for the radial deformation which is obtained by combining (7), (8), and (12) can be reduced to a system of two first-order equations. The first equation is obtained by substituting (44) into (12) and then making use of (45),

$$\frac{d\lambda_r}{dR} = \frac{\dfrac{\lambda_r}{\lambda_\theta}(\tilde{\sigma}_{\theta\theta} - \tilde{\sigma}_{rr}) - \dfrac{\partial\tilde{\sigma}_{rr}}{\partial\lambda_\theta}\dfrac{(\lambda_\theta - \lambda_r)}{R} - \dfrac{\partial\tilde{\sigma}_{rr}}{\partial R}}{\dfrac{\partial\tilde{\sigma}_{rr}}{\partial\lambda_r}}. \tag{46}$$

Note that the right-hand side of (46) is a function of λ_r, λ_θ, and R. The second equation of the system is given by (45).

One boundary condition arises from consideration of the continuity of $r(R)$ at $R = 0$,

$$\lambda_r = \lambda_\theta, \qquad \text{at } R = 0. \tag{47}$$

The second boundary condition arises from (14),

$$\tilde{\sigma}_{rr}(\lambda_r, \lambda_\theta, R)\big|_{R_0} = 0. \tag{48}$$

Let λ_0 denote the common unknown value in (47). For each value of λ_0, (45) and (46) can be integrated numerically by, say, the Runge–Kutta method. Let the radii at which the numerical solution is obtained be denoted by R_i, $i = 1, 2, \ldots, N$, where $R_1 = 0$ and $R_N = R_0$. The values of λ_r and λ_θ at these radii are then denoted, respectively, by λ_{ri} and $\lambda_{\theta i}$. Boundary condition (48) becomes

$$\tilde{\sigma}_{rr}(\lambda_{rN}, \lambda_{\theta N}, R_N) = 0. \tag{49}$$

The value of λ_0 is adjusted by iteration until (49) is satisfied.

Once the numerical solution has been obtained, the radial deformation function (43) is obtained by the relation $r = R\lambda_\theta$. The twisting moment and axial force are obtained by using the calculated values λ_{ri} and $\lambda_{\theta i}$ in the following numerical approximations to (35) and (36):

$$\frac{M}{4\pi\psi} = \sum_{i=1}^{N} \lambda_{\theta i}^2 [W_{1i} + \lambda_{ri}^2 W_{2i}] R_i^3 \beta_i , \tag{50}$$

$$\frac{N}{4\pi} = \sum_{i=1}^{N} [W_{1i} + (\lambda_{ri}^2 + \lambda_{\theta i}^2) W_{2i} + \lambda_{ri}^2 \lambda_{\theta i}^2 W_{3i}] R_i \beta_i . \tag{51}$$

In (50) and (51), W_{ij} denotes the value of W_i at radius R_j and β_i represent weighting coefficients in the particular numerical formula used to approximate the integrals. Finally, volumetric strain V is determined by evaluating

$$V = \frac{(R_N \lambda_{\theta N})^2}{R_N^2} = \lambda_{\theta N}^2 - 1. \tag{52}$$

The values of the twisting moment, axial force, and volumetric strain determined from the numerical solution are denoted, respectively, by $M^{\text{calc}}(\psi; \mathbf{c})$, $N^{\text{calc}}(\psi; \mathbf{c})$, and $V^{\text{calc}}(\psi; \mathbf{c})$.

5.4. Determination of Constants. Let the angles of twist per unit length at which measurements are made be denoted by ψ_j, $j = 1, 2, \ldots, j^*$. An error between the measured and calculated quantities can be defined by

$$E(\mathbf{c}) = \sum_{j=1}^{j^*} [\alpha_1(M^{\text{meas}}(\psi_j) - M^{\text{calc}}(\psi_j; \mathbf{c}))^2 + \alpha_2(N^{\text{meas}}(\psi_j) - N^{\text{calc}}(\psi_j; \mathbf{c}))^2 + \alpha_3(V^{\text{meas}}(\psi_j) - V^{\text{calc}}(\psi_j; \mathbf{c}))^2]. \tag{53}$$

In (53), α_1, α_2, and α_3 are positive constants which weight the contributions of each term to the total error.

At the actual strain energy density function, the error is zero. For the form of W assumed in (41), constants $\mathbf{c}$ are to be determined by minimizing $E(\mathbf{c})$ so that its value is as close to zero as possible. There are two approaches which can be used to carry out this process. In the first, the system of equations is derived which represents the conditions for a minimum of $E(\mathbf{c})$, namely $\partial E/\partial c_i = 0$, $i = 1, 2, \ldots, 6$. These are solved numerically. The second approach uses direct optimization methods. One such method was used by Wineman *et al.* (Ref. 13) and it is the intention to use a similar one in the present case.

Let $\hat{\mathbf{c}}$ denote the set of constants at which the value of $E(\mathbf{c})$ is less than some specified value. A strain energy density function and the corresponding radial deformation have thus been determined so that the equilibrium equations and boundary conditions are satisfied, and the measured and calculated quantities agree to within a specified value. In this sense, the W for the material has been identified.

In making reasonable estimates of the coefficients, one should note that the values of the volume change of the cylinder are expected to be small, as noted by McKenna and Zapas (Ref. 11) and observed by Duran and McKenna (Ref. 9). For an angle of twist of $\Psi R_0 \approx 0.1$, the volume strain is of the order 10^{-3}. There is also reason to expect some physical meaning of the parameters if one keeps in mind that $W_1 + W_2$ is the shear modulus and that this may depend on pressure for polymers. Also, it is plausible that the value of W_3 seems to be related to the bulk modulus, although how this is so is unclear.

6. Summary

The problem of torsion of a cylinder of a compressible isotropic elastic material has been discussed and a method of determining the strain energy density function based on the method of material identification has been presented. In future work, twisting moment, axial force, and volume change data using the NIST torsional dilatometer will be analyzed by this method and the strain energy density function so determined will be presented and discussed.

References

1. RIVLIN, R. S. and SAUNDERS, D. W., *Large Elastic Deformations of Isotropic Materials, VII, Experiments on the Deformations of Rubber*, Philosophical Transactions of the Royal Society, Vol. A243, pp. 251–288, 1951.
2. TRELOAR, L. R. G., *Stresses and Birefrigence in Rubber Subjected to General Homogeneous Strain*, Proceedings of the Physical Society, Vol. 60, pp. 135–144, 1948.
3. KAWABATA, S., and KAWAII, H., *Strain Energy Density Functions of Rubber Vulcanizates from Biaxial Extension*, Advances in Polymer Science, Vol. 24, Molecular Properties, Springer-Verlag, Berlin, Germany, pp. 89–124, 1977.
4. BECKER, G. W., *On the Phenomenological Description of the Nonlinear Deformation of Rubberlike High Polymers*, Journal of Polymer Science, Vol. C16, pp. 2893–2903, 1967.
5. RIVLIN, R. S., *Large Elastic Deformations of Isotropic Materials, IV, Further Developments of the General Theory*, Philosophical Transactions of the Royal Society, Vol. A241, pp. 379–397, 1948.
6. PENN, R. W., and KEARSLEY, E. A., *The Scaling Law for Finite Torsion of Elastic Cylinders*, Transactions of the Society for Rheology, Vol. 20, pp. 227–238, 1976.
7. RIVLIN, R. S., *A Note on the Torsion of an Incompressible Highly-Elastic Cylinder*, Proceedings of the Cambridge Philosophical Society, Vol. 45, pp. 485–487, 1949.
8. WANG, T. T., ZUPKO, H. M., WYNDON, L. A., and MATSUOKA, S., *Dimensional and Volumetric Changes in Cylindrical Rods of Polymers Subjected to a Twist Moment*, Polymer, Vol. 23, pp. 1407–1409, 1982.
9. DURAN, R. S., and MCKENNA, G. B., *A Torsional Dilatometer for Volume Change Measurements on Deformed Glasses: Instrument Description and Measurements on Equilibrated Glasses*, Journal of Rheology, Vol. 34, pp. 813–839, 1990.
10. PIXA, R., LEDU, V. and WHIPPLER, C., *Dilatometric Study of Deformation Induced Volume Increase and Recovery in Rigid PVC*, Colloid and Polymer Science, Vol. 266, pp. 913–920, 1988.
11. MCKENNA, G. B., and ZAPAS, L. J., *The Time Dependent Strain Potential Function for a Polymeric Glass*, Polymer, Vol. 26, pp. 543–550, 1985.

12. IDING, R. H., PISTER, K. S., and TAYLOR, R. L., *Identification of Nonlinear Elastic Solids by a Finite Element Method*, Computer Methods in Applied Mechanics and Engineering, Vol. 4, pp. 121–142, 1974.
13. WINEMAN, A., WILSON, D., and MELVIN, J. W., *Material Identification of Soft Tissue using Membrane Inflation*, Journal of Biomechanics, Vol. 12, pp. 841–850, 1979.
14. SPENCER, A. J. M., *Continuum Mechanics*, Longman, Inc., New York, New York, 1980.
15. LEVINSON, M., *Finite Torsion of Slightly Compressible Rubberlike Circular Cylinders*, International Journal of Non-Linear Mechanics, Vol. 7, pp. 445–463, 1972.
16. BEATTY, M. F., *Topics in Finite Elasticity: Hyperelasticity of Rubber, Elastomers and Biological Tissues-with Examples*, Applied Mechanics Reviews, Vol. 40, pp. 1699–1734, 1987.
17. BAKER, M., and ERICKSEN, J. L, *Inequalities Restricting the Form of the Stress Deformation Relations for Isotropic Solids and Reiner–Rivlin Fluids*, Journal of the Washington Academy of Sciences, Vol. 44, pp. 33–35, 1954.

Index

Complete Series Listing

Below is a complete listing of the volumes in the *Mathematical Concepts and Methods in Science and Engineering* series.

1 **INTRODUCTION TO VECTORS AND TENSORS, Volume 1: Linear and Multilinear Algebra**
- *Ray M. Bowen and C.-C. Wang*

2 **INTRODUCTION TO VECTORS AND TENSORS, Volume 2: Vector and Tensor Analysis**
- *Ray M. Bowen and C.-C. Wang*

3 **MULTICRITERIA DECISION MAKING AND DIFFERENTIAL GAMES**
- *Edited by George Leitmann*

4 **ANALYTICAL DYNAMICS OF DISCRETE SYSTEMS**
- *Reinhardt M. Rosenberg*

5 **TOPOLOGY AND MAPS**
- *Taqdir Husain*

6 **REAL AND FUNCTIONAL ANALYSIS**
- *A. Mukherjea and K. Pothoven*

7 **PRINCIPLES OF OPTIMAL CONTROL THEORY**
- *R. V. Gamkrelidze*

8 **INTRODUCTION TO THE LAPLACE TRANSFORM**
- *Peter K. F. Kuhfittig*

9 **MATHEMATICAL LOGIC: An Introduction to Model Theory**
- *A. H. Lightstone*

10 **SINGULAR OPTIMAL CONTROLS**
- *R. Gabasov and F. M. Kirillova*

11 **INTEGRAL TRANSFORMS IN SCIENCE AND ENGINEERING**
- *Kurt Bernardo Wolf*

12 **APPLIED MATHEMATICS: An Intellectual Orientation**
• *Francis J. Murray*

13 **DIFFERENTIAL EQUATIONS WITH SMALL PARAMETERS AND RELAXATION OSCILLATIONS**
• *E. F. Mishchenko and N. Kh. Rozov*

14 **PRINCIPLES AND PROCEDURES OF NUMERICAL ANALYSIS**
• *Ferenc Szidarovszky and Sidney Yakowitz*

16 **MATHEMATICAL PRINCIPLES IN MECHANICS AND ELECTROMAGNETISM**
Part A: Analytical and Continuum Mechanics
• *C.-C. Wang*

17 **MATHEMATICAL PRINCIPLES IN MECHANICS AND ELECTROMAGNETISM**
Part B: Electromagnetism and Gravitation
• *C.-C. Wang*

18 **SOLUTION METHODS FOR INTEGRAL EQUATIONS**
• *Edited by Michael A. Golberg*

19 **DYNAMIC OPTIMIZATION AND MATHEMATICAL ECONOMICS**
• *Edited by Pan-Tai Liu*

20 **DYNAMICAL SYSTEMS AND EVOLUTION EQUATIONS**
• *J. A. Walker*

21 **ADVANCES IN GEOMETRIC PROGRAMMING**
• *Edited by Mordecai Avriel*

22 **APPLICATIONS OF FUNCTIONAL ANALYSIS IN ENGINEERING**
• *J. L. Nowinski*

23 **APPLIED PROBABILITY**
• *Frank A. Haight*

24 **THE CALCULUS OF VARIATIONS AND OPTIMAL CONTROL**
• *George Leitmann*

25 **CONTROL, IDENTIFICATION, AND INPUT OPTIMIZATION**
• *Robert Kalaba and Karl Spingarn*

26 **PROBLEMS AND METHODS OF OPTIMAL STRUCTURAL DESIGN**
• *N. V. Banichuk*

27 **REAL AND FUNCTIONAL ANALYSIS, Second Edition**
Part A: Real Analysis
• *A. Mukherjea and K. Pothoven*

28 **REAL AND FUNCTIONAL ANALYSIS, Second Edition**
Part B: Functional Analysis
• *A. Mukherjea and K. Pothoven*

29 **AN INTRODUCTION TO PROBABILITY THEORY WITH STATISTICAL APPLICATIONS**
• *Michael A. Golberg*

30 **MULTIPLE-CRITERIA DECISION MAKING**
• *Po-Lung Yu*

31 **NUMERICAL DERIVATIVES AND NONLINEAR ANALYSIS**
• *Harriet Kagiwada, Robert Kalaba, Nima Rasakhoo, and Karl Spingarn*

32 **PRINCIPLES OF ENGINEERING MECHANICS**
Volume 1: Kinematics—The Geometry of Motion
• *M. F. Beatty, Jr.*

33 **PRINCIPLES OF ENGINEERING MECHANICS**
Volume 2: Dynamics—The Analysis of Motion
• *M. F. Beatty, Jr.*

34 **STRUCTURAL OPTIMIZATION**
Volume 1: Optimality Criteria
• *Edited by M. Save and W. Prager*

35 **OPTIMAL CONTROL APPLICATIONS IN ELECTRIC POWER SYSTEMS**
• *Edited by G. S. Christensen, M. E. El-Hawary, and S. A. Soliman*

36 **GENERALIZED CONCAVITY**
• *Mordecai Avriel, Walter E. Diewert, Siegfried Schaible, and Israel Zang*

37 **MULTICRITERIA OPTIMIZATION IN ENGINEERING AND IN THE SCIENCES**
• *Edited by Wolfram Stadler*

38 **OPTIMAL LONG-TERM OPERATION OF ELECTRIC POWER SYSTEMS**
• *G. S. Christensen and S. A. Soliman*

39 **INTRODUCTION TO CONTINUUM MECHANICS FOR ENGINEERS**
• *Ray M. Bowen*

40 **STRUCTURAL OPTIMIZATION**
Volume 2: Mathematical Programming
• *Edited by M. Save and W. Prager*

41 **OPTIMAL CONTROL OF DISTRIBUTED NUCLEAR REACTORS**
• *G. S. Christensen, S. A. Soliman, and R. Nieva*

42 **NUMERICAL SOLUTION OF INTEGRAL EQUATIONS**
• *Michael A. Golberg*

43 **APPLIED OPTIMAL CONTROL THEORY OF DISTRIBUTED SYSTEMS**
• *K. A. Lurie*

44 **APPLIED MATHEMATICS IN AEROSPACE SCIENCE AND ENGINEERING**
• *Edited by Angelo Miele and Attilio Salvetti*

45 **NONLINEAR EFFECTS IN FLUIDS AND SOLIDS**
• *Edited by Michael M. Carroll and Michael A. Hayes*